CLASSICS FROM PAPYRUS TO THE INTERNET

Ashley and Peter Larkin Series in Greek and Roman Culture

CLASSICS FROM PAPYRUS TO THE INTERNET
An Introduction to Transmission and Reception

JEFFREY M. HUNT
R. ALDEN SMITH
FABIO STOK

Foreword by Craig Kallendorf

University of Texas Press
AUSTIN

Copyright © 2017 by the University of Texas Press
All rights reserved
Printed in the United States of America
First edition, 2017

Requests for permission to reproduce material
from this work should be sent to:
Permissions
University of Texas Press
P.O. Box 7819
Austin, TX 78713-7819
http://utpress.utexas.edu/index.php/rp-form

♾ The paper used in this book meets the minimum requirements of
ANSI/NISO Z39.48-1992 (R1997) (Permanence of Paper).

LIBRARY OF CONGRESS CATALOGING-IN-PUBLICATION DATA

Names: Hunt, Jeffrey Michael, author. | Smith, R. Alden, author. | Stok, Fabio, author.
Title: Classics from papyrus to the internet : an introduction to transmission and reception / Jeffrey M. Hunt, R. Alden Smith, Fabio Stok; foreword by Craig Kallendorf.
Other titles: Ashley and Peter Larkin series in Greek and Roman culture.
Description: First edition. | Austin : University of Texas Press, 2017. |
Series: Ashley and Peter Larkin series in Greek and Roman culture | Includes bibliographical references and index.
Identifiers: LCCN 2016055397
ISBN 978-1-4773-1301-5 (cloth : alk. paper)
ISBN 978-1-4773-1302-2 (pbk. : alk. paper)
ISBN 978-1-4773-1303-9 (library e-book)
ISBN 978-1-4773-1304-6 (nonlibrary e-book)
Subjects: LCSH: Classical philology—History and criticism. | Learning and scholarship—History. | Communication in learning and scholarship—Technological innovations. | Communication and technology—History. | Paleography, Greek—History. | Manuscripts, Greek (Papyri) | Written communication—History.
Classification: LCC P96.T42 H866 2017 | DDC 002—dc23
LC record available at https://lccn.loc.gov/2016055397

doi:10.7560/313015

Uxoribus nostris carissimis,
Jenny, Diane, and Vinzia

CONTENTS

PREFACE ix

FOREWORD *by Craig Kallendorf* 1

CHAPTER 1. Writing and Literature in Antiquity 5

CHAPTER 2. Grammar, Scholarship, and Scribal Practice from Antiquity to the Middle Ages 46

CHAPTER 3. Classical Reception from Antiquity to the Middle Ages 85

CHAPTER 4. Classics and Humanists 149

CHAPTER 5. Classical Texts in the Age of Printing 191

CHAPTER 6. Tools for the Modern Scholar 221

NOTES 241

BIBLIOGRAPHY 301

INDEX 325

PREFACE

"THREE CLASSICISTS WALK INTO A BAR" — THIS PHRASE perhaps sounds like the first line of a joke leading to a ribald punchline. Yet in this case, it truly happened, as at the suggestion of Craig Kallendorf of Texas A&M, Jeff Hunt, Alden Smith, and Fabio Stok sat down in a bar in Rome first to discuss how Stok's *I classici dal papiro a Internet* might be adapted into a new book specifically designed not only to convey the history of classical scholarship but also to speak broadly to the training and development of a new generation of classicists. As we set about the business of planning the new work, based on but distinct from the original, we decided to take the last part of the foundational work's title quite seriously and develop a webpage to go with the book, a site that we will continually update and expand, designed to abet the research and development of graduate students and advanced undergraduates, teachers of classics, and those engaged in related fields.

We envision the webpage that dovetails with the book to become the modern-day heir of the pioneering work of many scholars, from Conrad Gessner, to Jules Marouzeau, to J. A. Nairn, to Maurice Platnauer. While the last chapter looks toward the website, the preceding five chapters, indebted to the important works of Rudolf Pfeiffer and L. D. Reynolds and N. G. Wilson, among others, are meant to provide background, so that the reader who is studying philology or is already a philologist may understand the grand contributions of those who came before. The road to becoming a modern scholar began a long time ago and has been built by the labor of many generations.

In addition to the works just mentioned, recent contributions have been vitally important. The *Introduction to Manuscript Studies* of Raymond Clemens and Timothy Graham is a superb work for budding paleographers. That book provides a basic overview of manuscript production alongside helpful and detailed examples of numerous scripts. Its lengthy list of abbreviations that can be found in Latin manuscripts is also very useful. It is not a study of textual transmission per se, as its aim is obviously paleographical.

Another solid contribution is the *Handbook for Classical Research* by David Schaps, which is a fine introduction to classics and its subfields. It offers instruction about reading a published papyrus or inscription, listing and explaining many resources now available for research, and even suggests how a bibliography might be assembled. Schaps's didactic focus is notably on the nuts and bolts of the present state of affairs in research methods and it is thus outside his purview to offer a history of classical scholarship. Our approach enlarges on the connection between the modern techniques outlined in Schaps's book with the monumental works of scholarship that gave rise to them. Nonetheless, the webpage that will complement the current volume is necessarily indebted to Schaps's fine contribution.

We have numerous people to thank for making this volume either possible or better. We begin with those who have underwritten this project in various ways. Truell Hyde, Baylor's Vice Provost for Research, has encouraged our work, and his office has generously financed much of the travel to make research for this volume possible. We are grateful, too, to Jennifer Good, Director of the University Scholars Program in Baylor's Honors College and to the dean of that college, Thomas Hibbs. We also want to thank Dean Lee Nordt of Baylor University and the administration of University of Rome Tor Vergata.

Other friends contributed in various ways to this enterprise. Professor Michael Beaty, chair of philosophy at Baylor, facilitated a semester at St. Andrews for Jeff Hunt, which experience allowed him access to the fine University Library and interaction with the faculty there. In particular, Hunt wishes to thank Mark Elliott, St. Mary's Head of School at the University of St. Andrews, and the entire Department of Classics at St. Andrews, and especially Jason König, chair of Classics.

Our colleagues at Baylor—Simon Burris, Jeff Fish, Kevin Funderburk, Dan Hanchey, David White, and Brent Froberg—were all helpful with various stages of the work. We wish to express our gratitude, too, to colleagues at Tor Vergata. Other friends provided us opportunity for interactions for this book. In addition to Craig Kallendorf, who gave this collaboration impetus, we thank Gianni Profita (La Sapienza), Piergiacomo and Annamaria Petrioli (Brown University in Bologna), Peter Arzt-Grabner (University of Salzburg), and Baylor librarians Janet Sheets, Eileen Bentsen, and John Bales. We also thank David Konstan, Lee Fratantuono, Peter Knox, and Rachel Sanders, and many of our students at Baylor, including Wes Beck, Kelsey Bell, Keller Bright, Joshua Conatser, Samantha Elmendorf, Jacob Imam, Joseph Lloyd, Kara Kopchinski, Cynthia Liu, Gabriel Pederson, Kelsi Ray, Madeleine Sullivan, and Jamie Wheeler. Further, we want to acknowledge the unflagging support of

Thelma Mathews and Charmaine Dull. Diane Smith and Jenny Hunt, too, were truly helpful in the final stages of this project, and we here thank them *ex corde*. How can we begin to thank Mary Claire MacDonald, the superb artist who rendered so faithfully the many images used in this book? Needless to say, opening to any illustrated page will amply evidence to even the most random of surveyors that Mary Claire's work is first-rate.

We wish to express sincere thanks to Jim Burr of the University of Texas Press for his professionalism and the support that he and his staff rendered to us throughout this project. It has been a pleasure to work with them, and they have been remarkably helpful.

Finally, we are deeply thankful to the reviewers of this volume, whose insights and criticisms have improved it. The authors are, of course, responsible for the final project, which we hope will provide a point of departure for inquiry into what constitutes classical philology, what it means to interpret inherited knowledge, and ultimately, therefore, to some extent what it means to be human.

CLASSICS FROM PAPYRUS TO THE INTERNET

FOREWORD
Craig Kallendorf

THIS IS AN EXCITING TIME TO BE A CLASSICIST—INDEED, we would have to go back as far as the Renaissance to find a similar degree of change in how scholarship in Greek and Latin is undertaken. It was then, under the guidance of Francesco Petrarca (Petrarch) and his intellectual heirs, that the philological method was developed, in which a careful study of the classical languages might allow later readers to recover how a work would have been understood within the culture in which it was originally produced. Gradually, thanks to the work of textual scholars like Angelo Poliziano (Politian) and his followers, scholars devised a way to trace a text back through centuries of erroneous scribal and editorial interventions to something that reflected the intentions of its ancient author.[1] This method was refined in the middle of the nineteenth century by Karl Lachmann and a group of scholars, mainly German, working around him,[2] so that for more than a century, classicists have been able to use the techniques described in books like L. D. Reynolds and N. G. Wilson's *Scribes and Scholars*[3] to follow in the footsteps of Niccolò Machiavelli, who declared that he had gained access to the ancient world, unimpeded by modern life; as he put it, "[I] step inside the venerable courts of the ancients ... where I am unashamed to converse with them and to question them about the motives of their actions, and they, out of their human kindness, answer me. ... *I absorb myself into them completely.*"[4] To be sure, now and again a classicist would note that passages and themes from the classics were reused again and again by later writers. This led to books like Gilbert Highet's *The Classical Tradition*,[5] which set out in painstaking detail how Ovidian material resurfaced in Shakespeare's plays and how Milton's epic poetry drew from Homer and Virgil. The term "classical tradition" suggests that the process by which Greek and Latin material was handed down from generation to generation is largely a passive one, and since the real goal was to interpret a work within the culture in which it was originally produced, this subfield remained a marginal enterprise.

Within the last two generations, however, this venerable model has been undermined on several fronts. The so-called theory revolution in literary studies has challenged the idea that there is an objective vantage point from which any past culture can be seen or that a later reader like Machiavelli could completely immerse himself in the mindset of antiquity. Stated forcefully, this principle has led to the argument that "our current interpretations of ancient texts, whether or not we are aware of it, are, in complex ways, constructed by the chain of receptions through which their continued readability has been effected. As a result we cannot get back to any originary meaning wholly free of subsequent accretions."[6] If we accept this premise, then everything else changes as well, at least to a certain extent. We still need to construct texts, but the original intentions of a classical author remain unrecoverable, especially for material as old as this. Not to worry, though; we can get close, but more important is what we can learn from the way in which the text was passed on through the ages. As Jerome McGann has argued, texts are socially constructed, so that the interventions of editors, censors, and generations of readers are of value in and of themselves.[7] Classical texts, it is now argued, are never simply handed over, but are transformed as they are passed along—indeed we now have a new word, "reception," that emphasizes the active cooperation of later readers in helping to create the meaning of classical texts.[8]

Classics as a field is very much in flux right now, so that not every classicist feels comfortable with every one of these changes. As a result, hybrid models currently exist in which, for example, texts are constructed by an editor with an eye on authorial intention while the same scholar might also write an article that rests in an active, reception-based appropriation of classical material in post-classical culture. But it is a brave new world out there, one which demands a new treatment of how classical texts have been passed from generation to generation and is compatible with the developments that are transforming classical studies as a field. This volume provides that treatment.

Let me give a couple of examples of why this book marks such a timely intervention into the evolution of the field. Under the traditional philological model, commentaries have served as a key way to recover the original meaning of a literary work: when a skilled scholar proceeds through a poem word by word, clarifying each meaning through comparison with other usages of the same word and citing parallel passages, the careful reader should be able to reconstruct from this what the text meant at the time when it was written. Commentators often relied to an extent on their predecessors—they did not always acknowledge their debts, although that is a different story—but there was generally little need to consult, say, a commentary from the Renaissance, since anything worth preserving would have made its way into a modern commentary, and modern scholarship has undoubtedly made some observations from five hundred years ago obsolete. But if, as noted above, "our current interpretations of ancient texts, whether or not we are aware of it, are, in complex ways, constructed by the chain of receptions through which their continued readability has been effected," then each of these early commentaries becomes valuable again, as a record of one moment in the interpretive chain for the text it is explicating. And to recognize this, we have to consult the early editions of the classics, for this is where we find these commentaries, and read them to get an idea of what struck a reader as important at that moment. We also have to do exercises in printing history, to see how many times a given commentary was printed, on the assumption that often-reprinted commentaries had an outsized importance in the reception of the text they accompanied. Once we know, for example, that Christoph Hegendorff's commentary to Virgil was printed thirty-nine times before 1600, while Richard de Gorris's was printed twice and Germain Vaillant de Guélis's only once, we will be in a better position to understand the Renaissance Virgil.[9]

There is another reason, as David Scott Wilson-Okamura puts it, that if "we want to read the same classics that our poets read, and especially if we want to make arguments about word choice or even verb tense, we need to read them in the bad, old, beautifully printed, sometimes horribly corrupt editions that Ariosto and Ronsard would have owned and studied."[10] This is because the early printed editions often carry the only evidence we have of how our classical texts were received. For example, the Renaissance editions regularly contain indexing notes, words or phrases like "courage" or "simile" that stand in the margin next to a marked-off passage. These notes show that the book in question was prepared for commonplacing, a process in which the indexing notes serve as rubrics and the marked-off passages were listed below as examples of the category in question. The commonplace book in turn served as a source book for the compositions of later writers, whose works show an intertextual relationship with the texts they read and broke apart. This explains a phenomenon often remarked upon, that Renaissance literature strikes many a modern reader as tissues of passages woven together from previous books, but our understanding of what is going on here depends on our seeing the early printed books that served as the sources for the commonplace books of the age.[11] Another example of how reception depends on manuscripts and early printed books comes from other physical signs of intervention in those books. To take Virgil as an example again, the commentary of Josse Willich was regularly censored in Catholic countries, so that passages containing gratuitous swipes against the Church in Rome were inked out and made illegible. The commentaries of Philip Melanchthon, Christoph Hegendorff, and Étienne Dolet were regularly subjected to the same fate, which resulted at the very least in their names being obscured, but in extreme cases entire sections of the book containing their glosses were removed.[12] Here, to bring into focus this link in the chain of reception, we have to see both what was printed and what was removed in the editions of the time.

In other words, any vision of classical studies that takes reception seriously should have a material foundation to it, such that to clarify our own understanding of a classical text, we must also study the manuscripts and printed books that have brought it to us. This book has been designed to help us in that endeavor. In a series of engaging, detailed chapters, it explains how the texts of Greek and Latin authors were transmitted, edited, and commented upon, from late antiquity through the computer age. Traditionally minded classicists will find this book a useful complement to *Scribes and Scholars*, covering much of the same ground but from a different perspective and with comparatively little repetition. And for those classicists who have embraced the new model for the field, or are in the process of doing so, the book will serve as essential reading as classical studies moves into the third millennium. In it, subfields like textual criticism, the history of scholarship, and the classical tradition merge into a broader, deeper vision of classical studies, one that continues to value language and context but that rests in an understanding of reception that moves a materialized text through time and space from ancient Greece and Rome to us.

1 / WRITING AND LITERATURE IN ANTIQUITY

THE OLDEST EXTANT WORKS OF GREEK LITERATURE, THE Homeric epics, famously begin *in medias res*. This expression most directly refers to the action of the plot, which for the *Iliad* begins near the end of the Trojan War and for the *Odyssey* at the end of Odysseus' wanderings. Homer's epics as literary works are also *in medias res* in the sense that they are a continuation of existing traditions and an incorporation of influences from Eastern literature. Yet they are often considered a beginning for Greek literature because of the remarkable, sustained cultural influence they held over ancient Greece and Rome. In fact, they are both. The *Iliad* and *Odyssey* could not exist without the established tale of the Trojan War and epic traditions of earlier cultures, but Homer's work was not therefore an obligatory or replaceable part of the greater literary tradition.

Homer's place in the literary tradition provides a fitting analogy for the development of Western thought, which was neither inevitable nor *ex nihilo*. On the contrary, prevailing currents of thought always stand in relation to the literature of the past, whether in emulation or rejection of preceding theories. It is often the case that old or ancient texts inspire fresh approaches to problems with which people have grappled for centuries, or even millennia (as recently as 1990, Derek Walcott penned his epic *Omeros*, a work whose name alone reveals its debt to Homer).

This text is primarily concerned with examining literary transmission in its broadest sense. Particular manuscripts will at times come into the discussion, though our focus will center largely on how attitudes toward texts develop over time, especially in reaction to changes in the physical form of the book. In so doing we will touch on a number of topics, including the origins of numerous scholarly disciplines that continue to shape our understanding of the past and the determined effort required to keep the literary tradition alive. We hope these analyses will collectively provide a window into current methods and attitudes toward texts, which, as at all points in time, are neither inevitable nor incontrovertible.

Our study of literary transmission must, like Homer's epics, begin midstream, as it were, as we commence with an exploration of the origins and development of writing in ancient Greek and Roman societies. To do so efficiently, we shall pass over millennia of development that led to a watershed moment when the Greeks began to write, or at least to write what we commonly call ancient Greek. We do not bypass so much material arbitrarily; we do so only because the capacity to have produced the first written form of the Homeric epics represents a defining moment in the intellectual history of the West, a moment from which many current conventions and attitudes toward books can be traced. Thus, while we acknowledge that the Greeks and Romans, for all their originality, owe much to other cultures and themselves benefited from the transmission of ideas, we begin our discussion *in medias res*, starting with the slender but important evidence that we have for the development of books and writing.

Writing, which ultimately became a hallmark of Greek culture, made its way to Greece relatively late and to Rome even later. The alphabet was most likely introduced to Greece by Phoenician traders sometime around 800 BC. Greek writing predates the introduction of the alphabet, as evidenced by clay tablets found in the remains of Mycenaean palaces. These tablets are written in Linear B, a script derived from the older, but as yet indecipherable, Linear A, which was already in Crete by ca. 1700 BC.[1] The Linear B tablets indicate the presence of a syllabary in use for Greek (primarily for record keeping) as far back as 1400, but its use seems to have been limited to the Mycenaeans themselves as no trace of it is found following the decline of that civilization around 1200.[2]

The transmission of the Phoenician alphabet to the Greeks is clear, for the names, forms, and order of the letters *alpha* through *tau* display remarkable similarity to corresponding Phoenician letters.[3] Perhaps the most significant alteration made by the Greeks was to assign each letter to a single sound; many forms from the Phoenician system represented open syllables (i.e., a consonant sound paired with any vowel, thus allowing for multiple sounds from a single letter, the correct one to be determined in context). The Greeks represented vowel sounds with unique characters, a practice that was nascent in Phoenician writing but not yet developed. Also significant is the Greek tendency to use *boustrophedon* writing (see example in fig. 1.1) instead of the retrograde style characteristic of the Phoenicians.[4] The bidirectional back-and-forth flow of text loosely mimics the motion followed by an ox plowing a field.

The transmission of the alphabet to the Greeks is placed by most scholars at about 800 BC, largely on the basis of the earliest appearance of Greek writing, which dates from the middle to late eighth century. Even at this early

FIGURE 1.1. Boustrophedon *inscription from Apollonia. Sixth century BC.*

stage, however, the Greek alphabet demonstrates significant differences from its Semitic source, leading some scholars to posit an undocumented phase of transmission in the ninth century or earlier.[5] The precise place of transmission is a more open question: Herodotus credits Cadmus with bringing the alphabet to Greece and identifies Boeotia as the point of transmission.[6] Few scholars take Herodotus' claim seriously, though Euboea, which is geographically close to Boeotia and was connected to it by trade, is a strong possibility, given the early date of inscriptions found both in Euboea and in colonies established by the Euboean cities of Chalcis and Eretria.[7] Other sites of transmission have been proposed, however, including the Greek trading city of Al Mina,[8] as well as Crete, Rhodes, and Cyprus, all of which are islands along the trade route to the East.[9]

That Greece acquired its alphabet from a single point of transition is largely agreed upon, despite the well-known diversity of numerous local (or "epichoric") scripts. Throughout Greece, regional scripts show consistent variations from their Phoenician source, including the same shift of certain Phoenician consonants to Greek vowels—the division of the Phoenician *wau* U into the Greek *digamma* F (which took *wau*'s place as the sixth letter in the Greek alphabet) and *upsilon* Υ (which was placed after *tau*, the final Phoenician letter). Greek scripts also show a consistent confusion in their presentation of Phoenician sibilants that is otherwise inexplicable.[10] The nearly ubiquitous Greek additions of *phi* Φ, *chi* Χ, and *psi* Ψ following *upsilon* Υ (although not always in the same order or with the same phonetic value) also suggest a single point of transmission.

Differences in the order of *phi*, *chi*, and *psi* and in their pronunciation help distinguish between broad categories of Greek scripts. The nineteenth-century scholar Adolf Kirchhoff assembled the various epichoric Greek scripts into a handful of categories and marked them on a color-coded map.[11] The groupings are still often identified by the colors Kirchhoff used in his influential study. Eastern Greek scripts, including those of Attica, Corinth and her colonies, and Asia Minor are known as "blue" scripts, which display the familiar order and pronunciation *phi* [ph], *chi* [kh], *psi* [ps]. The western Greek "red" scripts, which occur throughout much of the Peloponnese, Boeotia, and Euboea and her colonies, instead preserve a different order and pronunciation: *chi* [ks], *phi* [ph], *psi* [kh]. It was from these western Greek "red" scripts that the Etruscans and other Italic peoples would acquire their alphabets, which in turn would influence the Latin alphabet used by the Romans. Kirchhoff identified a third — "green" — group of scripts used on the islands of Thera, Crete, and Melos, in which *phi*, *chi*, and *psi* do not appear. The cause of variation between red, blue, and green scripts is unknown, although the distribution of the scripts suggests they were spread along sea routes.

The diffusion of the Greeks' recently formed alphabet must have happened rapidly, as Greek writing dating to 770 BC has been found as far inland on the Italic peninsula as Gabii.[12] This diffusion, however, certainly does not imply a high rate or level of literacy. Of the relatively few inscriptions from the eighth and seventh centuries, most simply identify an object's owner.[13] However, it is noteworthy that evidence of literacy can sometimes defy expectations; for example, an Attic abecedarium (inscription of the alphabet) from before 500 BC found inscribed on a rock in pastureland suggests that even a shepherd could possess a rudimentary level of literacy.[14]

The oldest reference to writing in literature occurs in the *Iliad*. Book 6 contains an account of how Bellerophon, a victim of slander, is sent by King Proetus to deliver a message to the king of Lycia. Unbeknownst to Bellerophon, the tablet he carries instructs the Lycian king to kill him.[15] Lines 168–169[16] refer to a folded tablet (*pinax ptuktos*) engraved with marks that convey Proetus' instructions (*semata lugra*, "woeful signs"). Clearly this passage demonstrates an awareness of tablets and writing, but its significance for Homer and ancient Greece has been debated. The passage seems to indicate that writing was normative in the eighth century,[17] not only for brief dedications and household use but also for more substantial messages that facilitated long-distance communication between cities. Folding tablets, such as that in the Homeric passage, were nothing new in the Mediterranean world of the time, and this passage reveals that Homer was certainly aware of them. The degree to which writing was known and used in Greece, however, is debatable. Homer's reference, for example, to *semata* (signs) instead of *grammata* (letters) may suggest that writing per se was not that widely diffused in eighth-century Greece.[18]

Inhabitants of Sicily and southern Italy soon encountered the Greeks' alphabet, and, perhaps influenced by Greek colonists from Pithecusae, the Etruscans adapted it to fit their language. Epigraphic evidence for Etruscan writing dates to the seventh century. Unlike Greeks and Romans, the Etruscans did not distinguish between voiced consonants (sonants) and voiceless consonants (surds), and so had no need for the letters *B*, *D*, *K*, and *Q*.[19] As a result, the Romans initially used *C* for both [k] and [g] sounds (presumably acquiring *B*, *D*, and *Q* from the Greeks or other Italic peoples, while *K* remained in use for a very few words). This practice can be easily seen in *praenomina*[20] on inscriptions, which retained the archaic *C* where *G* would be expected. The letter *G* begins to appear around 269 BC (the change is often attributed to Appius Claudius Caecus, though Plutarch credits a certain Spurius Carvilius).[21] It took the place of the Greek *zeta*, which was initially represented by the Latin *S*. The Greek *digamma* (F) became a Latin *F*. Already by the end of the seventh century, *Y* and *Z*, initially unused, had been introduced and placed at the end of the alphabet.[22]

In the first century AD, the emperor Claudius attempted to enlarge the alphabet from twenty-four letters.[23] Claudius invented three letters, one to represent consonantal [v], one to represent [bs] or [ps], and a new vowel with a value between [i] and [u]. Despite appearing in a few inscriptions, Claudius' letters were but short-lived.[24]

Along with the development of the alphabet presumably came the materials suitable for holding the written word. Precisely when imported material, such as papyrus, made its way to Greece is unclear. Many materials were used to receive writing, and a variety of factors could determine which medium was chosen. Those messages intended for permanent public display, for instance, would be inscribed on stone or metal, while reusable wooden boards allowed notices of immediate importance to be displayed and easily "erased" by whitewashing the board. The relative availability of materials also played a role at times in the chosen medium for writing. In other instances, the force of tradition dictated the use of particular materials in spite of the availability of more convenient media, such as papyrus.

There exists an extensive tradition among the Romans that their earliest books were written on the inner bark of trees (particularly lime trees, according to Pliny the Elder), and so the Latin word for "bark" (*liber*) came also to mean "book." Although the evidence is anecdotal, this scenario is thoroughly plausible. Pliny's further assertion that Romans once wrote upon leaves before making use of bark may also be taken seriously, given that palm leaves have long been used for documents in the East.²⁵ However prevalent wooden *libri* were in Rome, they were replaced at an early date by papyrus, whose superior quality swiftly allowed it to become the standard writing material for literary works. Papyrus played an outsized role in ancient writing and book production that will be discussed at length below. First, however, some other materials are worthy of note.

Among the principal materials that bore writing in antiquity was linen. It was produced using fibers from flax, a plant that was widely cultivated, though Egypt seems to have generated a particularly large flax industry.²⁶ Livy on several occasions notes the use of linen texts to preserve lists of magistrates, which were deposited in temples. In particular he identifies the temple of Juno Moneta, which was associated with coin production, as housing magistrate lists on linen. Another documented use of a linen book is mentioned by Livy, in his discussion of instructions for an old Samnite ritual recorded on linen.²⁷ The most famous surviving linen text, the *Liber linteus zagrabiensis*, is a fragment containing the longest extant Etruscan text (about thirteen hundred words). The date of the text is difficult to discern, though the script suggests it was written ca. 200–150 BC.²⁸ The text was found in the cartonnage of a mummy in Egypt, but the material is from Etruria and contains what appears to be a calendar of ritual events.

Not all texts produced by religious authorities were on linen; the *pontifices*, for instance, oversaw the publication of the *tabulae dealbatae*, whitewashed boards intended for reuse. Furthermore, Pliny recalls a period of transition during which different writing materials were used for public documents. While describing the Romans' shift from writing on bark to papyrus, he touches on other materials as well, noting that "later, public records began to be drawn up on leaden sheets, and soon even private ones on linen or wax."²⁹ The distinction between public documents on lead and what Pliny identifies as "private" documents recorded on linen or wax can obscure the meaning of the passage. Piccaluga is likely correct that "private" here does not involve something personal but rather refers to documents of social importance not intended for full public disclosure.³⁰ Thus many of the documents associated with linen could, at least early on, have been recorded on wax tablets as well.

Linen, in any case, certainly came to be associated with religious texts. Fronto notes that no corner of Rome lacked a shrine with linen books.³¹ That is not to say, however, that linen was reserved for religious use, for it seems on some occasions to have been used for the preservation of nonreligious texts.³² That is not surprising, since it was relatively inexpensive, did not require protective coatings of oil (as did papyrus), and was resistant to tears. In the period before heavy importation of papyrus, linen was itself cheaply imported and was an ideal choice for preserving important texts.

FIGURE 1.2. *Ancient Greek man with stylus (wax) tablet. Red-figure vase painting by Douris, about 500 BC. Staatliche Museen zu Berlin, Preußischer Kulturbesitz: Antikensammlung, F 2285. Courtesy of Pottery Fan.*

Another class of objects that commonly held writing was wooden tablets, which were used extensively for sundry purposes in Greece as well as Italy; the surviving examples come from Roman provinces. Wooden tablets can be divided into two types. The first is the stylus tablet, which was hollowed out and filled with wax, upon which writing was incised with a stylus. The second is the leaf tablet, which was a thin, polished piece of wood intended to receive writing in ink. Of these two types, stylus tablets have been preserved in far greater numbers: there are more than two hundred from Pompeii and Herculaneum alone.[33] This kind of writing board appeared in codex form, in which multiple leaves were bound together with leather thongs.[34] The method of inscribing the tablets is inferred from early fifth-century Greek vase paintings that depict an individual preparing to write on a tablet that he has opened (as in fig. 1.2).

Stylus tablets offered several advantages over writing in ink, perhaps most notably the ability to "erase" easily what one had written. This was accomplished by using the broad, flat end of the stylus to reshape the inscribed wax. The tablets would need a fresh coat of wax after extensive use.[35] The benefits of

FIGURE 1.3. *Wax tablet. Museo Gruppo Storico Romano. Courtesy of Michel Wal.*

erasure and reuse gave these writing boards great flexibility and longevity; they were still heavily used in the medieval period.[36] Paradoxically, they were used for both ephemeral writing, such as jotting down notes, and to preserve documents such as financial records and wills. The tablets from Pompeii, in fact, pertain to the finances of an auctioneer named L. Caecilius Iucundus. Stylus tablets found in Egypt tend to have been for legal use, and include birth certificates, wills, military diplomas, and census declarations.[37] In any case, stylus tablets were limited in their uses to either documents or short-lived writing.

In areas such as Egypt and the Greek East, which had ready access to papyrus, stylus tablets were less commonly used than elsewhere in the empire.[38] In places where papyrus was not indigenous, such as Rome, tablets tended to be preferred for documents or notes (see fig. 1.3). Pliny the Younger, for example, specifically mentions taking a *stylus* and *pugillares* (notebooks) on a hunting trip with the intention of bringing them back filled with writing even if he failed to catch a boar.[39] Though Pliny's hunting expedition may be no more than a literary fantasy, the situation Pliny imagines—an isolated location fit for creative composition—made the small stylus tablets an ideal writing material. Once finished, however, the final literary product would have been presented on a carefully written roll, not a tablet. The distinction is underscored in another letter,[40] in which Pliny exhorts himself to create a work that is worthy of being written on *chartae*, that is, papyrus rolls. Pliny's use of *charta* points to papyrus specifically over other possible materials, attributing to it a status befitting its lofty literary content.

Though stylus tablets were typically made from wood, other materials, such as ivory, could also be hollowed out to create tablets. In fact, in late antiquity it became traditional for consuls to present to the emperor and other important figures ivory *pugillares* with images of their accession to office carved on the front. These elaborate tablets, known as the consular diptychs, were most notable for the images displayed in bas-relief, but would also have the achievements of the newly elected consul written on the wax inside.[41] The copies that might be presented to the emperor were quite elaborate in their decoration and influenced later book covers.[42]

Though they do not survive in such quantities as stylus tablets, leaf tablets also played an important role in the ancient world. In 1973 over two hundred fragments from this kind of writing material, dating mostly from the early second century AD, were discovered at a Roman fort in northern England.[43] This find suggests a rather wide use for leaf tablets, which, being thin slivers of wood, suffer decay more readily than the thicker stylus tablets and so leave fewer examples. The location of the find is significant: although papyrus could be imported into Rome with relative ease, the same cannot be said of Britain. It is reasonable to expect that leaf tablets produced relatively cheaply—the process was still labor intensive—might have provided a substitute for papyrus for letters and other documents in provinces with an abundance of timber.[44]

EDUCATION AND LITERACY

Though writing yielded many benefits and became indispensable in antiquity, its introduction to society at large was not without controversy. The alphabet provided a convenient system for recording information, which had a dramatic effect on Greek and Roman governments, society, and culture. In Greece, from the Archaic period on, one can trace the increasing role of documentation and literacy, though this progression was far from uniform across the Greek-speaking world. Despite its thorough integration within public and private business, a high level of literacy was achieved by only a small percentage of any Greek population, including those at great cultural centers such as Athens, Alexandria, and Pergamum. Because achieving literacy required the leisure for school and the resources to hire an instructor, throughout antiquity literacy generally was an indicator of status, although it also obviously had practical advantages. Still, in the provinces especially it was rarely necessary for daily life, even among the upper classes.

As noted above, the earliest extant inscriptions are brief and limited in their uses; most show ownership, but a few religious dedications and inscribed tombstones survive.[45] By the close of the sixth century in Athens, the written word was integral to the newly instituted practice of ostracism. This term was derived from *ostrakon*, which was a potsherd incised with writing. Each *ostrakon*, it should be noted, need not—and typically did not—contain more than an individual's name, the inscribing of which would require minimal literacy. In addition pre-inscribed *ostraka* were available, as demonstrated by a collection of one hundred ninety-one *ostraka* written in only fourteen hands found on the North Slope of the Acropolis.[46] By the 480s, when ostracism required thousands of votes, a certain functional, basic reading knowledge may have gained traction among Athenians, while the high level of literacy required for reading and writing literary works was enjoyed by a select few throughout the Archaic period.

Writing was used from an early point to preserve literature; we do not know how or when Homer's epics were first written down, but we do know that Hesiod and other poets, such as Archilochus, Tyrtaeus, and Sappho, wrote rather than composed orally.[47] Pre-Socratic philosophy also seems to have provided an impetus for the written word as Ionian philosophers used writing to record and disseminate their ideas, though issues of circulation and readership remain entirely opaque.[48] Throughout the Archaic period, literacy continued to make inroads, expanding to reach more Greek cities and increasing among the residents of more populous urban areas.

The increase in readership, however, should be kept in perspective. Harris speculates that the most cultured towns of the Archaic period might boast hundreds of readers, while other Greek cities would have been home to literate populations in the dozens.[49] Additionally, the uses of writing remained largely unchanged through the end of the sixth century. It is tempting to infer widespread literacy from artifacts like inscribed *instrumenta domestica* (household articles), which suggest that the owners of these objects, and perhaps the craftsmen as well, were literate. Such brief inscriptions, however, give no indication of readership,[50] and Greece in this period remained primarily an oral culture.[51]

In the Classical period, the evidence for writing and literacy becomes more abundant, at least in the case of Athens. References in Aristophanes' *Clouds* and

Xenophon's *Oeconomicus* indicate that bookkeeping was a common practice, at least among the wealthy.[52] In legal matters, written testaments and documents of manumission became more prevalent, especially in the fourth century.[53] Writing featured heavily in managing the *polis*; in addition to financial accounts that were surely used to manage the tribute collected from Athens's allies, the epigraphical evidence demonstrates numerous new types of documents, including many decrees of the assembly, tribute lists, casualty lists, notice boards in the agora, and other matters written for public display, though how many Athenians could read them is questionable.[54] By the fourth century, complaints had to be submitted to courts in writing, even if many may have needed assistance to do so. Throughout the fifth century, oral complaints would be recorded by a court secretary.[55]

Demosthenes' speech *Against Timotheus* suggests that most citizens were unfamiliar with bankers' record keeping.[56] While the mere existence of writing does not necessarily indicate wide readership, one may reasonably infer that the level of literacy in Classical Athens must have been relatively high, for in order to manage the affairs of so large an empire, archons, polemarchs, and other officials depended not merely on oral communication but on the written word. By 405 BC there existed a state archive in Athens, with junior public clerks (*hypogrammateis*) first attested soon after.[57]

Despite the expanding role of writing in society, literacy was not deemed a necessity for normal social involvement—especially outside the city—except perhaps for those wealthy elite who aspired to high public office. Literacy continued to expand by the end of the fifth century and into the fourth, although its value was not universally appreciated. Eventually literacy came to be nearly the exclusive property of the upper class and thus slowly and unevenly mapped its place onto the ancient educational landscape.

Throughout the fifth century BC, formal training was divided into a dyad of music and physical training. Several sources evidence a preference for this traditional curricula and even demonstrate hostility toward the spread of literacy.[58] One can see this, for instance, in the case of a character named "Better Logic" in Aristophanes' *Clouds*, who ruminates on the "good old days" of the generation conversant with the Battle of Marathon, when, he says, students would visit the harpist (*kitharistes*) and the physical trainer (*paidotribes*). Such a dual curriculum is reflected also in Plato's *Republic*, where Socrates posits the ideal curriculum, which includes the broad categories of physical training and music.[59]

Letters and reading, referred to as *grammata*, were added as an additional component of ancient Greek education only in the early fourth century BC.[60] By the Hellenistic period, musical performance had been severed from the traditional subjects in favor of *grammata*, which eventually formed the core of a standard curriculum later referred to as the "all-encompassing education" (*enkyklios paideia*).

The incorporation of writing into the lives of average citizens, nascent at the end of the Classical period, exploded within the Hellenistic monarchies, especially in Ptolemaic Egypt.[61] Contracts involving monetary transactions in particular became normative, which required an individual participating in a transaction either to have a minimum level of literacy or to hire (and trust) a scribe who could explain the document. A handful of places, such as Teos, strove to achieve something like universal education, yet they were certainly exceptions and had little influence on learning elsewhere.[62] Alexandria, a city that provides much evidence for uses of writing and literacy, likely had a rate of literacy higher than much of the rest of the Hellenistic world.

The diffusion of literacy in Rome progressed differently from that in Greece. Writing spread from Greek colonists to the Etruscans and other Italic peoples from an early date. An archaic gravesite from Pithecusae (modern Ischia) in the Bay of Naples has yielded a find known as "Nestor's Cup," a drinking vessel bearing an inscription and dating to the end of the eighth century BC. In northern Italy, the introduction of the alphabet was facilitated by the Etruscans, but Italic peoples in the south, such as the Sikels and Oscans, acquired the alphabet directly from Greek colonists.[63] Etruscan texts date from the seventh century BC, many of them *instrumenta domestica* buried with their owners.[64]

Evidence for literacy in Rome is quite scarce until about the time of Plautus, at the end of the third century BC. Latin writing before Plautus includes some fragments of religious texts, such as the famous Lapis Niger, an inscription on black marble found near the *curia* and dating to the sixth century BC. The Roman legal text known as the Twelve Tables is ascribed to the fifth century BC, although the fragments that have come down to us are likely later modernizations of the text.[65]

By the end of the third or the beginning of the second century BC, substantial texts begin to appear that shed light on the nature of Roman literacy, which differed in some significant respects from literacy and its spread among the Greeks. As Woolf notes, the Roman aristocracy's reliance on slaves and freedmen as managers of estates and as representatives in other forms of personal business necessitated a high level of literacy among some in the lower classes.[66] This fact is supported by Cato's advice in the *De agricultura*[67] that an owner should audit written accounts of goods upon visiting his villa and leave behind written instructions for the manager.

These different types of literary activities as well as the need to communicate by letter suggest a more socially unified enactment of literacy among the Romans than among the Greeks.[68] Even if literacy was more diffuse among social strata, it was not necessarily more prevalent. Among the Romans literacy was probably achieved by only a small percentage of citizens. This is likely to be true despite apparent evidence to the contrary, such as Petronius' *Satyricon*, which features several instances of freedmen reading in a famous vignette known as "Trimalchio's banquet." Just as a Roman would need a literate manager for his estate (*latifundium*), it is not surprising that there should be a class of educated freedmen as Petronius depicts.[69]

Literacy occupies a particularly interesting place within Roman religion. From an early period, Roman religious colleges kept annual written records. The most famous of these, known as the *Annales maximi*, contained information regarding religious activities important to the *pontifices* (priests who managed the state cult and were authorities on religious law) as well as the business of the senate and *comitia*, including the passage of laws.[70] The Regia housed the *tabulae dealbatae*, which were used for the temporary display of the pontifical annals and which were the source for the *Annales maximi*.[71] The contents of the books of other colleges, such as the *augures* (who handled issues relating to divination), *quindecimviri sacris faciundis* (who kept and consulted the prophetic Sibylline Books), and the Arval brotherhood (responsible for the Dea Dia festival), are unknown, as is the manner of their use for religious or legal purposes.

Perhaps the purpose of such books was, in part, to accord authority to religious leaders, for only the priests would have had access to written information pertaining to public law and religion.[72] The strictness of Roman religion, in which a minor mistake could require repeating a ritual act, suggests that texts would be useful to those learning how to perform rituals with precision.[73] In an act disparaging of aristocratic power, Caesar as consul in 59 BC ordered the documentation and publication of senate proceedings.[74] If the theory of a largely singular form of literacy at Rome is correct, this tradition may indicate that literacy had a role in transferring some power to the people, even if the overall number of readers was quite low.

Oral publication of laws, read out by a herald, preceded their written form on stone, and formulae in inscriptions emphasize the importance of witnesses and oral expression of agreement. Perhaps implicit in legal inscriptions is the assumption, occasionally overtly expressed, that a reader would recite the text to an interested illiterate citizen.[75] Noteworthy, too, is the fact that not all laws or senatorial decrees were set up as inscriptions; written forms of legal publication were secondary to oral forms.[76]

As was the case in Athens, political and military uses for literacy increased with the growth of Rome's territory. Meanwhile private transactions could involve — but by no means required — documentation. Like those in Greece, Roman authorities relied upon writing for communication, record keeping, taxation, and various other functions. Undertakings such as the census would have been unimaginable without the aid of the written word.[77] In the imperial period, the Roman military made heavy use of writing: inscriptions recording rosters, pay records, and accounts of arms and provisions survive in abundance.[78]

The prominent role of writing in managing the Roman empire suggests that literacy was indispensable for magistrates and was also likely high among soldiers, perhaps disproportionately so relative to private Roman citizens. The publication of laws and senatorial decrees on stone or bronze was expensive and not used by Roman officials for widespread communication; rather, copies of documents would be deposited in local archives. Public inscriptions of documents on stone or metal were largely symbolic, as the official copies for consultation were those in the archives.[79] Public notices were communicated in writing using painted wooden tablets, which would be whitewashed and reused as needed.

Unfortunately our knowledge of writing and literacy in antiquity is sufficient only for drawing broad conclusions. Even so, the question of literacy will benefit from a more nuanced approach than is allowed by the simplistic division into categories of "literate" and "illiterate." The fact that not all individuals read with the same proficiency will provide an important context for the dissemination of literary texts. As we will see, the rolls and codices onto which texts were copied varied in quality, and therefore value and purpose. Like literacy rates, the readership of literary texts in antiquity should not be considered straightforward or static. Let us begin our foray into ancient literary texts by considering some additional materials and their associated fields.

Among the variety of materials used for writing, two in particular, stone and papyrus, have survived in substantial amounts. The former is highly durable and can preserve legible writing even when damaged or repurposed. Papyrus, by contrast, though durable, is prone to damage from moisture; arid climates have preserved great troves of texts on papyrus. Because texts on these two materials are so common, each has a field of study devoted to it; stone inscriptions are the principal object of study for epigraphy, while papyrology attends primarily to texts on papyrus. Nevertheless, not all inscribed objects belong to the field of epigraphy, as inscriptions on coins, for example, have their own field, known as numismatics.[80] It is sometimes difficult, then, to draw clear boundaries between the various disciplines concerned with ancient forms of writing. Both epigraphy and papyrology, concerned as they are with the materials of ancient writing, bear further consideration.

EPIGRAPHY

Epigraphy is most closely connected with inscriptions on durable materials intended to last indefinitely; some inscriptions not intended for posterity, however, such as graffiti from Pompeii, do fall to the epigrapher to study.[81] In the treatment of epigraphy here, our interest will center on the examination and interpretation of texts on stone.

Inscriptions served a variety of public and private purposes in antiquity, many of which can be grouped for convenience into general categories. The largest category, epitaphs (more formally known as *tituli sepulchres*), encompasses about 66 percent of known inscriptions.[82] Epitaphs vary between small stone memorials and elaborately decorated monuments, many of which can be dated according to the formulae used in their texts. The information about the deceased or surviving family members was variable; some Greek funerary inscriptions are no more than the name of the deceased, perhaps followed by the word for farewell (χαῖρε) or a patronymic. Others were in verse.[83] Roman epitaphs (and those in Greek from the Roman period) were also diverse in content, but it was typical for the dedication to include the name of the surviving family member who paid for the tomb and the name of the deceased, whose full name and father's name would be provided.

Another group of inscriptions are *tituli sacri*. These record dedications to a god or goddess and are often found in a temple precinct. Like epitaphs, dedications range from simple to very ornate. They might list no more than the divinity's name with the notion of dedication implied or could contain an elaborate explanation of the benefit the dedicator received. Dedications, especially small votives, are important sources of evidence, sometimes even helping to identify the deity to whom a nearby temple was dedicated. One such example is that of the temple of Hera at Paestum. This temple, known as Hera II, is the largest of the three in the city. While it was previously thought to have been dedicated to Poseidon, after whom the city's Greek name was fashioned, votive offerings found in the area suggest that the temple was in fact dedicated to Hera.

Inscriptions honoring men are differentiated from those honoring the gods. These *tituli honorarii* are found on Greek statue bases, most commonly with the dedicatee's name in the accusative case.[84] The Romans imitated the Greek practice of honoring men with statues, a debt evident in the fact that early Roman honorary inscriptions might place the honorand's name in the accusative case, though similar types of inscriptions use other cases, especially the dative.[85] Honorary inscriptions were at one time common in Rome, but most have been lost. The Forum of Augustus was lined with images of famous Romans, each of which was identified with a commemorative inscription known as an *elogium*, some of which are known from copies.

Similar to honorary inscriptions are those found on public buildings. These include many of the most visible inscriptions, such as those on temples, arches, theaters, and other public structures.[86] Inscriptions on structures often indicate who built or restored the monument, one of numerous possible displays of beneficence typical of Roman magistrates to commemorate their year of service.[87] Among the most famous is the brief inscription on the Pantheon in Rome, which simply states "Marcus Agrippa, son of Lucius, thrice consul, built it."[88] Some public monuments were constructed by individuals to honor a family member, and this intent would be recorded in an inscription on the edifice. Of course, numerous monuments, most notably the triumphal arches in Rome, were decreed by the senate in honor of the achievements of Roman emperors.

All the inscriptions described thus far are of a public nature, and have the specific purpose of explaining the function of the monument upon which they were carved.[89] Inscriptions could also be of a documentary nature, and, provided they are recorded on stone, metal, or a similarly durable material, they remain part of the body of evidence encompassed by epigraphy. Decrees, laws, and treaties are all forms of documents that might be published on stone. These documents were not created for record-keeping purposes, for which tablets better served, but rather as a means to commemorate an individual's civic service.[90] In such cases, the inscription was sometimes erected at the individual's expense. In other cases, local governments, especially in the provinces, ordered the creation of documentary inscriptions, because such public displays were an important means of demonstrating a relationship with Rome.

Historical interest in ancient, especially Latin, inscriptions dates back to the beginnings of the Renaissance, a point that will be examined in greater detail below. The earliest transcription of ancient monuments was undertaken as early as the ninth or tenth century AD. Published collections of transcriptions emerged as the result of observations made by traveling scholars. These collections were at first arranged regionally but were eventually displayed by type.[91] The former system of organization would become dominant with the publication of the *Corpus Inscriptionum Latinarum* (*CIL*), which marks the beginning of modern epigraphy. The history of the *CIL*, however, opens with the unsuccessful *Corpus Inscriptionum Graecarum* (*CIG*).

In a defining moment for epigraphy, August Boeckh undertook to gather all known Greek inscriptions into a single collection, known as the *Corpus Inscriptionum Graecarum*, whose first volume appeared in 1825.[92] Boeckh could not keep up with the pace of publication by epigraphers, and so the *CIG* and what is now known as the *IG* (*Inscriptiones Graecae*) remain incomplete.[93] Boeckh's work was important, however, because it provided a model for Theodor Mommsen's very successful *CIL*, which remains an extremely important epigraphical tool.[94]

It is the task of the epigrapher to record details about inscriptions, such as their provenance (findspot), dimensions, and the material on which they are inscribed, as well as to interpret the text. The initial publication of an inscription (known as the *editio princeps*) is expected to contain as much detail as possible;

often context is as important as content for dating and interpreting the monument. The issue of provenance, for instance, must take into account where an inscription was found and whether it was found in its original location or had been moved, assuming the findspot is known. The material itself also requires detailed analysis. For example, in some instances monumental inscriptions originally made with bronze letters can be reconstructed based on the remaining pattern of holes used to attach the original letters.[95] The surface of the stone may also show evidence of erasure, as is the case on the Arch of Septimius Severus in the Forum Romanum.

Another crucial aspect of epigraphic publication is the inclusion of an image of the inscription in question. This can be done, as it has been for centuries, using line drawings. Modern epigraphers are also able to take high-resolution photographs that provide a more accurate (though sometimes more difficult to read) representation of the inscription. Photographs are a useful component in an inscription's analysis inasmuch as they accurately record physical characteristics of the stone, such as its color, that are otherwise impossible to reproduce. Even the highest-resolution image has limitations, however. Some of the physical characteristics of an inscription can be better discerned through the practice of making a "squeeze" of the stone, that is, an impression of the inscription on paper.[96] In making a squeeze, specially treated paper is dampened and laid over an inscription. As it dries, it hardens and retains the three-dimensional image of the inscription into which it has been pressed.[97] Because the paper takes the shape of the incised surface, squeezes allow close analysis of inscriptions and are an important complement to photographic records.

Reading and interpreting inscriptions can present complex challenges. Depending on their date and dialect, Greek inscriptions may not conform to the Ionic alphabet, which came into widespread use in Greece in 403/2 BC and is familiar to most readers of ancient Greek. As noted above, the Greek alphabet was not uniformly adopted, a reality reflected in the differing alphabets and orthography found throughout the Greek world. Latin inscriptions are not necessarily as dialectically challenging as they are highly formulaic. Many formulae were condensed into standardized abbreviations. For example, DM, for *dis manibus* ("to the divine shades"), can be found on many epitaphs. In studying either Greek or Latin inscriptions, the epigrapher must contend with damaged or missing pieces of the stone while simultaneously attempting to decode errors in grammar or spelling.

In order to make the texts of inscriptions as accessible as possible, universal guidelines for proper publication have been established among epigraphers. These guidelines comprise the Leiden system, which was issued for the fields of papyrology and epigraphy in 1931. This standardized practice requires, among other things, that all abbreviations be written out, with supplied letters in parentheses.⁹⁸ There are similar rules for letters that are furnished as a reconstruction of the text, extraneous letters carved in error, and other situations that arise in establishing the text of an inscription.

The methods used to date inscriptions vary in reliability and are obviously dependent on the amount of detail provided on the stone. Such a date can be no more than a *terminus ante quem* (the latest possible date) or a *terminus post quem* (the earliest possible date). An epitaph bearing the words *dis manibus*, for instance, can be roughly dated in or after the middle of the first century AD.⁹⁹ The most reliable means of establishing a date is when a known historical reference occurs in a text. Though such a reference could denote a vital historical moment, most often it is simply the mention of the name of a magistrate or emperor.

The ancients themselves dated some inscriptions using the names of the leading magistrates for that year. Thanks to the discovery of a fragment of the *Fasti consulares*, which is a list of Roman consuls, and the fact that the lives of Roman emperors are well attested, it is possible to date many inscriptions that provide such indicators. In Greece the situation is more challenging. Each Greek city had its own magistrates — archons in Athens, ephors in Sparta, and so on — which limits the efficacy of historical knowledge for dating. For instance, the archons from Athens in the fifth and fourth centuries BC are relatively well attested, but this information provides no aid in dating inscriptions from the same period outside of Attica.¹⁰⁰ The Greeks made at least one attempt at a universal system for dating, which they based on the victors at the Olympic games. The system, found mostly in literature,¹⁰¹ is decipherable to scholars thanks to Hippias of Elis, a Sophist, who compiled a list of Olympic victors.¹⁰²

Another means of dating an inscription is by its inclusion of a reference to an identifiable individual (apart from eponymous magistrates or Olympic victors). This could be a historically famous individual, but often is someone known from other inscriptions. The number of different individuals named in inscriptions is tremendous, and it is not unusual for an individual to be attested in more than one place. The analysis and organization of this vast sea of information falls to the subfield of prosopography. The prosopographer compiles and collates names mentioned in inscriptions, interprets data, and classifies by type the information gathered. Examples of such classifications include Roman Epicureans, Roman senators, Roman women associated with a particular religious order, etc. Prosopography is an important tool for dating inscriptions that might mention an individual whose life cannot be dated from another source. The field also uncovers important details about ancient society generally, for example, the particulars of government administration or the lives and careers of even minor individuals.¹⁰³

Overlapping somewhat with prosopography is the field of onomastics, which addresses issues related to names in general rather than specific individuals. By compiling data on when and where names occur with frequency, scholars can approximate a date for an inscription or infer an individual's place of origin.[104] Since so many epitaphs are extant, they are a particularly useful source of data for prosopography and onomastics.[105]

Finally, approximate dates can be established on the basis of known formulae and letterforms, which falls under the field of paleography, to be discussed below.

It seems that the epigrapher has numerous tools available for dating and interpreting a text, but in fact the process can be quite difficult and the evidence contradictory. The well-known Themistocles Decree provides an excellent example of these challenges. The Themistocles Decree is an inscription on a marble stele that was found in Troezen, near Attica. The text gives instructions for the evacuation of Athens before the Persian invasion in 480 BC. The inscription includes a common formula and cites the famous Athenian general by name.[106] The location of the find, the formulaic phrasing, and the identification of a known figure all point to the fifth century. There are some problems, however. From a paleographical standpoint, the letterforms and spelling (orthography) on the decree appear to come from the fourth century, not the fifth.[107] The decree also contradicts historical evidence: Herodotus indicates that the Athenians evacuated Athens suddenly, following the Battle of Artemisium, when the Spartans failed to gather the Greek forces in Boeotia, where the Persian advance could be halted before it reached Athens.[108] The inscription, however, purports to have been passed (and the evacuation ordered) before the Battle of Artemisium.[109]

Is Herodotus wrong, or is the inscription a forgery? How should the competing evidence be understood? At this point the interpretive challenge of epigraphy becomes apparent, and requires the aid of history and paleography to advance, if not solve, the issue. The formula used in the inscription provides only a *terminus post quem* and does not guarantee that the stone is not forged. In fact, forged inscriptions are quite prevalent; more than ten thousand forged inscriptions were identified in *CIL* at the beginning of the twentieth century.[110] The letterforms date at the earliest to the fourth century, leaving open several possibilities: the decree could be a fourth-century copy of an original or an invented account of the evacuation. As to the historical evidence, Herodotus' account is not incontestable and could reflect a later impulse to depict the Athenians in dire straits[111] or simply be due to the vagaries of even the contemporary memories upon which Herodotus had to rely.[112]

The Themistocles Decree, then, is generally dated to the fourth century and not considered an original inscription recording the order to evacuate. The questions of whether it accurately depicts events from the fifth century, and the reason for its creation in the fourth century, remain open. It is true that the Themistocles Decree deserves special scrutiny because it addresses what was already in antiquity a momentous event in Athenian history, but the types of questions it raises are indicative of the process of dating and interpreting ancient inscriptions.

PAPYROLOGY

From epigraphy we turn now to papyrus and its eponymous field, papyrology. Papyrus holds a privileged place in the history of writing, which is perhaps best summed up by Pliny the Elder: "through papyrus especially does civilized society, and certainly memory, endure."[113] Pliny's comment highlights the incredible demand for papyrus in first-century Rome, but the truth of the statement extends far beyond Pliny's milieu. In fact, the importance of papyrus for writing dates back millennia even from Pliny's time, and was not, as Pliny almost comically observes in his *Naturalis historia*, owed to the conquest of Alexander.[114]

Papyrus was the writing material of choice for most purposes, apart from permanent public displays, which were engraved on bronze or stone. For most other writing, papyrus was likely to be used, and it falls to the papyrologist to read, describe, and publish extant fragments. In particular, the papyrologist details the provenance of the fragment, when possible, as well as its paleographical features. Before initial publication, an *apparatus criticus* is prepared.[115] The papyri contain a great variety of subject matter; most are documentary—that is, they record various aspects of private and public life, especially as pertains to governmental administration, legal matters, and private transactions. A much smaller group, perhaps about 10 percent of known papyri,[116] are of a literary nature.[117]

All ancient papyri fall, at least to some degree, within the purview of papyrology. Though the term "papyrology" suggests an exclusive concern with writing on papyrus, the specifically historical and cultural interests of the field encompass similar types of documents that happen to appear on other materials, such as potsherds and tablets.[118] The field of papyrology, then, extends beyond the physical material suggested by its name and, in addition to the translation and publication of texts, also involves interpretation. Nevertheless, the extent to which the role of the papyrologist is interpretive is generally limited to cultural, social, or historical aspects and thus often overlaps with other fields, especially history. The interpretation of literary papyri falls to philologists and literary scholars, since those texts are part of a literary tradition at the center of the field of philology.[119]

The divisions between fields, however, are problematic; due to the special

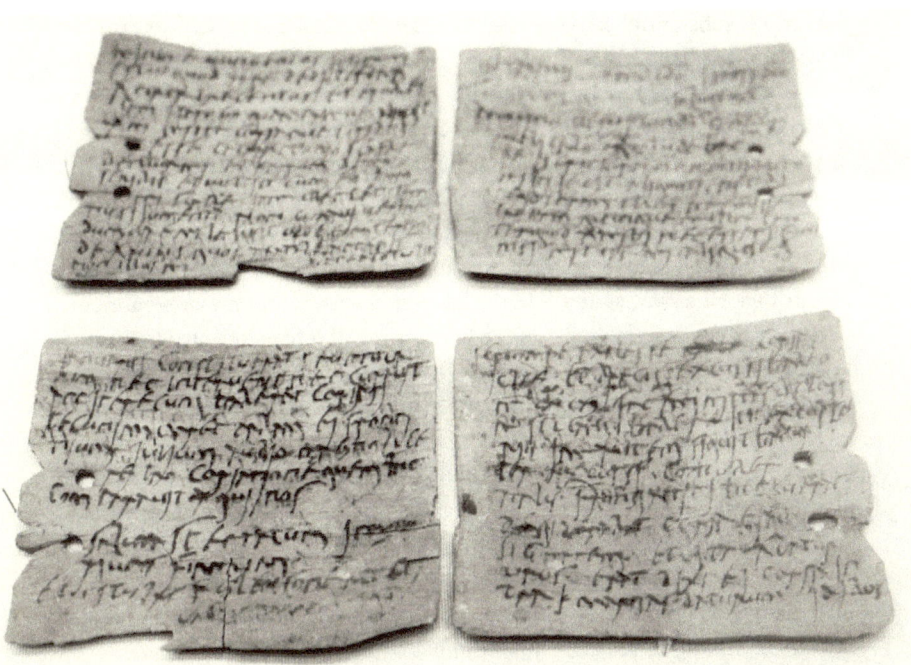

FIGURE 1.4. *Roman leaf-style writing tablet from Vindolanda in Northumberland. Tablet 343: Letter from Octavius to Candidus concerning supplies of wheat, hides, and sinews. First–second century AD. London, British Museum.*

training needed to decipher and properly understand papyri, the value of the evidence is not always fully accessible to all scholars. Even within the field of papyrology, the variety of languages in which the papyri are written can make it difficult for scholars without a command of Greek, Demotic, and Coptic, for example, to synthesize all of the papyrological evidence for a given place or time.[120] Despite these challenges, papyrology has vastly increased scholarly understanding of antiquity due to the thorough integration of papyrus into the workings of the Greco-Roman world.

Papyrus was in use as a writing material in Egypt at least as far back as 3000 BC, long before the Greeks used it as their preferred medium for the written word, and remained in use until about AD 1000.[121] Despite the considerable importance of papyrus in Greece and Rome for literary pursuits, governmental administration, record keeping, and private transaction, very little of the massive amount of papyrus consumed in antiquity remains due to its susceptibility to mold and decay. Like the wooden tablets from a Roman fort along Hadrian's Wall, Vindolanda (see fig. 1.4), accidents of preservation have led to the discovery of fragments of papyri, even in relatively humid climates. In fact, the earliest substantial papyrus discoveries were made from 1752 to 1754 at Herculaneum, where mud from the eruption of Vesuvius preserved fragments of rolls from a villa owned by Calpurnius Piso. At present, Piso's villa, known as the Villa of the Papyri, has been only partially excavated; the state of the fragments makes it difficult to estimate the number of volumes represented, but there could be as many as one thousand rolls.[122]

Discoveries in climates like that of Italy are exceptional, however, and most papyri have been recovered from the sands of Egypt, specifically from arid regions of the Nile river valley.[123] In the late nineteenth century, a large number of papyri from the Fayum area came to light, much of it from the city of Crocodilopolis (which was renamed Arsinoe by Ptolemy Philadelphus).[124] These papyri varied in date and language, though the majority of them were from the Roman period. Perhaps the most famous papyri find, however, is that of Bernard Pyne Grenfell and Arthur Surridge Hunt in 1897 at the ancient city of Oxyrhynchus, about two hundred miles south of Alexandria on the edge of the desert. A massive number of papyrus fragments were discovered in an ancient rubbish heap over the course of six seasons of excavation.[125] To date, a large majority of the many papyri fragments have not been published, even as the Egypt Exploration Society continues to produce new volumes of *The Oxyrhynchus Papyri*.

The abundance of papyrus finds from Arsinoe-Crocodilopolis, Oxyrhynchus, Herculaneum, and other places have done much to reconstruct the picture of daily life in ancient Greek and Roman society. The interpretation of the evidence is not without challenges, however, especially since the papyri that are discovered are disproportionately concentrated in place, such as Oxyrhynchus, and time, typically the Roman period, leading to concerns about how indicative the material is for Egypt, let alone ancient society at large.[126] Another challenge to interpretation is a lack of data: there is currently no cumulative record of published (or unpublished) papyri, which forces scholars to rely on rough estimates for the categorical distribution of the material.[127]

Despite its long history of continuous use, papyrus seems not to have made its way to Greece until relatively late—perhaps the early sixth century BC—and it likely did not arrive in Rome until even later, assuming, as seems plausible, that Greek colonists introduced papyrus into Rome.[128] No evidence exists for when papyrus first came to Greece, though references to an established book trade in Athens appear in, among others, Aristophanes and Plato,[129] suggesting that the material had already been in use in Greece for some time. Pottery from the sixth century depicting papyrus rolls as well as Greek contact with Egypt through the trading city of Naucratis further suggest, at the latest, a sixth-century date.[130] Alternatively, however, papyrus may have come to Greece at the same time the alphabet was introduced, making Phoenicia the source for both Greece's letters and the material on which it was written. In fact the Greek word for "book," *biblos*, conceivably derives from the name of the Phoenician city of Byblos, which may have served as the trading center from which the Greeks first obtained their papyrus rolls.[131]

Ennius was perhaps the first Latin author to write with the intention of publishing and circulating his works in book form. He divided his *Annales* into books, with the amount of text in each corresponding to the length of a papyrus roll. A fragment from the *Annales*[132] contains the technical term for publication (*edere*) and a papyrus roll (*charta*). Previously Naevius had not divided his epic poem, *Bellum Poenicum*, into books. Instead it was the grammarian Octavian Lampidio who, at the end of the second century BC, published Naevius' poem according to book divisions.

The papyrus roll was in common use in the age of Lucilius, who makes reference to it in a fragment from his *Saturae*.[133] Also on papyrus were the so-called Books of Numa, which were discovered on the Janiculum Hill in 181 BC. These were revealed to be forgeries (and burned), as they had not deteriorated as much as they should have if they had been buried for four hundred years.[134] This sort of observation presupposes a degree of familiarity with papyrus. The whole of subsequent literary production until the third century AD relied on papyrus rolls for circulation. Livy makes this point explicit when he groups the several *libri* that constitute his work into *volumina*—that is, papyrus rolls.[135]

As noted above, the vast majority of papyrus was used for documents rather than literature. Thus, for many individuals, the need to use papyrus was not contingent upon their ability to read or write.[136] On the contrary, any situation in which an individual might need to demonstrate ownership of property or the fulfillment of a debt would require writing, whether on the part of the individual or a proxy, such as a literate family member or scribe.[137]

Since the climates of Greece and Italy are generally unsuitable for the preservation of papyrus, little physical evidence remains from places like Athens and Rome to attest the deep reliance upon papyrus in many aspects of society. In these cities, much like those in Egypt, the demand for writing material among much of the population must have been considerable. Pliny the Elder provides some insight into the Roman need for papyrus when he reports that a crop failure caused a severe papyrus shortage in Rome under Tiberius, who appointed a special committee of senators to oversee distribution of the material.[138] Pliny's anecdote reveals not only the importance of papyrus, but its potential for shortfall as well.

Though there were some occasions when the availability of papyrus was limited, the material does not generally seem to have been scarce. The affordability of papyrus is difficult to determine with accuracy. It has been estimated to cost what a laborer would in earn in one or two days, a price of little consequence for the wealthy but a considerable burden for anyone on subsistence living, though such people would have had little need for papyrus and could use potsherds or other found materials as a substitute.[139]

Turning papyrus into paper was a labor-intensive process that yielded a sheet of papyrus (Greek, *kollema*; Latin, *pagina*) of varying quality, depending on the part of the pith used to make the sheet. Pliny the Elder describes the process of paper production and the grades of paper that result.[140] Pliny's description is not entirely clear, so much so that it has been speculated that Pliny never personally observed the production process but instead reports the results of his research on the matter. Nevertheless, Pliny's insights are valuable, and modern research on extant papyri has done much to shed light on Pliny's description.

In the fashioning of a papyrus sheet, the pith of the papyrus plant is divided into thin strips, of which those taken from the center produced the highest-quality paper; strips taken further from the center yield paper of progressively lower quality. Once the strips have been cut or peeled from the stalk, vertical strips are placed side by side to form the verso of a sheet. The vertical strips are covered by horizontal rows of papyrus to form the recto. Though both sides of the sheet could receive ink, the recto was the preferred side for writing; to write on the verso required writing across the vertical seams of the papyrus strips. Once laid out, the vertical and horizontal strips would then be pressed together, causing them to bond together without the need for additional adhesive.[141]

The papyrus sheets so produced were of varying quality, depending not only on the part of the papyrus pith used but also the technique used to make the sheets; poor manufacturing could lower the quality. A high-grade sheet was distinguished by its slim yet firm quality as well as its smooth texture and bright sheen. Further, each grade of papyrus had a standard width, and the better grades were wider: wider sheets meant fewer joins between sheets on a roll and thus a smoother writing surface. Pliny does not describe a standard height for different grades of papyrus, though the available data reveal that a high-quality roll fit for literary uses would average 25–32 centimeters in height.[142]

We owe to Pliny, too, the names of the various grades of papyrus. He labels the best *papyrus Augusta*, and this was followed by *papyrus Liviana*, in honor of the first emperor and his wife. The next level, *hieratica* (derived from the Greek word for "holy"), is so called, Pliny writes, because the paper had been used for religious writings. The next three grades of papyrus, *amphitheatrica* (or *Fanniana*), *Saitica*, and *Taeneotica*, were all named for their place of production, the first referring to the workshop in Rome of a certain Fannius, and the others for towns in Egypt. Finally, the lowest grade was called *emporitica*, which Pliny deemed unacceptable for writing. It was instead used for envelopes and wrappings for goods by merchants, from which it took its name (*emporus*, "merchant").[143]

Once the papyrus sheets were made, those intended to be used for book rolls would be joined together, usually using papyrus sheets of the highest grade, although there are exceptions extant.[144] A recent example of middling-quality papyrus, in this case *Fanniana*, comes from the Milan Papyrus, a long fragment of a Ptolemaic book roll containing epigrams of the Hellenistic poet Posidippus that had been reused as mummy cartonnage. The handwriting is skilled, but not excellent. The lettering is small and the intercolumniation is narrow, suggesting that this book was not intended to be a display piece.[145] Though the original size of this now-fragmentary roll is unknown, a typical length for a papyrus roll has been estimated at six to eight meters,[146] but they could vary considerably in this regard. Johnson goes so far as to assert that the notion of a "typical" length is a misnomer; in fact, a roll from the Herculaneum papyri has recently been estimated at twenty-one meters in length.[147]

The amount of text that a roll could hold depended on a number of factors, including its length, the size of the lettering, and the width of the space between the columns, but on average rolls would be long enough to contain between one and two thousand verses. Rolls of insufficient length could be extended easily by gluing more pages—or even another roll—to an existing roll. This task may have fallen to the retailer rather than the manufacturer, as implied by Pliny's insistence that rolls did not exceed twenty sheets in length.[148] The recent discovery in 2005 of a treatise by Galen entitled *On Grief* (*Peri alypias*) sheds further light on the typical length of a book. Galen says that he has forty-eight large books (rolls), some of which hold more than four thousand lines and need to be divided into separate volumes, some for public use and others for his own personal use.[149] The implication seems to be that the more usual size was a roll of two thousand lines, given Galen's inclination to divide the large rolls into two volumes, a size he considers "well proportioned."[150] Further, the rolls of four thousand lines seem limited to Galen's personal use rather than intended for circulation.

One theory regarding the length of rolls, championed by Birt,[151] is that literary works were designed to fit onto rolls of standard length. To take a limited example, one might observe that collections from the Hellenistic period, at least, appear to conform to a standard length. Hellenistic arrangements of Sappho (1,320 verses), Bacchylides (1,400 verses), and Pindar (*Olympians* [1,562 verses], *Pythians* [1,983 verses]) all would fall within the range of a single roll. Hellenistic authors' own works also typically fall within the limits of an average roll.[152]

Despite the circumstantial evidence, there are some problems with this theory. One is that it was a relatively simple matter to glue additional sheets to a papyrus roll, and so it would seem unnecessary to tailor a creative work to a particular length of roll. The presumption that authors designed their works to fit a certain length also assumes a rather odd priority for the structuring of creative works.[153] The length of a work surely came into an author's consideration, though more likely as, for example, a feature of the work's genre.

Aside from the dimensions of a typical roll, a few other features are worth noting. The first sheet of a roll was known as the *protokollon*. Since this was the outermost sheet and subject to the most wear and tear, it was often left blank as a protective cover, and could be reinforced with additional papyrus.[154] In fact, it would seem that the Milan Papyrus had its *protokollon* replaced, based on the manner in which it was attached to the following papyrus sheet.[155] Once assembled, a label (*sillybon*) would be added and, possibly, a protective parchment cover.[156] It is unclear how common *sillyba* were, and it is unknown whether they were added by the owner or the manufacturer of the volume; without them, one would have had to unroll the volume to learn its title.[157]

For rolls designed as high-value luxury items, rollers (*umbilici*) with ornamental knobs (*cornua*) could be added.[158] Whether such ornate texts were intended for reading or merely for display is unclear, but certainly more affordable, utilitarian rolls were available for avid readers. In addition to their aesthetic value, *umbilici* also offered additional protection to the fibers of a papyrus roll. Extant examples of *umbilici* are few, however, and it is difficult to say how commonly they were used.[159] The ease of reading a roll depended upon its size, and it is possible that the act of unrolling and rerolling would have made looking for particular passages rather difficult and tedious. For the purpose of skimming the text, the codex form would prove far superior.

Unlike wax tablets, which could be reused by reshaping the wax writing surface, papyrus was inscribed with ink and therefore could not be easily erased. Corrections were relatively easy to make during the writing process, because wet ink could be wiped away with a sponge. Once it dried, however, removing ink would require scrubbing the papyrus. Occasionally, a large amount of text was removed so a papyrus could be reused, resulting in what is known as a "palimpsest." Examples of palimpsests are extant but nevertheless rare. One reason for their paucity is simply that not all papyri were suitable for erasure; the papyri of the highest quality were too thin and fine to allow the ink to be scrubbed away.[160] Even if it were possible to erase a papyrus roll, the result would be a roll of lower quality, and thus reduced value, especially since it would be impossible to remove the ink completely from the original text. The process was thus impractical from a bookseller's standpoint.[161] Given the general affordability of papyrus, there was little need to go through the effort of erasing texts.[162]

Another more practical option for repurposing a papyrus roll was to write on its obverse. Such a roll is known as an "opisthograph" and, like palimpsests, is relatively rare among papyri finds.[163] One striking example of an opisthograph comes from the third-century estate of Aurelius Apianus in the Fayum. Heroninus, the estate's administrator, repurposed documents, some of them over a hundred years old, for bookkeeping.[164] While this incidence of writing on the obverse of a papyrus reveals that such recycling was practiced, the high number of non-opisthographic rolls found in Oxyrhynchus must caution against the assumption that opisthographic writing was a standard practice.[165]

Martial provides evidence that opisthographic writing in Rome was looked down upon, at least for literary works.[166] In one poem, he playfully addresses his book of poetry, telling it that if it meets with Apollinarus' approval, it need not fear being used as a wrapping for fish, but if it is rejected, it will go straight to the fishmonger's bookshelf for boys to scribble on its back.[167] It seems, then, that rolls intended as a work of literature, especially if it was to be sold, held writing only on the *recto*. Of course, the individual who owned a papyrus might well choose to write another text on the *verso*, perhaps for reasons of economy or availability.[168] If the papyrus was very thin, however, as was the case, according to Pliny, for literary-grade papyrus, writing on the *verso* could have been impossible, or at least difficult and unattractive, given the tendency of the ink to bleed through the page.[169] Examples of literary rolls repurposed to contain additional literary texts are extremely rare.[170]

Copying books was a time- and labor-intensive process, for which most of the evidence comes from the medieval period. The work of monastic scribes will be more fully addressed in the next chapter. Of scribal practices in antiquity, however, we hear almost nothing from the literary sources, but some conclusions can be drawn from observations of extant papyri. For instance, extrapolating from the types of errors found in the papyri and a comparison of texts' line and column lengths relative to their exemplars, it seems that ancient scribes typically made copies of text from dictation.[171] The uniformity of column width and intercolumniation, especially in the Roman period, indicates that scribes were well trained.[172] Despite the careful attention given to the size of columns, no attempt was made to avoid writing over the joins (*kolleseis*) between papyrus sheets.

A rather odd feature of ancient "professional" book rolls is the tendency of the scribes to write in slanting columns. From the first line of a column at the top of a roll, each subsequent line begins slightly closer to the left, causing the column to lean to the right. This tendency, now termed "Maas's Law," may have been a deliberate aesthetic choice, although it has also been proposed that the slant is the inadvertent result of scribes copying a text as it rest on their knees.[173] Marked guides for the scribes are rare in the papyri, but not entirely absent. A few papyri show a faint vertical line of dots along the left-hand margin of the column, which would have helped the scribe write a perfectly square column (not all high-grade books follow Maas's Law). The standards for quality book production in antiquity were high, and such items could not be produced by scribes without training, a point to which we will return below.

PAPYRUS TO PARCHMENT

The use of both papyrus and parchment date to an early period in Egypt. The earliest evidence of writing on parchment comes from an Egyptian hieroglyphic image from the third millennium BC.[174] For reasons not entirely clear, however, papyrus emerged as the preeminent writing material. Both parchment and papyrus are durable materials suitable for writing, yet parchment offers a number of advantages: its writing surface is entirely smooth, it lasts longer than papyrus, and the livestock needed to produce it were widely available. The papyrus plant, on the other hand, grows chiefly in Egypt and a few other regions of the Mediterranean basin.[175] Despite the comparability—some might argue superiority—of parchment, for centuries papyrus remained unrivaled as a material for writing.

Parchment is made from animal skin, typically sheep or goats (the finer material known as "vellum" is from the hide of a calf or kid).[176] The process for producing parchment was different from the tanning process; the hide was prepared by soaking it in a caustic calcium lye or other materials, such as egg, bran, or even feces, any of which could change the hide's alkalinity.[177] Such a change in alkalinity facilitated the removal of hair and fat, which were now easily scraped off.[178] The skin was then washed thoroughly before being tightly stretched while it dried. As this stretched skin dried it would shrink and turn into parchment.[179]

Roberts and Skeat speculate that the challenges related to the production of parchment gave papyrus the edge between the two writing materials; even as the production and distribution processes were becoming more refined, the established infrastructure of the papyrus industry made it increasingly more difficult for parchment producers to compete. By the time the process for parchment production had been worked out, an active papyrus trade already dominated the market. Its relatively low cost may have been a factor as well; still, there is no reliable evidence to establish a dependable comparison of the price of papyrus with a comparably sized sheet of parchment, and thus one cannot simply assume that parchment would have been more expensive.[180]

The preference for papyrus over parchment can be inferred from an anecdote related by Pliny the Elder on the authority of Varro. Pliny reports that parchment was discovered at Pergamum when Ptolemy, the king of Egypt, cut off their supply of papyrus rolls because of a rivalry he had with King Eumenes over their libraries.[181] That parchment was invented in Pergamum is certainly false, although it might be true that Egyptian papyrus rolls were not reaching Pergamum. The library at Pergamum was established under Eumenes II, during whose reign Egypt was invaded by Antiochus IV. Egypt's war with Antiochus (173–169 BC) may have curtailed the production of papyrus, causing a shortfall that forced cities like Pergamum to find a substitute material for writing. The solution may have been parchment, which was not suddenly discovered in Pergamum but rather would have come to prominence out of necessity. Soon after it was brought to Rome, where that same war may have also created a dearth of papyrus.[182] Pliny goes on to suggest that the use of parchment remained heightened even after the papyrus shortage subsided.[183]

Nevertheless, parchment did not truly rival papyrus in popularity until much later, and the emergence of parchment as the dominant writing material followed only after a lengthy transition. It is true, however, that Pliny's anecdote about the discovery of parchment at Pergamum likely heralds a significant turning point for the material—Eumenes II's successor, Attalus, was responsible for exporting parchment and likely headed a delegation that brought parchment to Rome during the papyrus shortage in 168 BC.[184] That Pergamum became an important center for parchment distribution is suggested also by the Latin word for parchment, *pergamenum*, which doubles as the adjective of the city Pergamum.[185] In any case, papyrus remained the preferred material for centuries, even after the Pergamene "discovery" of parchment.

The protracted transition from papyrus to parchment coincided with another significant change, that of roll to codex. Though the two developments are not causally related—a growing preference for the codex form did not initiate a preference or need for parchment—they do overlap.[186] As mentioned above, the codex form had existed alongside papyrus rolls in the form of tablets. A leaf tablet (fig. 1.4) was constructed from multiple sheets, while a stylus tablet had a leather strap along a single hinge (fig. 1.3).

The Romans' awareness of parchment may have increased after Attalus' embassy in 168 BC, but that led neither to a dramatic increase in the use of parchment nor to its acceptance as a material superior or even equal to papyrus. By the Augustan period, parchment notebooks existed alongside wax tablets, both in codex form. In his *Ars poetica*, Horace advises Piso's son to share whatever he writes with certain critics, then put his work, described as *membranis*, away for nine years on the grounds that if he chooses he can still destroy an unpublished work.[187] *Membrana* is another Latin word for parchment and here refers specifically to parchment notebooks. As the context makes clear, these *membranis* are for drafts, not final work, and thus are akin in use and status to wax tablets.[188]

Quintilian clarifies the connection between codices in parchment and wax tablets: for him, wax tablets are best for composition, for they are easily erased. Parchment notebooks (*membranae*) may be preferred for someone with poor eyesight since they are easier to see, but the constant need to add ink to the pen (generally a reed pen, known as a *calamus*) delays writing and disrupts the chain of thought.[189] Quintilian further advises leaving a few pages blank regardless of the material used in order to allow for additions to the text, reinforcing the informal role Horace attributed to the parchment codex. First-century Romans thus had ample experience using parchment codices, but apparently only as notebooks. They would not be deemed suitable for holding a published literary work until later.

Parchment notebooks appear to have been uniquely Roman, as may be inferred from the fact that the term *membrana* is incorporated into the Greek language as a loanword from the Latin.[190] Aside from Martial, who mentions what seem to be "pocket-sized" versions of literary works on parchment codices, there exists little evidence for publications of the first two centuries AD on any material other than papyrus.[191] Nevertheless, there is some evidence that parchment codices could hold very valuable works, despite not being in a format fit for publication. Among the possessions Galen lost in the fire of AD 192 were pharmaceutical recipes on *diphtherai pyktai* ("parchment tablets," which, as Nicholls notes, by the late second century seem to have a proper Greek name to replace the Latin *membranae*).[192] Such notebooks were not intended for initial drafts but instead would have contained a finished work, albeit one of a functional rather than literary nature.

The reason for using parchment codices is entirely unclear. Nicholls speculates that the paginated book form was more convenient for gathering and adding recipes from disparate sources. All of the three notebooks that Galen lost were either bought from an owner or given to him; they were not published, but kept instead as private collections of recipes.[193] Perhaps, then, along with the convenience of compilation proposed by Nicholls, one may further speculate that the function of the text was a factor. If one intended to concoct a drug from a recipe in the text, it would surely have been far easier to find it in a paginated notebook (especially as recipes must have been assembled in an ad hoc fashion, incorporated into the collection merely as they came into the owner's possession). The codex form would then be highly valuable both because of the wealth of knowledge it contained and because of its functionality, which further connects its use to that of the parchment notebooks mentioned by Horace, Quintilian, and others. Nevertheless, the notebooks that Galen lost were worth a considerable sum, setting them apart at least in terms of their monetary value from most parchment codices, but whether this signals a change in the status of such codices is debatable.

THE EMERGENCE OF THE CODEX *QUA* LITERARY EDITION

Both literary and material evidence has been adduced for studies of the early codex. As discussed above, the bulk of our early-period papyri, and to this we may now add parchment, come from Egypt, and a limited area of Egypt at that. Great care must be taken, therefore, not to extrapolate too broadly about trends in antiquity from such a narrow and limited sampling. Further, dating the material evidence can prove difficult; the dates for many texts can only be placed within a range of one or two centuries, sometimes more.[194] It is also important to consider that most of the evidence comes from the Roman period, making it challenging to draw firm conclusions about when and how frequently codices and parchment were used. The details of the transition to the codex are therefore quite hazy, but some general conclusions can be drawn.

From literary evidence it is clear that parchment codices were used as notebooks as early as the first century BC. The material evidence from Egypt, however, suggests that the codex did not become a rival to the roll until centuries later. According to statistics compiled by Roberts and Skeat on Greek literary texts, the codex failed to attain parity with rolls until about AD 300. Rolls comprised 80 percent or more of Greek literary finds from the first to third centuries AD; in the fourth century, rolls represent about a quarter of Greek literary finds and subsequently suffer a marked decline.[195]

The reason for this change in the fourth century is uncertain. Accompanying the growing presence of codices is an increasing use of parchment. Parchment codices existed alongside papyrus codices, and in fact papyrus codices at first far outnumbered those made from parchment, which do not have a significant presence until about the fourth century.[196] The reasons are once again obscure, and the data from papyrological discoveries in Egypt suggest that there is no clear connection between a book's form and the material used. As literature transitions to the codex, papyrus seems to have remained the material of choice despite the precedence of parchment notebooks in codex form. One can neither trace a development of parchment codices from *membranae* nor assume that papyrus codices were the special provenance of Egypt, the center of papyrus production.[197]

No single event or feature seems adequate to explain the gradual turn toward codices and parchment. The numerous advantages of the codex format—it did not require rerolling after reading and its physical form made searches for particular passages easier—likely played some part.[198] Moreover, a codex could hold more text than a roll on a comparable amount of material, which may not have affected the cost much but did allow longer works to be copied into a single book.[199] Conversely, for the copyist, rolls would have been easier to use than codices. Rolls offered continuous writing surfaces that could be augmented. Greater care was required for writing in a codex, since it could not be enlarged with the same ease as a roll.

Though no direct evidence is available, paleographers generally believe that codices were assembled after loose sheets had first been written upon by a scribe. This would place a greater burden on the scribe, who would have to keep the sheets in order and accurately estimate the number of sheets needed.[200] Leaves could be added, but only with considerably more difficulty than adding sheets to a *chartes*, which entailed only pasting a new sheet to the end of the roll. Unlike papyri, which were sold in blank rolls, there is no evidence for the sale of blank codices. Such an argument from silence, however, is far from conclusive. Rather, it seems likely that small, blank parchment notebooks were available for sale, since they represented an acceptable alternative to wax tablets.

Examples of papyrus codices have been found that bear leaves with *kolleseis*, indicating that the codex was constructed by cutting a papyrus roll into sheets.[201] It is not necessarily the case that all papyrus codices were made using rolls, although some certainly were. Presumably it was possible to obtain a papyrus codex that was not marred by *kolleseis* running down the page, but, given the absence of evidence for the sale of blank codices, one would likely have to have provided the materials for it as special commission. A separate influence, and perhaps the more important one, was the Christian preference for the codex.

The preponderance of pagan literary texts up to the fourth century are on papyrus rolls. For Christian writings, the situation is quite different. Roberts and Skeat estimate that of 172 fragments of Christian material from the first through the fourth centuries, only fourteen can be identified as coming from rolls, and some of these are counted twice because they are opisthographs.[202] Of the forty-two known fragments of the Gospels that can be dated prior to the seventh century AD, all come from codices. While that may not mean that the Gospels were written exclusively in codices, the codex was clearly the preferred form. The reason may be due in part to the capacity of the codex to hold all four Gospels, something impossible for a roll, which would become too unwieldy to use if it held that much information.[203]

Based on material evidence, we have a sense of the significance of the codex (and parchment) relative to rolls (and papyrus) in the first four centuries AD. As already discussed, Martial, writing in the late first century and early second century, recommended the use of parchment notebooks as novel gifts or portable travel volumes, but his interest in a literary codex seems to have been an isolated case. On the contrary, questions about the validity of the codex as a legitimate format for literary works persisted even after Martial.

According to the early third-century jurist Ulpian, the material does not pose an issue in determining what constitutes a book, but the form does: "in the term 'books' are included all volumes (rolls), whether on papyrus or parchment or any other material.... But let us see whether books in codices of parchment, papyrus, or even ivory or another material, or wax tablet notebooks should be included."[204] Roberts and Skeat summarize Ulpian's stance well: "it is clear that for Ulpian only the roll was fully and unquestionably a 'book'; but it is equally clear that the codex will not long be denied its place."[205] Ulpian's hesitation is perhaps bewildering to a modern reader, but in the early third century the literary roll was still far more prevalent than the new codex form. The material and literary evidence is patchy, but taken together they allow us to sketch an outline of when the codex began to rival the roll and the circumspection of those faced with a new form of book.

LIBRARIES

Whether libraries existed in antiquity as we typically think of them — that is, as physical structures that house books for consultation — is a rather hazy issue. The Greek word for "library," *bibliotheke*, has a range of related meanings, including "bookcase," "book collection," and, eventually, "library."[206]

A discussion of the roots of ancient libraries may reasonably begin with Aristotle, who, according to Strabo, first collected books and taught the Ptolemies how to organize a library. Aristotle possessed a collection of books that was handed down to his pupil Theophrastus. The books then came into the possession of a certain student of both Aristotle and Theophrastus, Neleus, who took the collection from Athens to Scepsis in Asia Minor.[207] Neleus' departure was likely not anticipated by Theophrastus, who presumably intended that the books remain in Athens, accessible to the philosophers of the Lyceum.[208] Theophrastus' *bibliotheke*, then, despite its initial availability to a philosophical community, was ultimately a private possession subject to the whims of its owner.

The most famous library in antiquity was the Alexandrian library organized under the Ptolemies. Despite its fame in facilitating the first steps toward modern philology and scholarship, few details about this library can be asserted with confidence. The problem begins with Strabo, who states that instruction from Aristotle helped shape the library.[209] This is impossible in a literal sense, as Aristotle was already dead in 322, only a year after Alexander the Great's death and before the library could have been organized. Another source, a letter by a certain Aristeas (who may be fictional) to Philocrates, and dated to the second century BC, contains numerous apocryphal accounts about the library, including an unlikely partnership between Ptolemy II and Demetrius of Phalerum, who supported Ptolemy II's rival at the time of his succession. Unlikely, too, is Aristeas' claim, embellished in other sources,[210] that the library housed two hundred thousand books in the time of Philadelphus.[211]

It is unclear whether the Alexandrian library consisted of a single structure for the housing of its books or if space was simply allotted in a structure already built for a separate purpose.[212] Christian Jacob proposes that the library was not its own structure but rather a part of the Mouseion.[213] This suggestion accords well with the need for poets and scholars in the Mouseion to have easy access to texts. Another library in Alexandria was later established under Ptolemy III, this time in a temple dedicated to Serapis.[214]

That the concept of the library was not fully developed in the early Hellenistic period does not necessarily confute the increasing importance of texts or the concerted efforts under the sponsorship of Hellenistic kings to gather them; libraries at the time were at the very least collections of books available to a select group.[215] By the third century, gymnasia have begun to house texts as they become centers of learning, and not only places for physical training. By ca. 117 BC, a graduating class of Athenian ephebes was required to donate a hundred books to the Ptolemaeum gymnasium.[216] The appearance of book collections in gymnasia coincides roughly with developments in the structure of an educational curriculum, which, from the Hellenistic period on, focused on *grammata* (letters, and by extension, reading and writing) and *gymnastike* (physical training) to the exclusion of the once-canonical *mousike* (musical training).[217]

In addition to its plentiful holdings, the Alexandrian library likely contained space for scholarly activity. Indeed, the detail and erudition of early Hellenistic works would not have been possible without access to a substantial number of texts.[218] The logistics of the collection, storage, and maintenance of such a large number of texts were undoubtedly rife with challenges, especially regarding their organization, of which little is known. Unfortunately, even in the case of the Alexandrian library, the organizational roles claimed for Aristotle and Demetrius of Phalerum are too confused to confirm actual Peripatetic influence.[219]

Callimachus' important role in the history of that library, however, can be asserted with more confidence. Beyond his many poetic works, Callimachus compiled a catalogue of the works housed in the library, from which grew his more ambitious work, titled *Pinakes*, which was a catalogue of all Greek literature in 120 rolls.[220] Unfortunately, we have no indication of how one might have used Callimachus' catalogue to locate a particular book in the library, but the need for such an undertaking speaks to the size of the collection in Alexandria (probably tens of thousands of rolls). In any case, Callimachus' work represents a turning point for libraries, as the need for organized book storage, especially on a large scale, came to be acknowledged.

Just as the great Hellenistic libraries of Alexandria and Pergamum were compiled at the behest of kings, it would take the influence of powerful individuals to spur similar projects in Rome. Among the Romans, a pivotal year for the collection of texts was 168 BC, when Aemilius Paullus defeated the Macedonians and seized a large collection of books, which he had brought to Rome among his other spoils of war. This collection was likely of a size unknown among the Roman aristocracy and represented the potential for Rome to become a literary center.[221]

In the following century, Sulla, during the fall of Athens (86 BC), seized the library of the Peripatetic Apellicon of Teos, including Aristotle's and Theophrastus' aging and damaged libraries, which he had acquired from the descendants of Neleus. Though many texts had suffered from neglect, Apellicon sought to emend the damaged sections and publish Aristotle's works. Strabo quips that, because Apellicon was more interested in books than philosophy, he made many errors in his corrections.[222]

In approximately 70 BC, Lucullus brought Mithridates' library to Rome. This collection featured books on multiple subjects, including philosophy and science. This and other book collections brought as spoils were used by Cicero and other leading figures of the time, making the libraries both a symbol of status and functional scholarly tools from a relatively early period in Rome. After the destruction of Carthage (146 BC), twenty-eight books of a treatise on agriculture written by the third-century Carthaginian Mago were taken to Rome; the senate commissioned a translation of the work into Latin. The plundered books, together with others bought in Greece and the East, enriched private libraries: we have reports of relatively vast collections of books owned by Cicero, Atticus, Varro, and some others. Public libraries would not appear in Rome until shortly after Cicero's death.

The first public library emerged in the 30s, when Asinius Pollio established a library in the *Atrium libertatis* using spoils from his campaign in Illyrium. The construction of several other major libraries in Rome followed on Pollio's. In 28 BC the Palatine temple to Apollo was dedicated by Augustus and soon after included two libraries, one for works in Greek and the other for those in Latin.[223] Sometime between 23 BC and 11 BC, another pair of libraries housing Greek and Latin texts was constructed in the *porticus Octaviae* and dedicated to Marcellus. Subsequent emperors continued to build libraries in Rome; the Temple of Augustus, constructed under Tiberius but dedicated by Gaius in AD 37, contained a library.[224] Vespasian financed the construction of another library in the Temple of Peace in AD 75, and the Forum of Trajan, dedicated in AD 112, contained a pair of libraries as well.

Beginning with Pollio's in the *Atrium libertatis*, Roman libraries likely differed from their Greek predecessors in several respects. Several sources, most notably Pliny the Elder and Suetonius, discuss the intent to make the libraries public (*publicare*).²²⁵ Whether the notion of "public" meant that their contents were available to anyone or should be taken in a more restrictive sense is unclear. In either case, Romans, even if only the elite, seem to have had greater public access to texts than Alexandrians did; the Ptolemies' library may have had its usership restricted to members of the Mouseion.

Books stored in Greek *gymnasia* were likely intended for educational use. Roman libraries, on the contrary, were monumental in scope, often with elaborate facades. In fact, the monumental nature of Roman libraries seems tied precisely to their public purpose.²²⁶ Pliny credits Asinius Pollio as the first to adorn a library with bronze portraits, which he considers an important cultural achievement.²²⁷ Suetonius asserts that under Tiberius, the immense statue of Apollo Temenites, a work so large that Verres was unable to pilfer it from Syracuse, was placed in the library of the Temple of Augustus.²²⁸ In short, Roman libraries were built to house more than just books.

In addition to displays of plastic art, libraries often hosted public business and poetic recitation. Augustus is known to have held senate meetings and hosted foreign delegations in the Palatine library, and the *porticus Octaviae* was similarly utilized.²²⁹ Notwithstanding the size and ornamentation of the structures, their book holdings were precious resources that attracted scholars like Galen, who remarks that he rented space in a warehouse in the Forum Romanum to hold his books and writing materials (among other paraphernalia), presumably because he was thus able to keep his research tools close to the collections in the Temple of Peace, the Domus Tiberiana, and the Temple of Apollo.²³⁰

Libraries were thus important sources of information for scholars and authors in antiquity. Of course, private libraries continued to be gathered, giving rise to the ancient book trade. How it functioned is not entirely clear, especially the book trade in Greece. The available evidence, however, presents some significant insights into the transmission of texts.

THE ANCIENT BOOK TRADE

A book trade of some sort was in place by at least the end of the fifth century BC. The first reference to the commercial sale of books comes from the comic poet Eupolis; unfortunately the reference appears in a small fragment consisting only of the phrase "where books are sold."²³¹ In Aristophanes' *Frogs*, Dionysus mentions reading Euripides' *Andromeda* to himself on a ship, and his doing so raises a number of questions.²³² The *Frogs* was produced in 405 BC, and the play Dionysus was reading, Euripides' *Andromeda*, was produced in 412, a seven-year difference. How long after the play was produced would it take for written copies to circulate? How long would a play circulate in written form? How was the first copy made and how did it reach the book market? Unfortunately, we have few details, only the fact of the availability of written plays. The scholiast remarks that the *Andromeda* was among Euripides' best plays,[233] so it may have lingered in the social consciousness and perhaps circulated in written form more widely than most plays. Furthermore, the plot of the *Andromeda* contains humorous parallels to that of the *Frogs*, which surely influenced Aristophanes' decision to have Dionysus read that particular play.[234] The image of someone casually reading a play may thus be less typical than the scene leads one to believe.

Later in the *Frogs*, the chorus affirms the audience's capacity to understand clever arguments because, they claim, now everyone has a book.[235] This comment should not be taken literally, of course, given the low level of literacy throughout antiquity, but the chorus at least suggests that books are not uncommon. The type of book to which the chorus refers is unclear; the scene portrays the playwrights Aeschylus and Euripides, characters in the comedy, as having a poetic contest in the underworld, which means the book that supposedly everyone has could be a technical manual on poetry or perhaps simply a play, or something else altogether.[236] While it is risky to conclude too much about the ubiquity of book circulation from a comedy, at the very least one can infer that there was a demand for books. Noteworthy, too, is that books are listed among the cargo of ships and must therefore have been an item of ancient commerce, if only limited in scope.[237]

Plato offers additional evidence for a book trade. In the *Apology*, Socrates mentions that the teachings of Anaxagoras can be bought for a drachma, which he considers a high price.[238] Socrates' comment could give the impression of a bookseller holding a stock of copies of Anaxagoras, much like a modern bookseller. Yet not all accounts suggest such plenitude. Strabo recounts that when Neleus took Aristotle's and Theophrastus' combined libraries to Scepsis, Aristotle's Peripatetic school was left with only a few esoteric books unsuitable for practicing philosophy.[239] If Strabo is correct about the dire straits of the Peripatetics, then it was likely the case that acquiring texts even in ancient Athens was more onerous than simply finding funds to cover the purchase. Suffice it to say that, though its beginnings are unknown, a book trade was well in place by the end of the fifth century.

Evidence from Rome on how texts circulated is more substantial. The book trade came to Rome relatively late and does not seem to have found acceptance until the first century AD.[240] Of course, Greek texts must have been reasonably available in Rome by at least the late third century as evidenced by the propensity of writers of Roman comedy, such as Plautus (ca. 254-184), to adapt Greek models. Paradoxically, books containing Greek plays were likely easier to obtain than contemporary Latin ones.[241]

Eschewing the nascent book trade, readers circulated most texts through a closed system of personal contacts, beginning with requests for comments and revisions and, afterward, expanding to a larger, but still intimate, group of acquaintances. Each individual who was asked to comment on the work would receive a copy from the author. The time and expense of making multiple copies of a work in progress underscore that literary activity was the domain of the upper class and an appropriate use of their leisure time (*otium*).[242]

The initial presentation of a finished text reflects the private nature of books and libraries; one might recall that an important early influx of books to Rome occurred when Aemilius Paullus brought home Perseus' library in 168 BC, which remained a personal possession of Paullus. Similarly it was through his personal relationship with Sulla and his son, Faustus, that Cicero was able to use the library of Aristotle, which, as we saw, had come to Rome through Sulla's intervention.[243]

Texts that were sent to friends for early feedback would be returned with passages marked with red wax.[244] Alternatively, criticism could be gathered from an oral reading of a new work, which, when undertaken, always took place among a small group of close friends.[245] The case seems to have been similar for poets. In one of his elegies, Propertius announces that a work "greater than the *Iliad*" is being created, a reference to Virgil's *Aeneid*, of which he must have seen or heard some portion made available by the author.[246] There is some physical evidence for such recitations. For example, the *auditorium Maecenatis*, located on the property of Maecenas on the Esquiline is likely to have been used by Augustan poets for private readings.[247] In Pompeii, the Odeum may have served for public recitations as well.[248]

The ancient notion of "publication" differed from that of later ages. Van Groningen has demonstrated that publication (*ekdosis*) should be understood as the author putting his finished work at another's disposal.[249] This does not mean that the work was submitted to a bookseller or anyone intending to profit from the text; in fact, the text was unlikely to reach a bookseller at all, at least in the Republican period.[250] The finished text would first be distributed by the author as a gift to his friends.[251] This represents the text's publication, from which point it was in the hands of the recipients, who were free to make more copies or loan them out to others to be copied, producing, so to speak, new "editions" of the text in a process known as *diadosis*.[252] The point of distinction between *diadosis* and *ekdosis* is that the former is done without the author's consent while the latter is the author's (or editor's, in the case of scholarly editions) intentional act of making a text available. One should bear in mind, too, that publication included oral performances, which continued alongside the dissemination of written texts.[253]

The closed system of textual publication and circulation had some significant consequences. First, it hindered the book trade in Rome. Unlike fifth-century Athens, where Aristophanes and Plato give the impression of a relatively thriving book business—some books would have been available, although how they were selected is unclear—Cicero suggests that finding a desired book in a bookshop was quite difficult. Cicero writes to his brother, Quintus, that he would like to exchange books with him so they can each make copies and fill their libraries, but he does not have anyone available to do the work. Further, Cicero comments that the books he desires are not for sale and require skilled labor to copy.[254] In fact, the primary means Cicero used to supplement his library was to borrow and copy books from friends.

Some sense of the quality of literature that was for sale comes from Catullus, who intends to seek revenge for bad poetry he has received from Calvus by returning bad poetry (which he calls "poison") that he finds at a bookseller.[255] It seems unlikely that booksellers were only repositories of works of unknown or bad authors, but on the other hand, the elite class that wrote and appreciated literature neither made their texts available for sale nor looked for a learned work from a bookseller without first seeking it from a friend. The availability of new literary texts, at least, must have been quite limited on the market.

Though few books were likely available from a bookstall, the bulk of the booksellers' business may have derived from custom orders. It was possible to hire the services of a scribe (*librarius*) to copy books or write out documents. The term *librarius* in fact refers to both a bookseller and a scribe, suggesting that a single shop provided both services.[256] In one of his epigrams, Martial attributes errors in his text to the haste of a scribe to count the number of verses he copied so he knew how much to charge for his services.[257] Many of the papyri from Herculaneum also contain stichometric counts (line counts) at the end of the volume, suggesting that they were copied by a professional scribe.[258] Certain allusions of one author to another by book and verse indicate that lines were apparently counted and numbered at times even by authors.[259]

A few centuries after Martial, in AD 310, Diocletian's price edict established tariffs for books and documents per every hundred lines of text.[260] Having one's own slaves to produce texts on demand, as Cicero's friend Atticus did, could hardly have been the norm, as Johnson notes. Thus, there likely existed a vigorous book trade run by professional scribes vested in the business, who either had a small stock of texts to copy or could acquire them from a library.[261] If the book trade was in fact focused on made-to-order books, these shops would have existed alongside the closed system of circulation for new texts.

The idiosyncratic nature of literary production in ancient Rome also limited opportunities for would-be authors. Authors were generally not paid directly for their work, and were not, in any case, given royalties by booksellers. Booksellers in Rome did not see profits from a wide distribution, as they merely sold their books locally, which furnished them with but a slender profit margin.[262] Martial states tellingly that although his writing is known as far away as Britain, it brings him no profit.[263] Only writers of works of popular entertainment, such as mimes, might earn a wage for their writing, from officials who needed to put on shows for festivals. Making money from such subliterary writing, however, was met with scorn.[264] With few exceptions, Roman authors tended to be of the equestrian or senatorial class, both of which met a property qualification of sufficient means for self-support. Though poets in Rome could have patrons, they were not supported with a stipend as the Alexandrian poets under Ptolemy are believed to have been, nor were they ever paid for commissioned work. Instead, poetry, and literature in general, was a proper pastime of the upper classes and, for those at the bottom of this elite group, a means of ingratiating themselves and gaining other forms of recompense, such as lucrative military positions.[265]

By the second half of the first century AD, the book trade seems to have gained some acceptance among elite authors. Pliny the Younger claims to know from a bookseller that his work has found an audience.[266] Quintilian, too, entrusts his work to a bookseller.[267] The reason for the book trade's increased importance is unclear; Starr offers two proposals: first, that authors may have begun using the book trade as a means of reaching a broader audience (that is, elites outside of Rome), and second, that outsiders looking for opportunities from patronage increasingly used the book trade to make themselves known.[268]

The notion of an expanded book trade must still be approached with caution. Pliny the Younger, for instance, claims to be surprised that there are bookshops in Lugdunum (modern Lyon) and further that his work is sold there.[269] How should Pliny's comment be understood? It seems from Pliny's own letter that he gave his work to at least one bookshop to sell.[270] Is Pliny truly surprised that Lyon has bookshops and, further, that his work has traveled so far from Rome? There is no evidence for a system of mass copying and distribution; rather, texts seem to have been copied and sold locally. Martial, too, claims that his work is known outside of Rome, though how it circulates is not specified. In epigram 7.88, Martial says his poems are read by everyone in the town of Vienne (in Gaul). In epigram 11.3, in which Martial laments that he earns no money from his poems, he says that his poems have been carried to Getica (just west of the Black Sea, in what is now Romania and Bulgaria) and Britain by soldiers.

What role the book trade played in spreading Martial's fame is uncertain. In any case, with authors such as Pliny, Quintilian, and Martial putting their works into the hands of booksellers, the quality of books available to the public certainly would have improved since Republican times, and so there would have been more opportunities for increased circulation. Even so, private borrowing and copying likely remained the most efficient means of acquiring texts. Even the emperors Marcus Aurelius and Theodosius II are known to have done so on occasion.[271]

Not all texts would be available from booksellers. As we saw above, Galen stated that he bought his books of drug recipes from a private individual, not a bookseller. Texts of a specialized nature, such as lexica and other reference works, likely circulated within a closed system of friends who loaned books to be copied rather than through the book trade.[272]

CONCLUSION

The written word came to Greece and Rome relatively late, but soon became transformative for those societies. Writing became integral, with lasting and far-reaching effects, to public administration, legal proceedings, and military endeavors, to name but a few. For those living within highly literate societies, such as the community at Alexandria, literacy became necessary for social engagement, such as making contracts or appealing to public authorities. So important was literacy that the traditional pedagogical paradigm was adapted to include it. Additionally, the increased flow of communication enabled the effective administration of large empires, far beyond what would be possible for an oral society; the world became a much smaller place.

The prominence of writing gave rise to new possibilities and challenges. Unfortunately, losses accrued. The depletion of the great library of Alexandria was more likely caused by neglect than the fire of 48 BC. The effort and expense required to maintain the holdings of that and other libraries could not be sustained, and as a result, many works were forever lost. Other threats arose as well. Forged works appropriating the names of popular authors were prevalent, against which authors had little recourse.

Yet all was not lost, as the desire to preserve cultural heritage is reflected in the creation of critical editions of texts established through scholarly methods that would ultimately give rise to the field of philology. In the following chapter, we will examine some of the conventions that arose with the establishment of writing and continue to trace the development of the codex from late antiquity to the Middle Ages.

2 / GRAMMAR, SCHOLARSHIP, AND SCRIBAL PRACTICE FROM ANTIQUITY TO THE MIDDLE AGES

TEXTS ENABLE NOT ONLY THE PRESERVATION AND DISSEMINA- tion of ideas but also their organization and codification. This chapter will examine the rise of textual scholarship in antiquity and its long-standing effects, such as the systemization of punctuation and grammar. The practice of inscribing literary texts will also come into consideration, under the field of paleography, as well as other matters of scribal import, including the use of reading aids, such as scholia, and the conditions of medieval *scriptoria* (monastic copying rooms). As will become apparent, many modern practices that may seem self-evident were not so to early readers; rather, current reading habits, far from inevitable, can be traced to processes that have evolved since their beginnings in antiquity.

EARLY PUNCTUATION AND *SCRIPTIO CONTINUA*

Writing in Greece was frequently done without spaces between the words, in a style known as *scriptio continua*. It is tempting to imagine that the practice developed to conserve precious space, but this can hardly be the case, because surviving papyri reveal no consistent effort to limit the size of letters or margins. Whether for aesthetic or other reasons, the choice to write in *scriptio continua* was not guided by mere frugality.[1]

The Romans, possibly due to Etruscan influence,[2] used interpuncts, which are raised dots, to distinguish clearly between words in texts. Seneca the Younger cites the use of interpuncts in differentiating Roman practice from that of the Greeks.[3] By the end of the first century AD, however, the Romans had chosen to do away with these aids and wrote instead in the more challenging *scriptio continua*, presumably in imitation of Greek writing style.[4] The use of *scriptio continua* in Latin texts long endured, abating only when monks in Ireland, faced with an unfamiliar language, introduced spaces between the Latin words. This technique was common by the end of the seventh century.[5]

Of other types of ancient Latin punctuation there is but slight evidence. The Romans seem at times to have used blank spaces, as did the Greeks, to indicate the end of a sentence or paragraph. What remains of the work of the early Roman grammarians has little to contribute to the matter; as we will see below, the Greek tradition is much fuller, though also fragmentary. By late antiquity, Roman grammatical sources demonstrate familiarity with Greek practice in punctuation. It may be that the Romans had developed a system of lectional signs before late antiquity, but the dearth of evidence for such a system precludes anything more than speculation.[6]

Reading words in *scriptio continua* would have been as difficult for an ancient reader as it is for a modern one, and some of our earliest inscriptions reveal efforts to facilitate reading. The famous inscription on "Nestor's Cup," from as early as the late eighth century BC, includes marks similar to a modern colon separating the words at irregular intervals. The first line, widely believed to be a version of an iambic trimeter,[7] has a double-dot sign between each word. In the two following lines, in dactylic hexameter, the same sign seems to indicate an apparent early demarcation of metrical units, specifically the caesura.[8] Although the precise intent behind the divisions is difficult to ascertain with certainty, it is clear that the author desires to facilitate reading, and takes special care to distinguish for the reader important metrical units.

From the late sixth century BC until the end of the fourth, a new, highly stylized form of writing known as *stoichedon* was used.[9] In *stoichedon* inscriptions, the letters are arranged in *scriptio continua* on a grid with perfect horizontal and vertical alignments. By the early fifth century BC, nearly all letterforms of *stoichedon* inscriptions had broadened to fill the square space that makes up their position on the grid.[10] While the punctuation of Archaic inscriptions suggests the priority of the reading, a much different concern is evident in *stoichedon* inscriptions, whose identically spaced letters and unusual word-breaks produce a pleasing visual effect at the expense of the legibility of the text.[11] Lettering on stone, including *stoichedon* inscriptions (fig. 2.1), was influential for contemporary chirographic (handwritten) letterforms.[12]

Reading inscriptions in Classical Greece became, perhaps surprisingly, more difficult than it had been in the Archaic period, just as it would later when the Romans did away with their interpuncts. Punctuation was commonly used in some areas of Archaic Greece, such as Attica and Argos, but in other regions, like Boeotia, Thessaly, and Corinth, it was much rarer.[13] In Archaic Attic inscriptions, words were most typically separated by three dots in a vertical

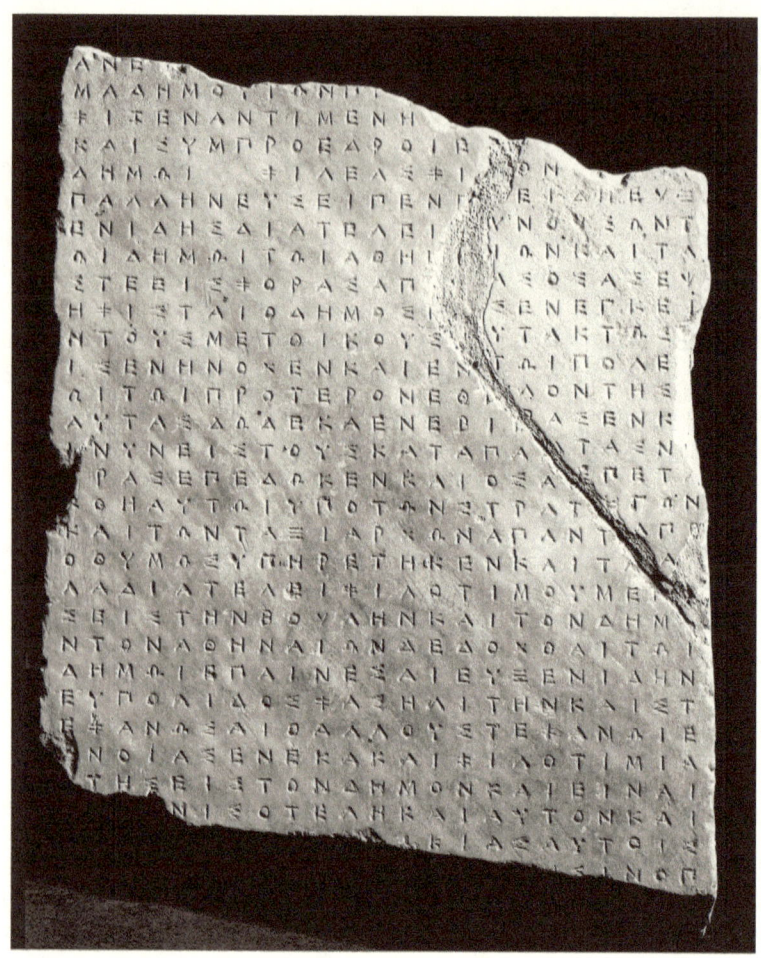

FIGURE 2.1. *Fragment of inscription with Attic decree written in stoichedon. Athens, National Archaeological Museum. De Agostini Picture Library/G. Nimatallah/Bridgeman Images.*

column,[14] but two dots, arranged much like the modern colon, were also used. Punctuation had disappeared from Attic inscriptions by the end of the sixth century BC and would not return until the Roman period.[15]

That unpunctuated *scriptio continua* would become standard for inscriptions may be owed to the fact that inscriptions on public monuments were not primarily intended for reading so much as to serve an aesthetic purpose. Inasmuch as scarcely more than 10 percent of the population was literate, the most effective means of disseminating information was through oral communication, whether by public proclamation or informal reading.[16] Thus, although monuments bore inscriptions—some of which are quite lengthy—they were not necessarily intended for a large readership; in the case of some inscriptions, particularly those of a magical or religious nature, the very acts of inscribing the text and placing it in a particular location were thought to effect desired outcomes.[17]

The privilege given in Greece to beauty and form over legibility in the arrangement of letters was a deliberate one and not limited to public inscriptions. In fact, many features of ancient literary texts aimed at form over function. An awareness of the beauty of texts is evident in the attention paid on the one hand to the materials used, particularly the grade of papyrus and the use of parchment coverings to protect rolls of high value, and on the other hand to the appearance of the written words on the page. Literary rolls typically contain texts written in lines of equal length, with care taken to keep the right-hand margin even.[18] This is all the more significant given that literary texts generally adhere to strict rules concerning word divisions at the end of a line. The margins tend to be ample, with upper and lower limits usually measuring 3–7 centimeters, far more than needed to keep the frayed edges of the papyrus from reaching the text.[19]

Johnson, noting that most literary papyri had text on only 40–70 percent of the front side of a roll, rightly cites the importance of aesthetics in ancient book production.[20] Books were thus beautiful objects, but in the case of *scriptio continua*, only at the cost of legibility. The ancients themselves must have recognized this, since their educational practices provided many aids now taken for granted. To meet the challenge of reading what must have initially appeared as no more than a string of letters, students from an early point in their education made use of such tools as texts that included word divisions, lists of mythological names divided into syllables, and other guides to identifying words.[21]

At the secondary level, practice readings prepared by the *grammatikos* contained aids for reading aloud and formed a sort of forerunner to punctuation. Even an accomplished reader might mark a text to indicate pauses and points to take a breath, concerns additional to the initial task of identifying words and phrases.[22] In short, reading a text required time and effort.[23] Books were works of beauty and, as Johnson puts it, "instantiated what it is to be educated."[24]

VOCALIZED AND SILENT READING IN ANTIQUITY

Despite the loss of interpuncts and similar markings used in Greek writing from an early period, other forms of punctuation persisted through the Classical period. The advent of scholarship in the Hellenistic period led the scholars of the Mouseion in Alexandria to undertake new, systematic approaches to lectional aids. Yet the signs developed to facilitate reading were specifically intended to assist in oral recitation, which was common in antiquity. Both ancient literary sources and the architectural design of ancient libraries corroborate the predominant practice of vocalization while reading.

While scholars generally accept vocalization as typical of ancient reading practices, there has been some debate as to whether there were any occasions for which silent reading was preferred.[25] Augustine's surprise at Ambrose's silent reading of a text and Cato's outrage at Caesar's silent reading of a personal letter have been taken as evidence for the rarity of the practice in antiquity.[26] Other evidence, such as Horace's remark that he reads silently and the lenient punishment meted out by Augustus of making guilty *equites* stand before a crowd while each silently read a notice of his crimes, suggests that the phenomenon was not entirely extraordinary.[27] Of course, reading to oneself would have been advantageous in some circumstances: personal documents and brief notices or inscriptions, either for privacy or convenience, would surely have been better read in silence, at least by an accomplished reader.

Texts written in *scriptio continua*, or even those separated by interpuncts but otherwise lacking punctuation, must have been easier to read by sounding the words out, at least on an initial reading. In this case it is easy to imagine a reader mumbling the words to himself, an act quite different from the full vocalization usually implied by reading "aloud." If nothing else, silent reading must have been the mark of a skilled reader.[28]

Reading aloud, however, was the norm, and to do so competently required practice. Scribes often did not add sufficient — or sometimes any — punctuation in the process of copying a text. They sometimes used spaces to indicate significant breaks,[29] but it was not unusual for readers to mark their own texts to aid reading. Early efforts to facilitate reading aloud include the *paragraphus*, a horizontal line that served as a form of punctuation that predates the more systematic efforts devised in the Hellenistic period. The *paragraphus* most often appears to the left of a column of text and, depending on the text, has several possible meanings. For example, the *paragraphus* may indicate a change of speaker in dramatic texts, the beginning of a new poem in a collection, divisions of metrical groups, or the end of a sentence.[30]

That same sign served also as a visual cue indicating where a reader should pause, perhaps to engage an audience, and then where to resume reading. Its value for vocalized reading is apparent from the fact that it is often used redundantly with a space or other mark in the text;[31] in fact, Isocrates even orders a clerk to begin reading "from the *paragraphus*."[32] Punctuation, then, was in the first case designed to facilitate reading aloud, the performative aspect of which was essential to being a skilled reader.

Nevertheless, one should not construe the use of the *paragraphus* and other

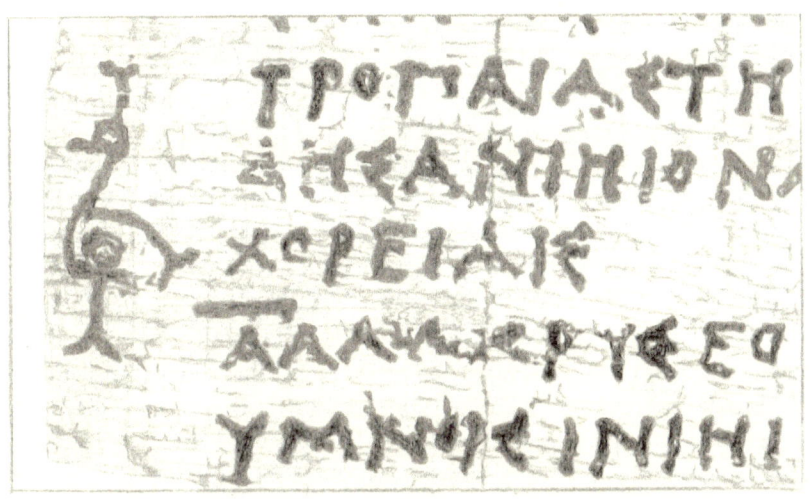

FIGURE 2.2. *Fragment of the Timotheus Papyrus containing the earliest known example of a* coronis (left); *a* paragraphus *appears as the slash after the third line. Drawing by Mary Claire Russell MacDonald.*

early signs (*semeia*) too narrowly. Punctuation also helped make sense of the text and indicates scholarly activity. The signs used were not initially systematized, and so various forms of punctuation could be used in the same capacity as the *paragraphus*. *Dicola* (rarely *tricola*) are occasionally used as stops (*stigmai*), but most commonly the *dicolon* indicated a change of speaker (a function the *paragraphus* could also serve).[33] Like the *paragraphus*, a mark known as the *coronis* designated a break, and is often found at the end of a roll, though it may also indicate the end of a scene in a dramatic text, the end of a poem, or the conclusion of a chorus (but not the end of a sentence).[34] The *paragraphus* is found in two of the oldest extant literary papyri, a fragment of Timotheus' *Persians* (fig. 2.2) and the Derveni Papyrus, an allegorical explanation of an Orphic poem. Both papyri date from the middle of the fourth century BC, with the Timotheus papyrus perhaps a bit earlier.[35]

The *coronis* appears less frequently than the *paragraphus* and is more variable in form. In the Timotheus papyrus, its earliest known appearance, it takes on a shape that bears some resemblance to a bird, which led Wilamowitz to propose tentatively that the likeness was intentional, especially since the Greek word *coronis* is orthographically similar to the Greek *corone*, which means "crow."[36] Some *coronides* are embellished forms of the *paragraphus* while others resemble a stylized backward "S" (Ƨ). Meleager apparently had some such image in mind when he compared the *coronis* to a snake.[37] Examples of *coronides* appear most often in texts that received scholarly attention.[38] Interestingly, they are especially infrequent in Christian texts and fell into disuse by about the fourth century.[39] In terms of the origins of the *paragraphus* and the *coronis*, it is impossible to say when or how these early marks developed, but they very likely predate our earliest examples, going back at least to the fifth century.

Although most signs used in texts were designed as reader's aids, some were not. Symbols also helped readers navigate the increasingly complex scholarly interest in textual criticism. A relatively early example is the use of marks to direct a reader to a commentary (*hypomnema*), a practice predating but closely associated with the scholarly commentators of the Hellenistic period,[40] an age that featured a turning point in textual criticism. Although it would still be some time before punctuation and other *semeia* became widely systematized, Hellenistic scholars took up the question of editorial practices in treatises with a range of results, some of which are reflected in existing papyri. In addition to pondering existing notions of punctuation, these textual critics devised new signs predicated specifically on the desire to create authoritative critical editions of texts. Perhaps the best example of the new scholarly impulse and its effect on scholarly *semeia* is the Hellenistic reception of Homer.

HELLENISTIC SCHOLARSHIP

Hellenistic scholars made significant intellectual contributions that would direct subsequent approaches to literary texts. Many of the works they produced, however, such as commentaries and lexica, did not originate with them. In fact, Homeric *glossai* (collections of obscure or rare words, especially from epic) were circulating already in the fifth century and possibly, to meet the needs of rhapsodes, earlier still.[41] Antimachus, a poet contemporary with Plato, is the first scholar whose name is associated with an edition of Homer.[42] By the Hellenistic period, such works were undertaken more methodically and systematically, a feat enabled by the vast scholarly resources housed in the Alexandrian library.

Philetas, an influential poet and scholar of the late fourth century, wrote the *Ataktoi glossai*, which became immediately popular and remained authoritative for interpretations of Homer for over a century.[43] Another Hellenistic scholar active early in the third century, Zenodotus, served as the first head of the great Alexandrian library, a position that included service as the royal tutor.[44] It is likely to Zenodotus that we owe the division of Homer's *Iliad* and *Odyssey* into twenty-four books each. He was also the first known *diorthotes*, or corrector, of a text. Such correction or emendation represents an effort to restore the text as much as possible to its original state, the result of which was an edition (Greek, *ekdosis*; Latin, *recensio*) that incorporated the scholar's changes.[45]

Though his methods are subject to speculation, in his capacity as corrector,

Zenodotus likely gathered numerous copies of Homer's text and, using those he deemed the best as a foundation,[46] made emendations derived from what he considered better readings found among the different copies. Like the modern philologist, Zenodotus would have had to choose between disparate readings of the same line, determine which lines were interpolated (added later by someone other than the author), and determine whether the lines were correctly ordered, among other challenges. The grounds for his editorial decisions are unknown, and, unfortunately, there is no evidence that he wrote a commentary to explain his methods. It may be that increasingly detailed notes incorporated into *ekdoseis* produced on Zenodotus' model led to advances in the commentary tradition.[47] In any case, the fruits of his labor were highly influential and mark a significant step in the history of scholarship.

In producing the first scholarly edition of a text, Zenodotus also introduced some important editorial practices. One, apparently beginning with Zenodotus,[48] consists of the use of an *obelus* (a cross-shaped, spit-like mark) to mark a presumedly spurious line of poetry. The introduction of the *obelus* shows considerable care and forethought on Zenodotus' part. When he undertook the task of creating an authoritative edition of Homer, Zenodotus was faced with a text that was rife with alternate readings and interpolations. Rather than excise lines that he disapproved of, some of which did not even fit metrically, Zenodotus simply marked them and preserved alternate readings. This practice allowed subsequent scholars, like Aristophanes of Byzantium and Aristarchus, to use their own judgments regarding questionable passages when they created their critical editions.

Scholars arranging texts did not concern themselves with punctuation, which remained the purview of scribes and readers.[49] Signs like the *obelus* were aids to understanding questions or problems with the text. The *obelus* was not the only new sign to help readers navigate scholarly editions. The Greek letter *chi* (χ) was sometimes used to direct the reader to a commentary, though in texts of Homer the *diple* (>) serves this purpose.[50] Another *diorthotes* of Homer, and the fourth head librarian, Aristophanes of Byzantium (mid-third century to early second century) prepared a critical text and, like Zenodotus, did not write his own commentary to accompany his edition. However, he augmented the number of scholarly signs to explain the reasoning behind his editorial choices. Some of the signs he likely introduced are the *sigma* (C) and *antisigma* (Ɔ), to mark lines that are interchangeable, and the *asteriskus*, to mark lines repeated elsewhere in the text.[51]

To Aristophanes of Byzantium is also attributed the invention of signs for the three Greek accents. Although both signs and terminology for discussing accents were developed in the Hellenistic period and used consistently after, accents are not regularly written in manuscripts until the medieval period.[52]

Interestingly, the accents that do appear after the Hellenistic period do not seem to reflect contemporary pronunciation but rather the pronunciation prescribed by Hellenistic grammarians. Eventually, manuscripts regularly used accents, and so scribes did not need to know rules of accentuation to copy the marks already on the page.[53]

Two third-century scholars, Lycophron of Chalcideum and Alexander Aetolus, are credited with correcting tragic and comic texts.[54] Another, Eratosthenes of Cyrene, the third Alexandrian librarian, was a man of extensive learning and wide interests. He wrote lexica and a treatise on Old Comedy and was also interested in philosophy and science, for which reason he refused to call himself a grammarian (*grammatikos*) and instead styled himself as a philologist (*philologos*), to which he ascribed the meaning "a man of broad learning."[55]

The sixth and last head librarian,[56] and the greatest of the Hellenistic scholars, was Aristarchus (216–144 BC), who did not produce his own *ekdosis* of Homer but did write an extensive commentary. He also largely adopted the signs introduced by Zenodotus and Aristophanes of Byzantium, but with a few modifications. From the *diple* he created the *diple periestigmene* (>:), to indicate his disagreement with Zenodotus' reading. He also combined the *asteriskus* and *obelus* to mark the repetition of misplaced lines.[57] Aristarchus was a prolific scholar and had numerous pupils who would influence the course of philology and grammar in their own right.

The accomplishments of Hellenistic scholars extended far beyond developing authoritative editions of Homer. Zenodotus began creating editions of Pindar and Anacreon, and wrote an alphabetically arranged *glossai* of Homeric words; Callimachus, most famous as a poet, wrote the *Pinakes*, an extensive work in 120 books that categorized Greek authors by genre, providing biographical details and titles of their works. Aristophanes of Byzantium would go on to supplement and correct the *Pinakes* about a half century later.[58] He was particularly diligent in creating critical editions; in addition to his work on Homer, he edited many lyric poets, such as Pindar, and likely Anacreon, Hesiod, Euripides, and Alcman as well.[59] In his new edition of lyric poets, Aristophanes added another important feature to the text: instead of writing the poems out in continuous lines, he divided them into shorter elements known as *cola*, which are metrical units of about twelve syllables that often correspond with pauses in a line. *Cola* are commonly found as metrical patterns in lyric poetry.[60]

The remarkably industrious Aristarchus, famous for his work on Homer, also wrote a commentary on the comedies of Aristophanes, as well as numerous monographs.[61] Moreover, he devoted attention to the works of Hesiod, Archilochus, Alcman, Anacreon, Bacchylides, Pindar, Aeschylus, Sophocles, and Euripides.[62] Nevertheless, Ptolemy VIII forced Aristarchus and his pupils to disperse after he supported the unsuccessful attempt of Ptolemy's rival to win the throne.[63] Some went to Pergamum, which, under the guidance of the Stoic Crates of Mallus, was already becoming the site of a philosophically oriented cultural center that rivaled Alexandria.[64] Crates, a contemporary of Aristarchus, was invited by King Eumenes II (197–158 BC) to Pergamum, where he was likely instrumental in the formation of the Pergamene library.

Crates described himself as a *kritikos* rather than a *grammatikos*, considering the latter term too limited in its concern with prosody and terminology. By the former he meant to signify a broader approach that encompassed philosophical knowledge.[65] Crates is perhaps best known for his adherence to *anomaly* over Aristarchus' *analogy*. In this context *analogy* refers to a systematic, rule-based approach to language that categorizes words according to their likeness. *Anomaly*, expounded first by the Stoic Chrysippus, recognizes that "similar things are often denoted by dissimilar words, and dissimilar things by similar words."[66] Among Crates' works, two titles are extant: the *Peri diorthoseos*, on textual criticism, and the *Homerika*, an exegesis of the Homeric poems.[67] From what remains of his works, it seems that Crates' interpretation of Homer included allegorical readings that uncovered Stoic cosmological ideas in the texts.[68]

Despite the dispersal of intellectuals from Alexandria, advances in Greek scholarship continued unabated. Dionysius Thrax (170–90 BC), a pupil of Aristarchus, fled to Rhodes at the accession of Ptolemy VIII. Like Aristarchus, Dionysius was a Homeric scholar but is famed for being the author of the first known Greek grammar, the *Techne grammatike*. If the attribution to Dionysius can be trusted, a point that remains a subject of debate,[69] then the text is also remarkable for being the only extant Hellenistic scholarly monograph.[70] The *Techne grammatike*'s popularity extends from late antiquity on, a fact that makes its emphasis on orality particularly noteworthy. In his work, Dionysius identifies six parts of grammar, the first of which is "practiced reading in accordance with prosody" (i.e., accents, breathings, pauses, etc.).[71]

On a different subject, Dionysius defines "reading" as the "flawless pronunciation of poems and prose writings."[72] The emphasis Dionysius places on the oral nature of reading is an important reminder that the Hellenistic period's apparent shift away from oral performances toward an interest in compositions designed to be read silently or individually, as well as the rise of philological activity, did not negate the cultural practice of reading aloud, a fact which is perhaps too often overlooked in analyses of Hellenistic poetry. The importance of orality for Dionysius' grammar extends to his discussion of punctuation, which he treats as lectional signs rather than the *semeia* of scholars. He distinguished three forms of stops (*stigmai*): the full stop (*teleia*), the "middle" stop (*mese*),[73] and the *hypostigme*, a pause after an incomplete thought. According to Dionysius, the difference between the three stops is the length of the pause, again indicating a focus on the oral nature of reading and punctuation. The duality of a full stop and an incomplete stop was the underlying structure for punctuation in antiquity, but it did not become immediately standardized. In fact, a later writer, Nicanor (second century AD), proposed a system of punctuation that included eight different kinds of stops.[74] Nicanor's system, however, was quite idiosyncratic and not widely used.

ANCIENT SCHOLIA AND POST-HELLENISTIC SCHOLARSHIP

Apart from Dionysius Thrax' *Techne grammatike*, the tremendous amount of scholarship produced during the Hellenistic period is in large measure lost, as are many other important ancient treatises from the Roman period on. What does remain has primarily been reconstructed from scholia, exegetical notes written in the margins of a text (as opposed to independent commentaries (*hypomnemata*) and interlinear notes on vocabulary (glosses).[75] How commentaries eventually became distilled into scholia is unclear, but the process was likely gradual, beginning perhaps as early as the fifth century AD. Since then, scholarly monographs and commentaries continued to be transferred to the margins of texts, and much of this likely occurred after the ninth century.[76]

One of the most famous texts to include scholia is the Venetus A manuscript, a tenth-century text containing marginalia in the hand of a single scribe. The scholia in Venetus A are derived from four ancient sources, and so the compilation of these sources has come to be known by the German term "Viermännerkommentar" (VMK). This set of scholia is highly significant because of the antiquity of its sources; the four works included in the VMK, as identified in a subscription by the scribe, are Aristonicus' treatise on signs, Didymus' compilation of Aristarchus' commentaries, Nicanor's treatise on punctuation, and Herodian's work on accentuation in the *Iliad*.[77] While these works date from the first or second century AD, Didymus' work on Aristarchus grants us some insight into the thinking of the greatest of the Hellenistic scholars. Not all scholia are as valuable as the VMK, but many provide an important bridge to ancient interpretations of texts. Without these aids, many passages of ancient texts would be indecipherable to the modern reader, who lacks the linguistic and cultural fluency of an ancient scholar.

Of the Greek scholars who came after the Hellenistic period, a few have already been mentioned as part of the VMK, while some others deserve attention. Apollonius Dyscolus, a grammarian of the second century AD, was prolific. Fortunately, a few of his treatises have survived. These works are more extensive than Dionysius Thrax' brief *Techne grammatike* and include the *Peri antonymias* (*On the Pronoun*), the *Peri epirrematon* (*On Adverbs*), *Peri syndesmon* (*On Conjunctions*), and the four-book *Peri syntaxis* (*On Syntax*).[78] Titles of other works are known through references in scholia, the *Suda*, and Apollonius' own works. Apollonius' grammatical work was highly influential, as was the work of his son, the grammarian Herodian.

Herodian, one of the VMK group, was born in Alexandria and traveled to Rome while Marcus Aurelius was emperor (161–180). Best known for his work on prosody and accentuation, he also produced an extensive range of grammatical works. He wrote a work entitled *Katholikes prosodias* (*On General Prosody*) in twenty books, dedicated to Marcus Aurelius, which dealt extensively with accentuation and may have provided rules for accenting as many as sixty thousand words.[79] His works on prosody in the *Iliad* and the *Odyssey* have been reconstructed from scholia, including the Venetus A.[80]

In addition to the grammarians, the lexicographers are also crucial to modern understandings of ancient texts. Harpocration, active in the second century AD, wrote a lexicon on terms used by orators. His work is notable both as a source of fragments and information on ancient orators and for its fully alphabetized entries, unlike other works that are grouped only by the first few letters of the word.[81] Perhaps, however, the most famous ancient lexicon is that of Hesychius, who lived in the fifth or sixth century AD. Hesychius provides insight into many rare words and even provides definitions for proper names.[82] His lexicon is not preserved in its entirety, but large portions of it were transmitted by Byzantine scribes.

One final, monumental work of scholarship crucial for our knowledge of ancient authors remains to be mentioned. In the late tenth century, a group of scholars compiled an enormous work, known as the *Suda*, that combined lexical entries from Hesychius' great lexicon with encyclopedic entries. The *Suda* contains about thirty thousand entries largely based on lost sources and therefore often provides information on authors and their works that is attested nowhere else. Strikingly, approximately five thousand of the *Suda*'s entries are derived from Aristophanes' texts and scholia, a proportion that is difficult to account for.[83] Despite the somewhat idiosyncratic nature of its sources, the *Suda* is an indispensable tool for modern studies of ancient history and literature.

In contrast to the developments in Alexandria, early scholarship in Rome does not seem to have been so prolific. Suetonius provides an insightful, if fanciful, anecdote regarding a visit from Crates of Mallus.[84] Suetonius relates that when Crates had been sent on an embassy to Rome (ca. 168 BC), he fell in a sewer hole and broke his leg. During his recuperation, he gave lectures in Rome and thereby provided a model for Romans to imitate. Suetonius, several centuries removed from the event he relates, appears to exaggerate its importance, neglecting the fact that by the time Ennius arrived in Rome (203 BC), the Romans were already quite taken with Greek culture and ideas.[85] Given the prevalence of philhellenism in the early second century, it seems unlikely that Crates would have introduced rather than simply advanced Roman interest in scholarship.

Furthermore, the dates of Suetonius' examples of the "imitators" of Crates are questionable, as are the particulars of their activities.[86] Suetonius credits Gaius Octavius Lampadio with dividing Naevius' *Punic War* (*Bellum Punicum*) into seven books, but the extent of his work on the text is unknown. He also comments that Quintus Vargunteius frequently read aloud (*pronuntiabat*) Ennius' *Annales*, though it is not certain how or if Vargunteius engaged with the text in any scholarly way.[87] Suetonius further reports that Laelius Archelaus and Vettius Philocomus worked on Lucilius' *Satires*. The vagueness of Suetonius' account provides little aid in understanding the state of philological activity in Republican Rome, but clearly the influence of Greek scholars was at work alongside early Roman literature.

What Suetonius' references lack in specificity, they make up for in numbers. It is thanks to his *On Grammarians and Rhetoricians* that we have the names of numerous Roman scholars, some of whom we know only from this work. There is, unfortunately, little to say about most, especially since Suetonius' own knowledge of the figures is often questionable. The lack of particulars notwithstanding, certainly the academic tradition of Alexandria, Pergamum, and Athens continued in Rome, which itself became an important cultural center.

Literary production was an appropriate pastime for the Roman senatorial class, and scholarly activity was regarded in the same light. It will suffice here to survey the beginnings of Roman literature and consider a few of the more famous Romans who pursued interests in grammar and scholarship. Although the Romans were making use of writing by the seventh century, written Latin literary and scholarly activity was a relatively late phenomenon. The ability to write did not make writing ubiquitous for all purposes, and culturally significant works — that is, literature — were often reserved for oral performance, even in cultures that possessed writing.[88] Oral culture was dominant in Roman society until about the end of the fourth or early third century BC, and continued to persist for centuries alongside the steadily increasing role of books and writing.[89] As written works of literature began to appear, the Romans looked to the Greeks for literary models, and, in fact, the first literary work in Latin was written by a Greek.

Traditionally, Latin literature is considered to have begun with Livius Andronicus, a Greek slave who was brought to Rome and who put on the first play in Latin in 240 BC. He also translated the *Odyssey* into Saturnian verse, a metrical form still poorly understood by modern scholars. Latin literature, of course, predates Livius Andronicus, as does Roman philhellenism, though earlier works exclusively preserved by oral tradition were lost already by Cicero's time.[90] Even Ennius' epic *Annales*, the masterpiece of Republican literature, owes a significant debt to the Greeks. It was a groundbreaking work, written not by a member of the senatorial class, but by the provincial Ennius (239-169 BC), who was invited into the entourage of Marcus Fulvius Nobilior and traveled with him during his campaigns.

Ennius was fluent in three languages, Latin, Greek, and Oscan, and his most lasting contribution to Latin literature was the application of his knowledge of Greek language and literature to his epic. His modification of Greek dactylic hexameter to Latin verse would become the norm in subsequent Roman epic. Ennius' *Annales* may well have been the first work to draw the attention of Roman scholars in the second century. Perhaps the influx of texts brought to Rome as spoils of war in 168 BC by Aemilius Paullus helped spur interest in grammar and scholarship; in any case, fragments of Ennius' *Annales* are extant thanks in large part to quotes preserved by ancient scholars.[91]

Marcus Terentius Varro (116-27 BC), a contemporary of Cicero and, like Cicero, of the equestrian order, led active intellectual and political lives. Varro studied first under the grammarian and (perhaps) *kritikos*, Lucius Aelius Stilo Praeconinus (born in the mid-second century BC), and later in Athens under Antiochus of Ascalon. The influence of both teachers is evident in his works. He allied himself with Pompey and held a number of offices, such as tribune and praetor.[92] As the civil war turned against Pompey, Varro was reconciled with Caesar, who later tasked him with organizing a public library.[93] This work was interrupted by Caesar's assassination in 44 BC, and Varro was subsequently added to the proscription list by Marc Antony. Varro's possessions were seized, but he survived the brutal purge that would claim Cicero's life.[94]

Varro was a prodigious scholar who wrote on a number of subjects. His approach to antiquarianism was apparently novel, as Cicero credits Varro with "explaining the age of the fatherland, descriptions of its times, the rights of its religions and priesthoods, the domestic and wartime practices, the location of its lands, the names, types, offices, and causes of all its secular and divine matters."[95]

Varro followed his teacher, Aelius, in attempting to determine which Plautine plays were genuine. Aelius first identified twenty-five such plays, a difficult task owing to his reliance almost entirely upon what he considered "true" Plautine style.[96] Varro, in his *On Plautine Comedies*, confirmed only twenty-one genuine plays, a list that was not immediately authoritative. The current canon of Plautine plays does, however, come from Varro, whose identifications of genuine plays remain extant in the form of a palimpsest from the third or fourth century AD.

The only fully extant work of Varro's is his three-book work on agriculture, the *De re rustica*, a technical manual on farming presented in the form of a dialogue. One may see in this work, too, Varro's interest in etymology. Another of Varro's works is the indispensable *De lingua Latina*, which is the earliest extant work on the Latin language in Latin.[97]

Another influential *grammatikos* and contemporary of Varro and Cicero was Nigidius Figulus. Although only fragments of his massive *Commentarii grammatici* (in at least twenty-nine books) remain, Figulus' work was influential. Gellius goes so far as to place him on par with Varro.[98] He was eventually seen as something of a mystic, as Lucan depicts him predicting future disasters by reading the stars, while Cassius Dio has him prophesying Octavian's rule at the moment of his birth.[99] Finally, even Julius Caesar concerned himself with grammatical works. While in Gaul, perhaps ca. 54 BC, he wrote his *De analogia* in two books.

The number of names and fragments of Roman authors who wrote grammatical works is extensive, but, unfortunately, little remains. Nevertheless, studies of grammar, syntax, etymology, accentuation, and the rest that began in antiquity remained influential for centuries, even as they were augmented by subsequent writers. But the material those writers wrote upon would soon change.

CONSTRUCTION OF THE CODEX

We have discussed already the process and terminology for constructing a roll. Yet the roll would cede its place to the less cumbersome codex by the end of the second century AD. A codex was made up of sheets (*bifolia*) folded in half at the midpoint to form leaves.[100] Each leaf could hold writing on the front and back and hence consisted of two pages. Two *bifolia* stacked together and folded would yield four leaves and eight pages. Most codices used a standardized set of four *bifolia* to form a *quaternio*, a Latin term from which we derive the term "quire."[101] A "quire," or "gathering," refers generally to any number of sheets, or even one large folded sheet, that forms a single grouping of leaves. The *quaternio*, by contrast, specifically denotes a group of four *bifolia*.

The fabrication of papyrus, with horizontal strips layered over vertical strips, produces a *recto* and *verso* for each sheet. Though the terms *recto* and *verso* most properly describe papyrus, parchment may be considered to have a *recto* and *verso* as well. The two sides of a sheet of parchment are the flesh side, which is brighter and smoother (*recto*), and the hair side, which has a yellow tinge and could appear spotted from the hair follicles (*verso*). Ideally, the codex would be constructed in such a way that, when opened, two facing pages would both be *recto* or *verso*, which would create a pleasing uniformity that masked the quality difference between *recto* and *verso*.[102]

Another difference among codices was whether they were composed of a single quire or multiple quires. Single-quire codices were generally limited in the number of leaves they could contain. Large single-quire codices could be made, but had disadvantages. Turner comments on such a codex, which contains the plays of Menander, observing that it was constructed of sixteen sheets stacked together and folded down the middle. The resulting codex was prone to breaking away from its spine. Further, when so many sheets were stacked together, the pages of the codex would not line up when closed.[103] The difference in width could be substantial, averaging, for example, 20 millimeters for the leaves of the Nag Hammadi codices.[104] The pages could be cut to form an even appearance, but those in the middle would then be noticeably smaller than the outer pages.[105] The text would likely not have been added until the codex was complete, meaning that careful planning would be needed since there would be no way to add leaves to the codex once it was bound, which was typically done using a pair of leather thongs strung through holes in the centerfold.[106]

Lengthy codices could be accommodated by stitching multiple quires together. In this case the text would have been written on the quires before they were bound together into a codex. It is important to note, however, that the multi-quire codex did not develop from the single-quire codex; multi-quire codices date from as early as the second century AD, and even outnumber single-quire codices.[107] The multiple-quire parchment codex eventually became the dominant form of codex, with quaternions as a standard quire size.[108]

Before a parchment could be used to hold text, it had to be prepared to receive ink. Since monastic communities did not produce their own parchment, they likely would have had to treat the material they acquired by smoothing it with pumice and coating it with chalk before it was inscribed.[109] Sheets were often ruled in preparation for writing. Horizontal rulings on the page were bounded by two vertical lines; the horizontal lines aided the scribe in writing straight lines across the page, while the vertical lines made it easier to write a straight column.

In the medieval period, pages were ruled with a hard stylus drawn across the parchment, which produced indentations. In the twelfth century, lead was used, which was then replaced by ink in the fifteenth century.[110] To produce perfectly straight horizontal lines, "prickings" were used. Prickings are holes in the parchment made in a vertical line on either side of the page; lines drawn between corresponding prickings produced straight horizontal ruling. When pricking first became standard practice is unknown, but it was in use by the mid-fifth century AD.[111]

The holes made by pricking were a convenient aid for the scribe since they could be made at the same time for multiple *bifolia* stacked together and were nearly invisible to the reader.[112] Early manuscript prickings appear within the space overwritten by text; in later manuscripts, beginning in the mid-fifth century AD, they appear in the outer margins.[113] Prickings were made in a variety of ways (with an awl of various shapes, with a compass, or perhaps with a wheel that produced punctures as it was rolled along the page),[114] locations on the page, and shapes (as holes or as slits, which could be horizontal, vertical, or angled). These variations were often particular to a time or place, and so are important tools in determining a manuscript's date and origin.[115] Since monastic scribes probably pricked and ruled their own parchment, sometimes it is even possible to identify particular scribes based on prickings.[116]

PALEOGRAPHY AND CODICOLOGY

The majority of surviving manuscripts date from well after antiquity, although, as we have seen, the roots of manuscript production and scribal activity reside in antiquity. The term "manuscript" most literally refers to books or documents written by hand. This definition, naturally, is so broad as to require some refinement. The study of Western manuscripts is typically the study of parchment codices, although this could include papyrus codices as well (papyrus documents and rolls fall to the fields of papyrology and, for literary texts, philology). Depending on the breadth of study, manuscripts can sometimes even be printed texts.[117]

Manuscript studies largely belong to the fields of paleography and codicology. Paleography in essence is the study of ancient handwriting, but is sometimes broadened to a more general study of the issues that pertain to writing in manuscripts. The origins of paleography are rooted in the seventeenth century, beginning with Jean Mabillon (1632-1707), who developed the science of paleography in his *De Re Diplomatica* of 1681.[118]

Mabillon divided chirography (handwriting) into several categories: Old Roman, Gothic, Lombardic, and Saxon. He initially subdivided Roman script into capitals, uncials, and a third script based on a single (now known to be forged) document. The discovery was groundbreaking, and has since been invaluable to scholarly understanding of ancient texts. Though his contribution in shaping the field of paleography was great, Mabillon's nomenclature did not go unchallenged. The science of paleography quickly drew scholarly attention, and new taxonomies that varied in complexity were proposed, including broad divisions into majuscule and minuscule texts, as well as numerous subdivisions of cursive.[119]

Codicology, a discipline related to paleography, examines the codex as a physical object, focusing on the particulars of how codices were constructed, including the materials used, the method of binding, and other pertinent features. Codicology has provided considerable advances in our understanding of how and why codices developed as they did, as well as enabling theoretical reconstructions of the contents of damaged codices based on the evidence of the remains. The first use of the term "codicology" occurs in a work titled *Les manuscrits* by the Hellenist Alphonse Dain in 1949. Dain's purpose was not to establish a new scholarly field, however, and his ideas about codicology were restricted to the preservation and distribution of manuscripts, information supporting the work of dating and localizing texts that was proper to paleography.[120]

Soon after Dain's book came out, François Masai challenged Dain's views on the scope of paleography and codicology, proposing some ideas of his own. He essentially identified the former as a subspecies of history and the latter as properly part of the field of archaeology.[121] The scientific nature of codicology was also being explored in German handbooks, such as those by Löffler and Kirchner, although the field was still nascent.[122] Of course, a proper understanding of a manuscript is likely to include both paleographical and codicological analyses, and the two disciplines are, in a sense, intertwined. Thus, scholars refer to the sum of information pertaining to the study of manuscripts, both paleographical and codicological, as "manuscript studies."

The study of scripts has contributed significantly to our understanding of ancient and medieval textual production. It is often possible for paleographers to assign dates and locations to manuscripts based on chirography, especially since in many cases large scriptoria, such as that at the monastery in St. Gall, developed their own scripts as a result of the intensity of scribal activity. In addition to assigning origin and provenance (history of ownership), paleographers also study other marks, sigla, and punctuation. These studies have revealed much about scribal practices in general, including on occasion the habits of individual scribes.

DEVELOPMENTS IN HANDWRITING

Well before formal book hands developed, the Greeks had begun writing with an instrument called a *calamus*, a stiff reed sharpened to a split point. The *calamus*, which functioned as a pen, was then dipped in ink consisting, in the earliest years, of soot and arabic gum dissolved in water.[123] This practice continued in the Western tradition, and by the fifth century AD the preferred writing instrument was no longer a reed but a quill made from a goose feather.[124] Use of the *calamus* existed in Alexandria alongside Egyptian practice, which employed a stiff, fine-point brush made from a rush. Egyptian ink was not dissolved in water, but rather the moistened brush was rubbed on the dried ink and then painted onto the papyrus.[125] The use of the *calamus* allowed for the creation of finer strokes and smaller lettering than possible with the brush. The composition of Greek ink changed by the second century AD from what we term "India ink" to mordant inks. Mordant inks contain metallic elements but, unfortunately, fade from black to brown over time, making them difficult to read.[126]

Broadly speaking, writing served two purposes. Its formal function entailed copying books and documentation. Conversely, individuals used it informally for such purposes as letter writing. Early scripts were derived from monumental inscriptions and therefore demonstrate some features particular to lapidary writing. Replicating epigraphical forms using a pen requires deliberate effort and additional time. It was easier and faster to write with the pen in continuous contact with the page, in a style that came to be known as cursive,[127] and as a result, features such as ligatures developed, that is, letters joined together by alteration of one or both letterforms. An example can be seen in the contraction of the Latin *et* into the still-used & ligature. While Greek scripts were highly conservative and retained many features of lapidary letters, as can be seen in extant papyri, writers of Latin took much greater liberties.[128]

Cavallo and Maehler propose that the predominance of oral culture as late as the early fourth century BC may have precluded the need for a large number of documents to be written quickly, thus deferring the development of a cursive script. The situation changed in the Hellenistic period, when the creation of the Alexandrian library generated large amounts of scribal work and likely led to the development of new scripts to accommodate a faster pace of writing.[129] Hellenistic scripts tended to retain their debt to the lapidary letterforms and kerning (proximity of letters). Greek cursive came into use only in the mid-second century BC. The emerging cursive style was characterized by abbreviations and ligatures, which suggest a rapid pace of writing.[130]

Roman inscriptions of the Republican period are in capitals, and are rather plain, with the strokes of the letters of equal width.[131] During the reign of Augustus, the increased availability of marble, which was superior to the tufa previously used for carving inscriptions, allowed for new practices with letterforms, and thus the *scriptura monumentalis*, also known as *capitales quadratae*, became prominent throughout the empire.[132]

The earliest examples of Latin cursive can be dated to about the first century BC, but earlier monumental letterforms clearly reflect their debt to the

FIGURE 2.3. *Building inscription, AD 487–488. From Sevastopol, Chersonesus. Drawing by Mary Claire Russell MacDonald.*

Etruscans and, before them, the Greeks. By the first century AD, one may find elements of cursive writing making their way also into inscriptions with increasing frequency.[133] In the third century AD, rounded capitals and uncials appear in an inscription in North Africa. The use of these letterforms elsewhere in the empire (e.g., the Chersonesus; see fig. 2.3) reveals their widespread prevalence. The encroachment of elements of handwritten scripts into carved monuments occurs in Greek inscriptions as well. Rounded *epsilon* (ϵ) and lunate *sigma* forms (c), as well as double-rounded *omegas* (ω) can be found from the third century AD.[134]

Different scripts developed for different purposes, and it is sometimes difficult to trace the line of development. As was the case for the Greeks, the Romans developed formal scripts for books (book hands) and more casual cursive scripts for documents and informal writing. Roman cursive is generally divisible into two types, majuscule and minuscule. "Majuscule" is a script in which the lettering is all the same size and fits within a band of writing. By contrast, "minuscule" letters are not all of equal height and some strokes may rise above or descend below the level on which the letters are set. Majuscule and minuscule forms of writing have continued to exist alongside each other, ultimately producing the upper and lower case in use today.[135] The older majuscule cursive is often referred to as Old Roman Cursive (ORC), and it was in use until about AD 300; the later minuscule cursive, known as New Roman Cursive (NRC), was dominant by the late third century AD and was a harbinger of the Carolingian minuscule of 800.[136] Examples of Roman cursive writing are, unfortunately, infrequent. An important exception is the collection of wood tablets found in Vindolanda, which preserves numerous examples of Old Roman Cursive.[137]

SCRIBAL PRACTICES IN ANTIQUITY AND THE MEDIEVAL PERIOD

In late antiquity, the book trade grew and gained legitimacy as texts spread throughout the Roman Empire. Of the ancient scribes little is known, other than that they were slaves or freedmen, and the evidence grows as the centers of writing shift from shops to the monasteries. By Augustine's time, public officials of sufficient rank could have books copied for themselves at a discount,[138] perhaps a continuation of scribal services performed by early booksellers. As early as the late second century, scribes were adding *colophons* (known in Latin as a *subscriptio*) to their work,[139] that is, notifications that identified the shop where the text was copied, and sometimes the date and the name of the scribe.[140]

Colophons are not ubiquitous, but when present they provide valuable evidence for a manuscript's history and scribes' attitudes toward their work. While many monastic *colophons* are devotional, others served as an opportunity for the scribe to complain about the challenging work of copying, call curses down on would-be thieves, or simply rejoice that the work was completed. One common rhyme that appears at the end of manuscripts is *Explicit hoc totum / Pro Christo da mihi potum* ("The whole work is finished by my quill / For Christ's sake give me a swill!").[141] Another theme found in *colophons* is a demand for payment for the finished work. One fourteenth-century manuscript reads *laus tibi sit, Christe, quonium liber explicit iste / detur scriptori merces equate labori* ("Praise be to you, O Christ, since the work is finished / May the scribe be given a payment equal to the labor").[142] Scribes sometimes apologize for errors or defiantly challenge the reader to do better: *O bone, non ride; vis melius scribere? Scribe. / Lauda scriptorem donec vides meliorem* ("O friend, don't laugh; do you want to write better? Then write. / Praise the scribe until you see someone better").

Finally, as noted previously, scribes often identify themselves. Scribes name themselves in a number of ways, sometimes directly, but sometimes in a more playful fashion, as in the case of the scribe called "Johnny": *Iohannes: Hic -nes est finis, -han medius, Io- quoque primus* ("The *-ny* is last, *-hn-* is the middle, and *Jo-* is first").[143]

The rise of Christianity, and especially self-sustaining monastic communities, created a new source of book production. Literacy was already an important part of monastic life in the fifth and sixth centuries. The *Rule of Saint Benedict* (early sixth century), which provides instructions for operating a monastery and the daily activities of the monks, assumes its members are literate, but makes no mention of book production. Even earlier, in 404, a passage from Jerome's translation of Pachomius (a fourth-century monk, who is considered the founder of cenobite monasticism) asserts that monks who cannot read must be assigned a teacher, and those who refuse to learn must be forced.[144]

Literacy played an important role from an early point in monastic life, but that is not to say that a high level of literacy was required of all monks; some likely remained largely illiterate, while others may have attained enough literacy to meet the minimum prescription of reading a single book in a year.[145] This attitude toward literacy corresponds to the initially modest aims of monastic scriptoria. Monastic communities in the first place created copies of texts required for their own worship, which was but one aspect of monastic life; others included agriculture, crafts, and any other tasks needed to meet the basic needs of the monastic community.[146] Among these activities would have been literate monks teaching their illiterate brothers to read, but also emphasizing the memorization of scripture and liturgy. In fact, learning to read was not part of the standard educational curriculum, but was a precursor to it, and in the sixth century was still more akin to manual labor and crafts.[147] It is in this context that one should consider the monastic scriptorium.

The term *scriptorium* typically refers to a room in a monastery dedicated to the activity of copying texts. Not all monasteries, however, had a room designated for solely that purpose; for some monasteries, the scriptorium may have been wherever the scribes found a suitable place for their activity.[148] Some *colophons* indicate, for example, that the scribe worked outdoors.[149] Most monasteries would not need space dedicated solely to book production, which waxed and waned as need arose.[150] For the textual endeavors of especially active monasteries, the scriptorium was a dedicated space, and could be quite large.

Our understanding of scriptoria must be largely constructed from written evidence like *colophons*, because the archeological evidence is extremely slight.[151] However, some sense of what a large scriptorium may have looked like comes from a ninth-century plan for the monastery at St. Gall (fig. 2.4). That monastery, founded in 720, had a functioning scriptorium by the end of the eighth century. Under a series of abbots who cultivated the work of the scriptorium, the holdings in St. Gall's library grew, necessitating the creation of a library catalogue, which included some 584 books by the end of the ninth century.[152] The plan for St. Gall, never fully realized, called for an enclosed scriptorium above the library (fig. 2.5); the center would have contained a large table, perhaps to store books when they were not being used to make copies.[153] The north and south walls of the room were to be lined with seven writing desks.[154] These desks could be shared by the monastery's scribes, who would work at different times during the day.

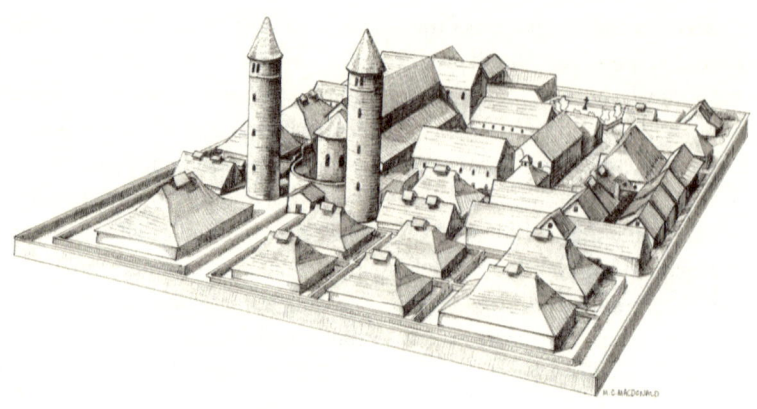

FIGURE 2.4. *Monastery of St. Gall. Drawing by Mary Claire Russell MacDonald.*

By one estimate, based on the number of unique hands found in manuscripts from St. Gall, there were as many as one hundred scribes working in the monastery in the ninth century.[155] It was surely advantageous to limit a scribe's work to only a few hours at a time, since arduous moiling[156] over a text for a sustained period would surely have diminished the quality of the work.[157] Even so, the labor was draining.

Book production was not limited to monasteries. The nuns of the early convent at Arles made their community an important center of book production as well.[158] It is important to note that, while men could play a role in copying and overseeing scriptoria at convents, the nuns were well educated and vigorously engaged in scribal activity.[159]

Great monastic libraries and scriptoria arose in many parts of the Christian world and formed important intellectual centers. Nevertheless, smaller monastic communities were limited by the availability of resources, especially parchment, and would have been less prolific. Much also depended on the monastery's leaders. The fourth-/early-fifth-century theologians Jerome and Rufinus greatly fostered the growth of their monastic libraries.[160] Similarly, in the mid-sixth century Cassiodorus founded a monastery known as the Vivarium on his Calabrian estate. He is credited with establishing the principles and methods of monastic scriptoria.[161]

Religious texts were not the only products of early medieval scriptoria. Rufinus of Aquileia, who established a monastery on the Mount of Olives along with Melania the Elder, had works of Cicero copied for Jerome.[162] The contents of monastic libraries and the products of their scriptoria extended beyond the communities themselves. Monks were allowed access to the holdings of other monasteries and were able to augment their own monastery's library by making copies. The early sixth-century *Regula magistri*, which contains guidelines for the organization of religious communities, acknowledges the work of scribes and instructs that all surplus material produced by the monastic community be sold outside the monastery at below market price.[163]

Prior to the invention of movable type, there is no evidence that a large number of copies of a single book were produced in order to create a stock of books to be sold.[164] As already observed, the most immediate concern of the scriptorium was to meet the needs of the particular monastic community. Copies made beyond those needs were individual requests, not large "print runs." Despite

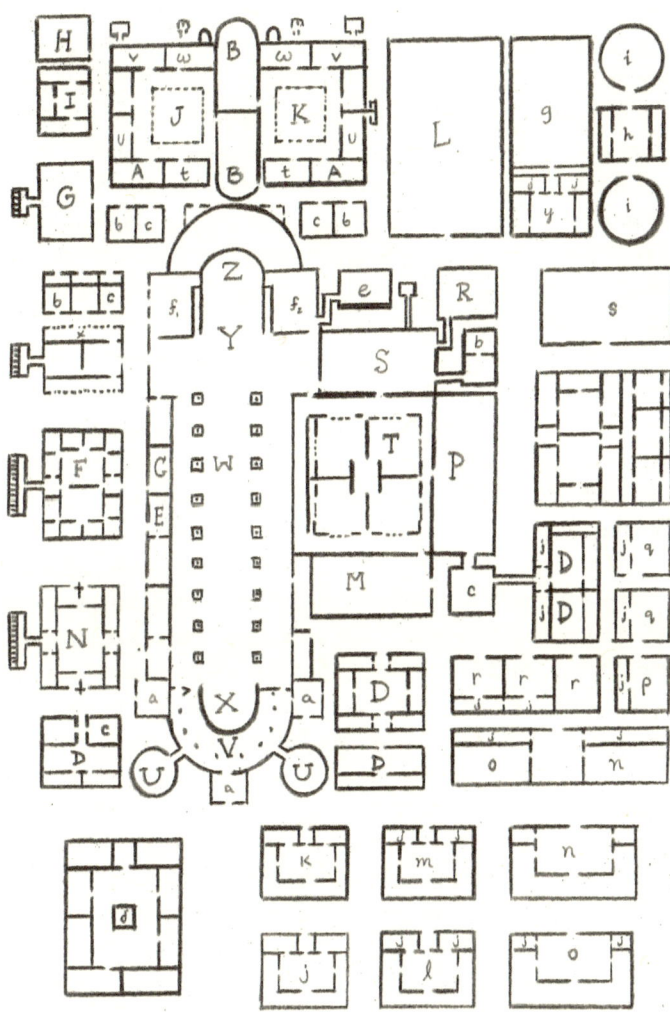

FIGURE 2.5. *The floor plan of St. Gall; Room f1 is the scriptorium. Drawing courtesy of Mary Claire Russell MacDonald.*

the large size attained by some scriptoria, they should not be considered sites of mass production or publication, as the prohibitive nature of the cost of materials militates against this notion. The amount of parchment needed to produce a single Bible, for example, would have been considerable. Those who wished to employ the scribes of a scriptorium to produce a book would likely have had to provide the material for it.[165]

The monasteries, of course, required books for their own purposes, which could be acquired by several means. If an *exemplar*—that is, a text from which a copy could be made—were present, the book could be copied in the scriptorium. Copies of other texts, particularly those sought by scholarly abbots who desired to increase the holdings of their libraries, might be requested from another monastery that had an *exemplar* in its holdings.[166] In either case, in-house copying at one or another monastery entailed the practical concerns of acquiring ink and parchment.[167] These materials had to be purchased, and several sources of funds were available: monasteries received income from services like book production, from the sale of surplus goods, and from donations. The generosity of just such a donor was the impetus for the only partially realized plan of the monastery at St. Gall. Donations were not limited to monetary contributions, and those of books were not uncommon.

Book production required more than mere copying, and the most ornate books required even more steps, such as rubrication and illumination. The amount of time needed to produce a book is difficult to determine, but such scheduling was an important component of the workings of a scriptorium. By compiling evidence from manuscripts, especially *colophons* in which scribes state when they began and finished a work, Michael Gullick has calculated that on average a scribe would write approximately one hundred and fifty to two hundred lines per day.[168] That figure can only be a guideline, of course, but it does give a reasonable sense of how arduous a task copying a manuscript was. For example, a small thirteenth-century Bible—containing 408 leaves and copied at the rather slow rate of 126 lines per day (at six days per week)—would have taken 450 days to copy.[169] Some scribes, however, worked considerably faster, at a rate of three hundred or more lines per day. The time required simply to copy a text, however, was variable.

Some of the variability in speed with which books were copied can be explained by the different capacities in which scribes operated. As noted, monks who copied texts, especially in the stricter orders, generally worked for only four to six hours per day. Books might also be copied by scholars for their own personal use or by professional scribes, that is, individuals who made a living by copying texts.[170] Professional scribes might be employed by a monastic community if their own scribes were inundated with texts that needed to be copied.[171] Unlike the monks, who also charged for their services but whose time at the writing desk was regulated, these professional scribes had no limits imposed upon the time spent copying. Like their ancient forebears, they did not charge by the hour, but rather their fee was based on the amount of text to be copied, and so they were well incentivized to work as quickly as possible.[172]

Sixty-four scribes were active in the scriptorium of St. Mary Magdalene in the latter half of the twelfth century,[173] of whom four seem to have been the scriptorium's master scribes and whose hands can be seen intervening in texts of unsatisfactory quality.[174] Scribes were expected to match their *exemplars* in quality, and a poorly made copy could be rejected by the customer. In the 1070s, for example, the archbishop of Canterbury, Lanfranc, sought a copy of Pope Gregory's *Moralia in Job* from a prior named Anselm. Anselm found a scribe for the work, but wrote back to the archbishop apologetically when it was determined that the scribe's handwriting was too poor and the project would have to be done again when a qualified scribe was found.[175] The scribe in question was likely a professional whose work was unmonitored.

When necessary, scribal labor could also be drawn from students in the monastery's school, though typically book production was left to the more experienced scribes. After all, copying a text entailed more than mere reproduction; the scribe would first have to learn to prepare a manuscript by studying and practicing the *exemplar* of his overseer, which also meant learning to write using a common script.[176] This he would have to do well enough to write with some speed, since the ink would pool if the pen were allowed to rest too long on the parchment. Lines were employed to ensure that he might write in organized, justified columns.[177] In addition, the scribe would also have to plan ahead by allowing space for decoration or illumination. Thus, the production of a book required forethought and active care for both copying and ornamentation. Accordingly, only talented monks were allowed to work on books. An individual skilled enough to be entrusted with the task of producing a book was called an *antiquarius*, while a novice, given tasks of lesser importance, was known as a *librarius* or *scriptor*.[178]

The act of transcribing a text was long and difficult, but was only one of the many activities that went on in the medieval scriptorium. Some of the work preparing quires for copying has already been noted, but other roles had to be filled to make a book of high quality. Some of these positions focused on decoration. One such task was that of the *rubricator*, the term for which comes from act of writing in red ink (the Latin *ruber* means "red").

The rubricator, like the *scriptor*, did more than merely copy text, however, and in fact rubrication was a task in which a scribe might specialize.[179] The rubricator used red ink for certain aspects of the manuscript, some of which may not have been in the *exemplar*. Rubrics could be used to write the author's name and title of the book, mark divisions in a text, provide chapter headings, or even offer a moral commentary.[180] Paragraph and verse initials were also rubricated. Since rubrication was added after the text was completely copied,[181] space for the rubrication, including leaving blank areas for initials, would be incorporated into the scribe's layout of the text prior to copying.

Another role of monks in the scriptorium was that of illuminator, the artist who designed the portraits, heraldic images, pictures, and other ornamentation that has become iconic in medieval manuscripts.[182] The illuminator could be a specialist, or could have handled all of the art for a manuscript, including the rubrication. Particularly skilled scribes were known to have copied, rubricated, and illuminated manuscripts by themselves.

To ensure the quality of the book being produced, proofreading and correction of manuscripts was needed. This task fell to the most learned monks of the scriptorium, and sometimes even to the abbot.[183] The corrector would compare the copied text to the *exemplar* and correct mistakes by adding notes in the margins.[184]

Once the quires had been copied, rubricated, illuminated, and corrected, they could be bound together into a codex. Various materials were used for book covers, including simple parchment. More typically, however, wood (normally oak)[185] was employed. Wooden boards were covered in fabric or leather, and could serve as a base for metalwork and jewels.[186] The two boards were joined using leather thongs that were pasted to a leather backstrip that served as the book's spine.[187] Before the ninth century, quires were sewn directly into their covers, a technique that persisted in the East.[188] In the West of the Carolingian period, however, quires began to be attached to supports of flax, which were essentially cords laced onto the wooden boards.[189] Attaching the quires to the sewing supports rather than joining them directly onto the boards was an important technique for adding strength, and accommodated larger books.[190]

Levels of literacy and book production must have varied significantly from one monastery to another. Furthermore, while monasteries played an increasingly important role in the Middle Ages as centers of education, libraries, and book manufacturers, they did not replace traditional publishers. Universities, too, played an important role, for university students copied texts under what is known as the *pecia* system. The growth of universities created a consistently high demand for a relatively few texts, which required a fast, efficient means of producing books. The process began with a professional scribe, who would use an *exemplar* to make a copy of a book that others could copy. The reproduction made by the scribe came to be regarded as the authorized copy after it passed the inspection of a committee of masters.[191]

Various methods of copying texts were available, but it is generally true that in the early Middle Ages a shift occurred, away from copying by dictation to copying from an *exemplar*. Of course, neither practice was ever completely absent in any time period, especially if it was necessary to produce books under extraordinary circumstances. For example, it is suspected that the instant popularity of Dante's *Divina Commedia* led to the quick production of copies by dictation to meet the demand.[192]

SCRIBAL ERRORS

Whether by dictation or visual reproduction, however, the act of copying always poses a great risk of incorporating errors into the copied text. One can sometimes identify whether these are due to an aural or visual lapse; that is, whether the mistake was made in copying from dictation or from line-by-line copying, but more often than not, the evidence is ambiguous.[193] In addition, it was not unusual for an *exemplar* to be changed intentionally, if the copyist or corrector identified an error, real or imagined, in the original. These changes in the copied text, whether intended or not, are known as "corruptions."[194] Corruptions can be simple, as in the case of a misspelled word, or complex, as when a scribe compounds the mistake of a previous scribe in attempting to make a correction and puts the text at an even greater remove from its original form.

The line-by-line method of copying popular in the medieval period involved reading and memorizing a sequence of text that was then transcribed, with the copyist's glance moving alternatively between the original (the *exemplar*, or, as it is also known, the *antigraph*) and the folio used for the copy (sometimes called the *apograph*). This process affords opportunities for various types of errors resulting from the copyist's visual perception or a mnemonic lapse. Among the possible errors, prominent are those that modify a single word, those that involve omissions of words or sequences of texts, and those that comprise additions of words or sequences.

In the case of single-word errors, the word transcribed by the copyist somehow comes to differ from the original, with modifications involving a mere letter, a syllable, or more elements. The word may be deprived of meaning as a result—for example, the occurrence of the nonsensical *narravertit* instead of *narraverit*. Such errors might have been due to negligence, but it should be kept in mind that not all scribes were equally fluent in Latin.[195] In cases where meaning is lost or unclear, it can sometimes be easy to infer the original word. Typically, however, scribes knew Latin and thought in terms of words and meaning rather than individual letters, and thus they were likely to mistake entire words.[196] The corruption of a single letter, conversely, could often yield a word of different but still comprehensible meaning. For example, *amnes* ("rivers") instead of *omnes* ("all") or *rapida* ("swift") instead of *rabida* ("frenzied"). Uncommon words were particularly susceptible to corruption into more common ones.[197] Sometimes the better of two alternate readings caused by the corruption of a single word cannot be securely determined; in fact, many variants that characterize the tradition of the Virgilian text are of this type.

Although corruptions of a single letter typically result in a meaningful Latin word, particular letters that hold traits in common with another letter were especially prone to change. In the first place, the appearance of any letter is obviously subject to the individual's skill in writing, and this, if deficient, can easily cause cases of ambiguity. Furthermore, the forms of some of the letters in various types of scripts (majuscule, minuscule, capitalis, Visigothic, Caroline, etc.) are particularly prone to confusion. In uncial script, to be discussed below, confusion between *i* and *e* is frequent, for example, *virum*, a noun meaning "man," in place of the adverb *verum* ("but") and vice versa. In Carolingian minuscule script, *s* and *f* are differentiated only by the horizontal stroke of the *f*, resulting in a high frequency of error in transcriptions, for example, *forti* ("to the brave one") instead of *sorti* ("by chance"), and the name *Fufidius* corrupted into *Fusidius*.[198]

Such a typology of corruption establishes an important clue for reconstructing the tradition of a text: if an error of a certain type is frequent in the text, one can hypothesize that the *exemplar* used by the copyist was written in a script prone to that particular form of corruption. Knowing the script of the *exemplar* allows one to postulate its origin (e.g., if a text shows errors usually caused by the Visigoth script, one can conjecture that the text was previously transcribed in the Iberian Peninsula).

Another practice that can generate errors is abbreviation, which was rare in ancient literary texts but became especially common in late medieval manuscripts.[199] Abbreviations were often devised for pronouns, conjunctions, prepositions, prefixes, and commonly used terms. At the extreme, terms of scholastic philosophy could be very highly abbreviated, such as ña for the word *natura* or rbīs for *rationabilis*.[200] More common (and less severe) examples of abbreviation include the enclitic conjunction *-que*, usually abbreviated *q*. The relative pronoun (*qui*, *quae*, *quod*), too, could be thus abbreviated but accompanied by a dash whose position (over, under, or next to the letter) indicates the gender of the pronoun. The Latin accusative ending *-m* when abbreviated is usually represented by a horizontal dash placed above the preceding vowel (the same abbreviation was used for a final ν in Greek).[201] The brevity and similarity of abbreviations made them susceptible to corruption, particularly for those manuscripts written in *scriptio continua*, which required the copyist to reconstruct the individual words, leading to such changes as *incipiet* ("he will begin") in place of *incipe et* ("indeed begin") or *invenere* ("they found") in place of *in Venerem* ("unto Venus").

Two other types of conceptual mistakes should be noted. The first, haplography, is an omission in which one or more letters or words that should appear twice are written only once (e.g., the reading ἄλλος instead of ἄλλοις ἄλλος in a manuscript of Pindar).[202] Another type of error, dittography, occurs when the opposite happens, and text that should only be written once is written twice, whether as letters in a word (e.g., in another manuscript of Pindar, the nonsensical ἑκόνκόντι instead of ἑκόντι) or as the repetition of an entire word.[203]

Mnemonic errors could also crop up during copying. The copyist would read the *exemplar* up to a certain word, memorize that sequence to copy into the *apograph*, and then return to the *exemplar*, starting again from the last word memorized. In so doing the scribe might allow a word to influence the orthography of another that immediately precedes or follows. In the case of the former, the phenomenon is a type of corruption known as "anticipation," in which the scribe changes a word to match the form of a word that follows (e.g., in the Codex Mutinensis Graecus 144, a Greek translation of the *Rhetorica ad Herennium* one finds ἐν χώρῳ ἐρήμῳ for ἐν χώρᾳ ἐρήμῳ).[204] When this phenomenon occurs in the opposite direction, it is called "perseveration."[205]

In addition to mistakes in orthography, scribes sometimes committed errors in the placement of text, often as a result of confusion occurring when returning to their place in the *exemplar* after copying text into the *apograph*. A common locational error involves the omission of a portion of text ranging in size from a few words to an entire line or a cluster of lines. Particularly vexing are instances in which the same word reappears in the *exemplar* a few lines apart: in these cases the copyist, searching for the memorized word at a glance, focuses on the second occurrence rather than the first, and takes up again the copying from that point, resulting in the omission of the portion of text between the two identical words. This phenomenon is identified by the French expression *saut du même au même*.

The transposition of individual letters, words, or even fuller portions of text often derives from an error of omission that had been corrected in the margin of the *exemplar*. The problem begins with the *exemplar*, which during its own creation suffered from an omission of a portion of text or a certain number of verses. The *exemplar* would have been subsequently corrected by having the missing text transcribed into the margin, but without clear indication as to where to insert it in the principal transcription. The subsequent copyist, not understanding what his predecessor was indicating, then inserts what he reads in the margin into the text at the wrong point. Another iteration of this type of error occurs when a copyist has at his disposal a disordered *exemplar*, in which the folios are no longer in the original order and he copies the text with extensive transpositions.

Finally, the necessarily frequent back and forth between *exemplar* and *apograph* could cause unwanted additions to the text. The copyist, in returning his gaze to the *exemplar*, might memorize a sequence that has already been transcribed and write it down it again. Another type of error caused by addition can occur when the copyist incorporates what was originally a marginal gloss or an alternate (often preferred) reading set between the lines of the text. This error is common in cases where the previous copyist had consulted other codices, collecting divergent readings.

Some of these errors, as we have seen, are not difficult to identify, such as when in the process of copying a word an incoherent or grammatically (or metrically) incongruous reading occurs. Errors of omission entail greater problems: if they pertain to a single witness or archetype, the error is nearly irreparable. Larger margins allow for errors of addition, of which duplication is most easily identifiable. At times, the copyist even introduces a missing preposition or adverb, such as *vel* or *alias*, offering alternate readings in the marginalia or between lines. Unfortunately, it is not always easy to determine which of the two variants represents the original reading.

Thus far we have considered numerous forms of corruption that can be attributed to one or another aspect of the mechanical process of copying a text, whether it be a mistake in recognition, memorization, or finding the correct place to resume in the *exemplar*. Another source of error derives from the copyist's preoccupation with something other than the text he is reproducing. It might happen that the scribe writes a word he is thinking of rather than the word he means to copy. These errors of preoccupation are challenging because their identification depends on a judgment of the copyist's thoughts rather than the more logical mechanical errors.[206] Nevertheless, such errors have long been recognized and, like mechanical errors, fall within a typology.[207]

The first of the errors of preoccupation are the so-called polar errors. The term "polar error" was first used by Douglas Young to describe an instance in which a scribe replaced the Greek word μέν with the semantically opposite word δέ.[208] In such errors the word to be copied is replaced by another word of opposite meaning (on the logical-semantic level, but also on the cultural level), for example, *bonum* in place of *malum*. Polar errors are less frequent than mechanical errors, making them all the more challenging to detect.

Another corruption caused by the inattention of the scribe is the alteration of a phrase to that found in another work of literature, quite often Virgil's *Aeneid*.[209] For example, one manuscript of Lucan that should read *audit temptare tenebras* instead has *audit temptare latebras*, most likely influenced by *Aeneid* 2.38 *uteri et temptare latebras*.[210] Similarly, Christian imagery sometimes works its way into texts of pagan authors. In one such lapse, the scribe wrote *cadit post Paulum gratia Christi* for *cadit post paulum gratia ponti*.[211]

Scribes also had to face difficulties caused by their own lack of knowledge. Particular problems were posed to medieval copyists by Greek words in the texts of classical authors, since Greek remained almost entirely unknown in the West until the fifteenth century. In the face of words written in Greek characters, then, medieval copyists tried at times to reproduce the form of the letters, with varied results, or otherwise did not transcribe them, leaving a blank space in the copy.

Other uncertainties were caused in the Middle Ages by linguistic changes in Latin in different areas of the former Roman Empire. For example, the transition away from the diphthong *ae* to the vowel *e* led to the widespread use of the vowel alone in place of the diphthong; analogously, authors reflect the phenomena whereby *v* becomes a fricative and the resulting confusion with *b*, and the oscillation between *t* and *c* before a vowel (e.g., in the form *vicium* for *vitium*).

Notoriously problematic was the reproduction of texts of a technical nature (rhetoric, medicine, etc.). Copyists not expert in the material naturally had an elevated risk of making errors in transcription. Analogous is the oftentimes erroneous transcription of mythological and geographical names in poetic texts.

In addition to unintended corruptions, scribes often altered texts intentionally. These deliberate interventions on the part of the scribe are known as "interpolations." Depending on the work and its context, some forms of interpolation might be expected or even appropriate. Dramatic texts were subject to frequent interpolations by actors, and lexica especially benefited from additions and refinements that could easily be implemented by the scribe.[212] Other genres stood to gain from interpolation as well, including commentaries, medical and pharmaceutical texts, to which prescriptions and remedies were frequently added, and legal texts, whose older accounts of Roman jurisprudence were integrated and updated in the age of Justinian and beyond.

Scribes would sometimes intervene in texts with the intention of correcting an apparent error or even to rectify a deficiency. Confusion over a correct *exemplar* reading occurred in a few manuscripts of Macrobius, which contained a quotation of Accius' *Annales*. The scribes were puzzled by the use of synizesis (treatment of two syllabically distinct vowels as a single syllable), and attempted to correct the meter by changing *eumque* to *huncque* in one manuscript and *cumque* in another.[213]

Intentional interventions in texts thus have several possible motivations. At times the scribe was attempting to clarify the text, and on rare occasions a text would be censored. Such is the case for some manuscripts of Herodotus that omit a passage on sacred prostitution. For one passage of Martial, we find a scribe exchanging the tamer *adultor* for the more scandalous *fututor*.[214] Unfortunately, many of the scribes of the Middle Ages were far removed from the philological methods of ancient Hellenistic scholars, who tended to place comments or proposed changes in the margins or commentaries rather than directly in the text.

ERRORS OF ATTRIBUTION AND PLAGIARISM

The practice of interpolation challenges the notion of recovering an "original" text and compels the philologist to attempt to define all the stages and alterations taken on by a text over time. This can be an extremely difficult task, since late antique and medieval exegesis is often the result of multiple stratifications, in which each commentator selects and revises scholia, glosses, and commentary. Only in a few cases is it possible to connect a stage of this continuous work in progress to a particular name. Then and only then can we speak of this or that commentator's work, bearing in mind that even in such cases that name obscures the contributions of unnamed commentators who came before.

The commentary of the early fifth-century Virgilian scholar Servius is an excellent example. The fullest edition appears under the name of *Servius auctus* (or Servius Danielis, from the name of his first modern editor, the Frenchman Pierre Daniel), completed in the late Middle Ages. The portion that was added later includes material that was actually very old (pre-Servian). This combination resulted in a circuitous investigation that rigorously tested twentieth-century editors of the text, as the failure of the Harvard edition undertaken in 1946 demonstrates. The result is that when a scholar refers to Servius, he or she often does so without distinguishing between the original fifth-century commentary and its later additions. In medieval codices of Virgil, some marginalia select and integrate various sections of the Servian commentary, with differing results; a famous example is that of the Virgilian codex owned by Petrarch. Petrarch's own notes are mixed with those by Servius and other sources.[215]

Beyond the accumulation of errors and interpolations, manuscript transmission is vulnerable to other alterations, both deliberate and inadvertent, that have consequences for the established relationship between the author and his work. The frequency in ancient literature of anonymous, spurious, and plagiarized works demonstrates the challenges to an author's control over his work in his own time and the instability of his oeuvre over subsequent generations. It should be emphasized that authorial control in antiquity was decidedly more insecure than it is now.

Some works come down to us without the author's name. One can imagine easily enough how a work may be separated from that important piece of information. The name of the author appeared only in the *inscriptio* located at the beginning of the work on the first folio of the codex, precisely the part of the book most likely to deteriorate or become lost or illegible. A missing name may already have affected a scribe's *exemplar*, and in that case the work would be transmitted without a named author. This is precisely the situation that occurred for the fragment of the *De bellis Macedonicis* and the papyrus *Carmen de bello Actiaco*.

On some occasions, anonymous works are received in their entirety. For example, the *Consolatio ad Liviam*, a composition of the early imperial period, was attributed by Renaissance printers to Ovid but had actually been handed down anonymously in the manuscript tradition; the codices of this work are no longer extant. In the case of the *Ilias Latina*, a compendium in hexameters of Homer's *Odyssey* dated to the Neronian period, the name of the author could have dropped out because the text as it was handed down in the medieval period was attributed to Homer.

Akin to the problem of anonymous works are those that are presumed to be spurious, that is, works incorrectly attributed to an author. Although occasionally a name is attached to a work as a supposition, as was the case for the *Consolatio ad Liviam*, likely written in or before the age of Nero,[216] some texts were written with the intent of falsification. Galen considers this phenomenon to have begun in the era of the rivalry between Alexandria and Pergamum, who "competed in the acquisition of ancient books," thus stimulating the initiative of the falsifiers.[217] Since plays were among the few literary works that could bring a profit, theater managers were probably the source of the attributions of a great number of comedies to Plautus; these comedies, as we saw above, were later excluded from Varro's canon. Suetonius (*Vita Iulii* 55) speaks of orations attributed to Julius Caesar. According to Suetonius, the authenticity of some of these orations, such as one in defense of Quintus Metellus, was questioned by Augustus, who thought the speech in its written form was the result of a stenographer's inability to keep up with the oral delivery.

On certain occasions, works were deliberately falsified to bolster a political or religious agenda. The spurious correspondence between Seneca and Saint Paul, for example, was a forgery produced in the Christian environment of the fourth century that aimed to justify reading the pagan philosopher in a Christian society. Another famous forgery, devised in the medieval period, is the "Donation of Constantine," a spurious document by which the emperor Constantine supposedly ceded temporal power to the Church.

Forgeries can sometimes be identified by using objective criteria, but the situation for some texts can be quite complicated. Dionysius Thrax' *Techne grammatike*, for example, may be a mixture of authentic text and interpolation. Even for those works that are not considered heavily interpolated, authenticity can be difficult to prove. Arguments often rely primarily on stylistic or structural evidence, requiring a distinction to be made between author and imitator; this can be extremely challenging in the case of a highly skilled imitator.

Sometimes, stylistic evidence can be sufficient to reach a scholarly consensus, as is the case for pseudo-Theocritus' *Idyll* 8—even though Virgil considered it authentic—as well as the *Onos*, an ancient novel attributed to Lucian, about which doubts were raised as early as the seventeenth century.[218]

For some works, the stylistic evidence is not clear enough to settle firmly the question of authenticity. For example, the authorship of the poems of the *Appendix Vergiliana* is still subject to debate. This small corpus is a collection of poems attributed to Virgil first mentioned only in the fourth century, and whose manuscript evidence dates from the tenth century.[219] The authenticity of the individual poems is still under scrutiny, while many scholars argue that the entire work is spurious.[220] Though debate persists, the *Appendix Vergiliana* is not likely to have been composed with malicious intent, as these poems attest the popularity of Virgil. The pseudo-Theocritean *Idyll* 8, moreover, is likely another example of an imitator following a popular stylistic model.

Incorrect attributions made some time after an author's death were normally not intentional falsifications but rather errors born out of confusion in the tradition. This is the case for the attribution to Cicero of the *Rhetorica ad Herennium*, attested in the late antique period and justified, it is said, by the fact that the work was handed down together with the Ciceronian *De inventione*. The authority attributed to Cicero on the subject of rhetoric strengthened the attribution. The source of the attribution of the *Octavia* to Seneca could be analogous; it may have been copied together with the tragedies of Seneca due to its Neronian setting—in fact Seneca is one of the characters in that play—and then at some point during its transmission along with the genuine tragedies of Seneca, the *Octavia* was attributed to him.

The more sinister act of falsification, which we now term "plagiarism," is found already in antiquity. Greek and Latin do not have specific terms to denote literary plagiarism. They use circumlocutions or generic terms like *klope* and *furtum*, which mean "theft" in Greek and Latin, respectively. In Latin, the verb *intercipio*, meaning "to intercept," but also "to take for one's self that which belongs to another," is used in the context of plagiarism. The phenomenon of plagiarism is known, but not well defined, and it is sometimes confused with characteristics that today would not be defined in such terms. The appropriation of another's work, which is by modern standards true plagiarism, would not have been difficult to accomplish in the system of manuscript reproduction; it is not hard to imagine a spurious poem being introduced into a corpus.

Pliny the Younger outlined an eventuality of this type to his interlocutor, Octavius, who decided not to publish his own writings. In inviting him to take action, Pliny warned that some of his verses were already circulating and could be falsely claimed by another.[221] If such were the case, the poet would have had little recourse in reclaiming his poems.

Acts of this sort were, of course, subject to censure, though they did not carry any sort of legal penalty. Vitruvius reports an instance of plagiarism in the introduction to book 7 of the *De architectura*: Aristophanes of Byzantium, librarian of Alexandria, revealed to Ptolemy that some poets who had participated in a competition had actually performed others' works, which Aristophanes was able to find among the books in the library. Ptolemy then, according to Vitruvius, had the men found guilty of *furtum* and sent away in disgrace. Suetonius tells of a similar case: the son-in-law of Aelius Stilo, Servius Clodius, appropriated (*intercepisset*) a work unpublished by his father-in-law; when discovered, his father-in-law disowned him and he left Rome in shame and grief.[222]

Sevius Nicanor, the first author to have become famous for his grammatical treatises, was suspected of similar behavior; Suetonius reports rumors of his treatises having been mostly appropriated (*intercepta*).[223] A distinction must be made, however, between Servius Clodius' plagiarism and Sevius Nicanor's copying. In the former case, Clodius claimed another's work in its entirety as his own. Nicanor, however, seems to have quoted passages without attribution, a practice that Pliny the Elder, in the introduction to his *Naturalis historia*, acknowledges as common while condemning it.[224] Because the notion of plagiarism was ill defined, unsuspecting authors like Nicanor could find themselves censured. The accusation of plagiarism could also be used as a means of attack on one's opponents. Philosophical schools on the rise in the fourth century BC especially sought to denigrate the founder of an opposing school, if not his ideas, through allegations of plagiarism.[225] Plato suffered several such accusations, including the claim that he plagiarized his *Timaeus* from a work by the Pythagorean Philoaus.[226] Aristoxenus even went so far as to accuse Plato of plagiarizing the *Republic* from Protagoras' *Antilogies*.[227]

Similar polemics characterized poetry, which was largely based on imitation and intertextuality. Aristophanes of Byzantium wrote a work (now lost) on the *furta* perpetrated by Menander against other authors; the same accusation was turned against Virgil by Perellius Faustus, author of a work on Virgilian *furta*, and by Octavius Avitus, who wrote at least eight books on the verses that Virgil reused from other authors. In the Virgilian biographical tradition, on the other hand, Virgil emerges as a victim of plagiarism perpetrated by his friend Varius Rufus with the tragedy *Thiestes* (but the account was devised to denigrate Varius).

The dangers of plagiarism and anonymity must have been long known to ancient authors. In the Augustan period (and even earlier, in Greek literature, in the Hellenistic period), authors sometimes identified themselves in their texts, a feature known as a *sphragis*, or seal. Such is the case, for example, at the end of the *Georgics* (4.563–566) in which Virgil cites his name, the earlier *Eclogues*, and the place (Naples) where he composed the work.

A further relevant issue of ancient publication is that of editions created without the consent of the author. Texts of this type were frequently copied from lectures, orations, and other types of oral speech. In a letter to Atticus in July of 58 BC, Cicero laments the circulation of the text of the oration *In Clodium et Curionem*, delivered in 61 BC, affirming that he wrote it in anger and considers it "shabbier than all the others." Cicero's concern over the speech's promulgation was triggered by his need to come to terms with Curio, who was consul-designate. Cicero even suggests that the ease with which works were forged might work to his advantage. Since the speech as published was more carelessly written and therefore unlike his others, he claims he could plausibly deny authorship (*quia scripta mihi videtur neglegentius quam ceterae puto ex se posse probari non esse meam*, Cicero, *Ad Atticum* 3.12.2). Though one hardly need imagine that many authors sought to disown their works, here we have a glimpse of the evidence that might be adduced to prove or disprove authorship, which is strikingly similar to the evidence that modern scholars use.

Even Cicero's philosophical works were not entirely safe from risk of undesired promulgation. As was typical practice during the Roman Republic, Cicero sent his works to a friend, Atticus, for review before making final revisions, though in the case of the *De finibus*, Cicero later discovered that unauthorized copies of the text had been made for some friends, an act for which either Atticus himself or his copyists were responsible. In a letter from the summer of 45 BC, Cicero complains to Atticus how much had happened, pointing out that the copy made public did not contain the last touches that he himself had given to the work. On the same occasion, Cicero praises the control exercised by his own copyists.[228]

In the imperial period an expanding book trade caused the publication of texts without the author's consent to become widespread, especially for famous authors. In the introduction to his *Institutio oratoria*, Quintilian laments the circulation of two books of rhetoric attributed to him that were prepared by students who attended his lectures.[229] He also reports that he previously published only one oration, while the others in circulation under his name were "corrupted by the negligence of stenographers eager for profit."[230] This situation highlights the vagueness of the notion of publication, as the difference between sharing lecture notes with a few friends and publishing them must have been remarkably blurry.[231]

Like Quintilian, Galen was concerned with the accurate attribution of his works, and so wrote the *De libris propriis*, which provides an authenticated list of his writings.[232] Galen provides a number of reasons for making an account of his own works, such as the fact that his books are read in many nations, but in mangled form. In particular, however, he records an anecdote about an encounter he witnessed at a bookshop on the Vicus Sandaliarius ("the street of the cobblers" in Rome, near the Argiletum, known for its bookshops and as a meeting place for intellectuals).[233] There he noticed a dispute on the authenticity of a book that bore his name, the falseness of which Galen allowed to be revealed by a *philologus*, rather than intervening himself. Galen sought to protect his reputation by swiftly producing an authoritative list of his works. This example shows how difficult it must have been to exercise control over one's own published work.

Just as modern books may be printed in multiple editions, variants of some ancient texts came about by the author's own hand in a new edition after initial publication. Unlike in the case of modern presses, however, the process in antiquity could be quite untidy. For example, while a text was being copied, an author could intervene with modifications and corrections, with the result that some copies included the corrections, but others did not. A case of this type pertains to the *Orator* of Cicero. From a letter to Atticus at the end of 46 BC we learn that Cicero realized that he committed an error in the draft of the work, attributing a verse to the comic poet Eupolis instead of Aristophanes.[234] In the letter he asks Atticus, who was the usual editor of his works, to have the error corrected not only in his personal copy but also in those that Atticus' copyists were preparing. The incident came to a good end, as the codices that we possess bear the name of Aristophanes. We can hypothesize that the diffusion of the *Orator* was still limited (the work was composed in the preceding months), and that Atticus still had control of the copies being produced.

In other cases involving the same individuals, however, the intervention was not successful. In 50 BC, writing to Atticus from Laodicea, Cicero informs his friend that he discovered, while reading Dicaearcus, that the place name *Phliuntii* used in the *De re publica* should be corrected to *Phliacii*, a correction that he had brought about in his own model, while asking Atticus to make the same correction in his copy (*Ad Atticum* 6.2.3). Yet the work must have already been made public, as the late antique palimpsest in the Vatican library bears the incorrect reading.[235] In each of these cases, the manuscript tradition is univocal, but one may hypothesize that in analogous cases both readings, whether original or subsequent, may be present in the manuscript tradition. Either of these authorial variants might be included in the text with full legitimacy.

A high likelihood of authorial variants is expected in cases of works whose author prepared multiple editions. In 45 BC Cicero published two editions of the *Academia* within a few months, the first of which was in two books composed between March and May and sent as usual to Atticus for publication. Dissatisfied with the individuals chosen for the dialogue, Quintus Lutatius Catulus and Lucius Licinius Lucullus, Cicero substituted Cato and Brutus for them, and also inserted Varro into the work, as he had expressed a desire to be included. To do so, Cicero modified the setting and expanded the dialogue into four books.

Though he promulgated the second edition in a timely fashion, the first edition was nevertheless not entirely supplanted, for the second book, dedicated to Lucullus, has survived; from a different manuscript tradition comes one of the books of the second edition, dedicated to Varro.

In many other cases of dual editions, however, the more recent version seems to have prevailed. So it is in the case of the *Amores* of Ovid, whose second edition remains extant, in three books, published around AD 1. No trace of the first edition in five books survives. In another much earlier case, after the poor initial reception for his *Clouds*, Aristophanes apparently rewrote much of the play (although he faults the audience for failing to appreciate the first performance), and put on the revised version to greater acclaim.[236] Though nothing is known of the circulation of the first edition and the question is complicated by the nature of dramatic performance,[237] it is clear that the revised version completely supplanted the original composition, of which scarcely five lines remain.[238]

Even in late antiquity, book production still occurred on a relatively small scale and publication remained a rather modest affair, reminiscent of the restricted circle for literary dissemination among the senatorial class of the Roman Republic. The process, as described by Jerome, begins with a copy made in shorthand from dictation, which, when recopied into a normal script, is known as a *schedula*.[239] The *schedula* would then be emended by the author and the whole work recopied into a version that incorporates the emendations.[240] When this initial process is finished, the author may be said to have produced an "edition" (*editio*).

Like the Greek *ekdosis*, the concept of the *editio* refers to the completed process of correction and emendation of the text.[241] A letter from Augustine to Firmus sheds some light on this process. Augustine offers a copy of his *City of God* to Firmus and instructs him to allow it to be copied, but not by too many people,[242] adding that pagans should be allowed to copy the text that they might benefit from it spiritually.[243] Yet the benefits were not merely spiritual, as we shall see.

3 / CLASSICAL RECEPTION FROM ANTIQUITY TO THE MIDDLE AGES

THE NOTION OF RECEPTION, INTRODUCED HALF A CENTURY ago by Hans Robert Jauss, enhances and qualifies the old-fashioned phrase "classical tradition."[1] The word "tradition" suggests, by its very etymology, an aspect of timeless cultural heritage with enduring value that is transmitted through different eras. Prior to Jauss, T. S. Eliot had commented that it is not "preposterous that the past should be altered by the present as much as the present is directed by the past."[2] While reception theory calls into question the assumption that the texts as transmitted through manuscripts "are stable and can delineate clear boundaries,"[3] it does not disqualify the importance of their content on that account. Rather, the notions of both tradition and reception point up that the way texts have been read over the centuries must be seen both within the context of historically and culturally determined reception and under the umbrella of abiding tradition. Reception is therefore the most important factor in the preservation of the textual tradition, what we might call the transmission of a text.

Practically speaking, the text, to be passed on, must have been reproduced in a sufficient number of copies to compensate for scribal error and the deterioration that can damage any individual copy. The number of copies corresponds to the level of interest in that text. The greater the interest, the lower the cost, which depends also on the medium used, whether papyrus or parchment, as well as the labor inherent in copy making. The high cost of the materials, particularly parchment, often prevented the reproduction of a text, resulting in a general trend toward making fewer copies—only as many as could be sold or were needed for textual preservation—and thus compounding the risks of natural and accidental loss. That the material was precious is evident in the tendency to reuse parchment in palimpsests, many of which can be dated to late antique monastic culture. The process of removing the original writing from parchment is both a blessing and a curse—the latter, because the original text was damaged, the former, because careful study of a palimpsest allows the scholar, in many cases, to reconstruct much of the text that was lost. A famous example is the *Schedae Rescriptae Veronenses*, housed in Verona's Biblioteca Capitolare, which preserves limited but important sections of the Virgilian corpus.

If interest in a particular author waned, that text was unlikely to persist, as it would be either destroyed, lost, or simply forgotten. The conditions of a text's reception, along with its interpretation, affected the likelihood of its replication. Changes introduced by copyists, discussed in chapter 2, were in part determined by the cultural context in which a copyist happened to work. Thus, the history of literary reception coincides with the history of literature itself. Each generation simultaneously benefits from the literature of its time and selections taken from those who had come before and had worked on the text in question. Such fresh assessments were, of course, compatible with the specific criteria of various sociocultural and political contexts, as well as prevailing taste. Accordingly, when we speak of the reception of a text, we mean the manner and means by which a work might be selected, transformed, and adapted to diverse cultural contexts at different times.

LATIN LITERATURE IN ANTIQUITY

It is likely that at least some of the Latin literature created in the third through the second centuries BC was lost not long after it was written, in some instances because the works were never produced in book form. Many texts, particularly those used in oratory or theatrical performance, were transcribed on impermanent materials, such as tablets. Even for texts in book form, access was limited by the low rate of literacy. Furthermore, surviving texts of Latin literature transcribed onto papyri might, after a brief period of circulation, be left in the archives of the families under which they were produced, thereby severely curtailing their distribution. Suetonius already gives evidence on the loss of texts as he recounts the activities of the second-century grammarian/philosopher Crates of Mallus and scholars of his age: "They were reading poems that had not been in circulation but were the works of deceased friends or anybody who might win a prize by producing an epistle, quipping on it, or publicizing it."[4]

The work of philologists of the second and first centuries BC no doubt played an important role in the preservation of ancient literature that eventually went missing but was rediscovered by the end of late antiquity. A striking example of the influence of early philological studies on the transmission of texts can be seen in the case of Plautus. As discussed in chapter 2, of the many comedies attributed to him, only the twenty-one that Varro considered genuinely Plautine have been preserved. This late antique collection, however, derived from only one list by Varro, although he elsewhere attributed other comedies to Plautus, which were passed down under another name.

Gellius cites one of these, attributed to Aquilius, and fully agrees with Varro's judgment when he says, "no one who regularly reads Plautus will have any doubt, if you will ponder those verses of the play as they are, saying that they are 'most thoroughly plautine' [*plautinissimi*], an expression befitting Plautus himself."[5] Yet neither Gellius' interest in identifying Plautine characteristics in other verses nor Varro's attribution of other comedies under other names proved influential enough to modify the tradition established by Varro's original selection.

In the Republican era, the philological activity that sought to preserve earlier literature is not well documented. Some early losses likely affected the production of Greek annalistic histories, while Latin translations may have supplanted the original Greek works.[6] Alongside the established tradition of the circulation of annalistic writing that was taking shape, solidified by the unbroken continuity of the genre until the late Republican period, oratory also emerged as a written and preserved literary form, even though it was no longer connected with politics to the degree it had been in the Republic. The preservation of a new style of speeches, with characteristics more of rhetorical exercises than of urgent speeches delivered before the senate in a time of crisis, would ultimately be entrusted to the archives of the individual orators' families. Still, even at the height of oratory in Cicero's day, it seems that the availability of speeches was limited.[7] Cicero avers that he had difficulty in tracking down copies of Cato the Censor's orations, although according to Cicero's own reckoning there were one hundred and fifty of them.

Another genre that suffered significant loss in the closing years of the Republic was theater, in part because not all tragic or comic texts enjoyed wide circulation in book form. The literary crisis surrounding theatrical production that began in the second century BC and continued into the Augustan age determined what selection of texts libraries might hold, based on canons that were elaborated upon during this era.[8] Circulation of archaic texts declined in the early imperial period in the wake of Caecilius Epirota's reappraisal of the scholastic canon, which significantly expanded the list of literature from merely Virgil and his contemporaries.

The divergence of existing tastes for ancient texts is well highlighted by Horace's famous judgment of Saturnian verse as *horridus* and his favorable appraisal of Greek influence on archaic literature.[9] The basis for such assessments was refined by experimentation in the Julio-Claudian age (AD 14–68) and then further tempered by the Flavian period's (AD 69–96) emerging classicism, by which time the work of Cicero, Virgil, and others came to be recognized as canonical. The judgments made by Quintilian in the tenth book of his *Institutio oratoria* reflect the assessment of literary history that prevailed in this age. Quintilian places Virgil on the same level as Homer, while calling Ennius "venerable" like "sacred groves venerated for their antiquity."[10] It is no coincidence that in his consideration of genres prominent during the height of Rome's archaic age, such as comedy, Quintilian deems Latin authors inferior to their Greek forerunners. Quintilian's literary evaluation adumbrates a system for the preservation of literary heritage, broadly speaking, wherein the library was the arbiter of textual transmission. This differs from the circulation of what the Romans would have regarded as classic works, whose preservation was ensured by their use in schools and by the production inspired by those classics.

The continued endurance of books was further complicated by the fact that they not only could fall out of circulation by the nature of their genre — such as those used primarily for declamation or performance, and not found in schools or frequently imitated by successors — but also could be the victim of totalitarian regimes under which they did not find favor. In fact it was not uncommon for oppressive governments to exert some form of control over the circulation of literary texts. Though merely anecdotal, the story of Peisistratus is representative of a ruler's authority over a text. Plutarch tells us that this sixth-century Athenian tyrant was said to have censored a verse of Hesiod and even to have interpolated a verse into the *Odyssey* to defend the prestige of Athens.[11]

Similar anecdotes occur on the Roman side as well. From Rome's regal period, for example, episodes of political censorship are reported by Plutarch, who speaks of the books of Numa being set afire.[12] In 186 BC the Bacchanalia were suppressed, an incident that Livy tells us also included the destruction of sacred books.[13] Such censorship recurs, too, during the imperial period, when the emperor's condemnation of his opponents was often accompanied by the degradation or banning of their works. A prime example is Augustus' burning in 12 BC of two thousand prophetic books in both Greek and Latin.[14] More egregiously, only a few years earlier, in 26 BC, at the opening of that same imperial regime, the *Amores* of the first Roman elegist, Cornelius Gallus, suffered a similar fate.[15] That ominous exercise of authority recurred later in Augustus' rule: in AD 11 the works of the rhetorician Cassius Severus and of the historian Titus Labienus were burned publicly and, although his works were not put to the flame, in AD 8 the poet Ovid was exiled to Tomis.

In the case of Ovid's banishment, only one specific literary motive is mentioned, the *Ars amatoria*. From Ovid's works published during his exile, we know that the *Ars amatoria* had been removed from the library known to have been in the Atrium Libertatis, yet it was never consigned to the flames.[16] He suffered a rather lighter sentence than other writers, as evidenced by his continued literary production during exile. Augustus' political control over the circulation of books is confirmed by Suetonius, who records that Augustus gave the order to Pompeius Macrus, the director of the Palatine library, not to make the writings of Julius Caesar available to the public.[17] Tacitus tells us that under Tiberius the historical writing of Cremutius Cordus was burned because in his work he celebrated Cicero, as well as Caesar's murderers, Brutus and Cassius. Even in this case, however, the work survived, thanks to his daughter,[18] and was republished during the reign of Caligula.[19]

In this same era, the hostility of Caligula against the works of Livy and Virgil had little lasting effect, although the emperor "came close to banishing their works and portraits from all libraries."[20] Other book burnings are reported well into the age of Domitian. In his *Agricola*, Tacitus mentions the case of elegies written by Arulenus Rusticus and by Herennius Senecio, which cost the two authors their lives.[21] In this case, the decemvirs were given the task of burning their works in the forum. Even in these instances, it was difficult for the repressive regime to account for all copies of a text within a cultural milieu characterized by an exclusive literary class and numerous private libraries.

The revival of Greek writers as well as archaic Roman authors that characterizes the era of the five good emperors (Nerva, Trajan, Hadrian, Antoninus Pius, and Marcus Aurelius) certainly played a role in the preservation of texts. In his *Noctes Atticae*, Aulus Gellius, who was born under Hadrian and died just after the reign of Marcus Aurelius, writes extensively about his own search for ancient books. Gellius provides a memorable description of a copy of the *Annales* of Fabius Pictor, which was "of good and authentic antiquity," offered for sale by a bookseller in the market of the Sigillari.[22] He also describes a copy of the *Annales* of Claudius Quadrigarius as an "ancient book" containing a prayer of Cato the Censor[23] that Gellius himself read in the library of Tivoli.[24]

Any one of these references may have been fabricated, of course, and some of the reports vaunted by Gellius sound improbable. One such claim that rings false is his boast to have seen a roll of *Aeneid* book 2 that belonged to Virgil. This roll was shown to him by the grammarian Fidus Optatus and was supposedly purchased for twenty gold coins at the market of the Sigillari.[25] Be that as it may, it can be assumed that the search for old books would provide opportunities for fraud on the part of booksellers and that Gellius and some of his friends might have been victims of such machinations. It is also possible that the roll from the bookseller was genuine, providing a direct and palpable connection with the past.

A reference to a copy of Ennius, emended by Octavius Lampadio (purportedly in Lampadio's own hand) and available to Gellius, offers an even more perplexing example.[26] Even if such a copy were in circulation, Gellius may be referring to the edition produced by Lampadio and not to a book that he had transcribed. Similar considerations apply to an edition that Gellius describes as being "of recognized authority" and containing Cicero's fifth Verrine "transcribed with precision and Tironian care."²⁷ In the case of the Cicero volume, however, it is unclear whether Gellius refers to a text that Tiro actually edited or to a specimen that Tiro copied. Though Zetzel provides ample caveats to accepting any such story wholesale, this and other accounts—such as a subscription possibly written a few years before Gellius by Titus Maximus Statilius, referring to an edition of the *De lege agraria* apparently prepared by Tiro—provide us valuable, if not consistently reliable, information.²⁸

However unlikely any one of these examples may be, based on the sum of them it is possible to establish a kind of Roman cultural authority discernible in Gellius' work.²⁹ The frequency with which Gellius gives the moniker "of venerable antiquity" to the memorable books that he or other characters in his narrative have tracked down is emblematic of an attempt to construct a cultural authority of rich Roman heritage, despite some uncertainty about his sources.

CULTURAL AND LITERARY DEVELOPMENTS IN LATE ANTIQUITY

Nearly a half century ago, Peter Brown identified the period of late antiquity as the span between antiquity and the Middle Ages.³⁰ The epoch of late antiquity is generally accepted as beginning in AD 180, the year of both Marcus Aurelius' death and the martyrdom inflicted on Christians in Africa at Scillum.³¹ The end of Christian persecution, decreed by the edict of Constantine in 313, opened a period of coexistence between Christian and pagan cultures that would extend into the next century with the Christianization of the empire and the evangelization of the Germanic peoples who had settled in the western regions. The close of the period of late antiquity, however, is less well defined, although the most widely accepted marker is the Arab expansion of the seventh century.

Late antiquity's first beginnings in the third century are characterized by a period of marked stagnation in Latin literature. Only a few minor authors go back to this period, and those that do are, for the most part, merely compilers, such as Gargilius Martialis, Censorinus, Solinus, and a few others. There is even a thirty-year period (254-284) to which no Latin text can be dated.³² Such a dearth is, in fact, surprising, because the political and military turbulence of the increasingly fragmented empire had not yet come to the fevered pitch it would reach in the years that followed.³³ Even so, when such upheaval did occur, it proved but a minor obstacle to the production of Greek literature. In the second and third centuries AD, the novelist Longus and the Neoplatonist philosopher Plotinus were active.

In Latin production, endogenous factors were also likely at work. Initially, there existed a stratified society whose cultural elite had aspirations to erudition and supported the social policies of the contemporary political authority, which had been at first well disposed toward literature and the arts. Later in that period, however, members of the imperial court, hitherto educated in second-century rhetorical schools, were displaced "by a barely literate ruling class, which itself had limited access to literature."[34]

Although the transmission of texts did not stop, book production was likely affected by the unfortunate political climate and barbarian invasions, which in this century occurred on an unprecedented scale. From the 230s through the next half century, Rome's leadership fell into complete disarray in what is known as the Crisis of the Third Century. Some twenty emperors would reign, many only holding power for brief periods. By AD 260 the army of the Alemanni had arrived outside Milan, and the Heruli sacked Athens soon after (267). While in the next century literary production slowly recovered, its rebirth was founded more on the idealization of traditional culture than on a thriving market for books. Thus the demand for books in the elite culture of late antiquity slowly began to narrow.

In spite of such cultural contraction, there arose nonetheless a good deal of innovation in various forms of script. Until the second century AD, the rustic capital was normally employed for writing in the *liber*, the dominant repository for writing of all kinds. Other writing materials, such as tablets, did obtain, but they tended to be chiefly used for the business of court scribes, who also wrote in rustic capital letters. For book production, however, cursive began to be mixed with the rustic letterforms. For example, a fragmentary witness to a *Carmen de bello Actiaco*[35] presents shabbily drawn shapes and letters tending toward the cursive form (fig. 3.1).

During the course of the third century AD, the Roman cursive, or minuscule cursive, emerges fully, consisting of rather small letterforms. Vertical lines diverge, "ascending" or "descending," from the body of each letterform, framed by two horizontal lines that touch the top and bottom of each row of letters. Each capital letter is suspended between the two horizontal lines. Examples of Roman cursive are attested in the writing of later centuries, not only in nonliterary documentation, but also in marginalia of editions of authors such as Terence,[36] the text of which, by contrast, is chiefly written in capital letters.

In late antiquity, rustic capitals continued to be used, but this script was limited to the sorts of luxury codices that normally contained canonical authors. Examples of these include three well-known codices of Virgil: the Codex Mediceus, the Codex Romanus, and the Codex Palatinus. An important palimpsest of Lucan housed in the Biblioteca Nazionale of Naples also preserves this font.[37] These examples reflect the sense of traditionalism pervasive in late antiquity. There were also various codices written in an elegant capital known as the "monumental capital," among which are two fragmentary codices of Virgil:

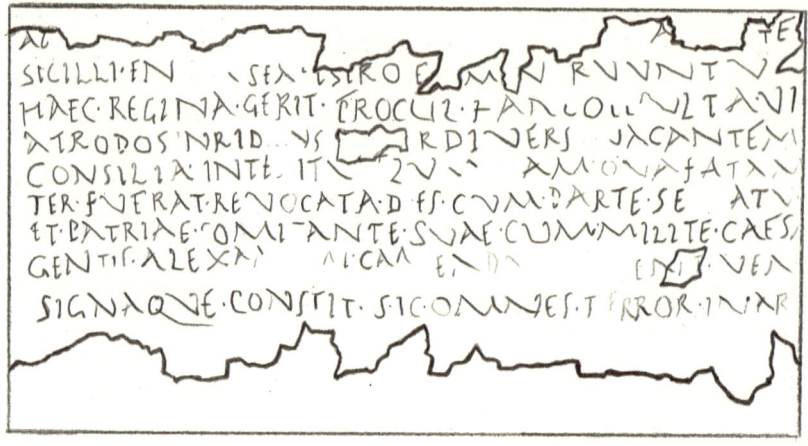

FIGURE 3.1. *Example of rustic capitals.* Carmen de bello Actiaco *on* P.Herc. *817. Drawing by Mary Claire Russell MacDonald.*

the *Augusteus*[38] and the *Sangallensis*.[39] By the end of late antiquity, however, writing in such capitals would be replaced by the script called "uncial," literally meaning "inch-high."[40]

During the fourth century, uncial writing in majuscules spread (fig. 3.2). The uncial majuscule style of calligraphy features the rounding of the letters, which became more spacious and heavy in contrast to rustic capitals. Its origin is uncertain, but it may have begun in Africa, as examples are detectable in African inscriptions as early as the third century AD.[41] Such majuscules employ the rounded configuration of letters that becomes typical of cursive writing.

The spread of the uncial style roughly coincides with the transition from the roll to the codex. Ultimately, the fortune of the uncial was closely connected to the affirmation of Christianity.[42] Although one does find some late antique codices of non-Christian texts written in uncials, such as the remarkable Ambrosian Palimpsest (which reveals Cicero's *De re publica*), copyists of Christian texts did not often use the majuscule. Instead, a form of minuscule book writing, derived from cursive and known as half-uncial, emerged in the Eastern empire in the third century. An example of the half-uncial's immediate precursor dates to the first half of the third century AD, as one finds in the case of Papyrus Oxyrhynchus 668,[43] an epitome of some books of Livy (fig. 3.3).

We find in that Livy codex the archaic half-uncial, also known as "oblique," featuring minuscule letters drawn with a chiaroscuro vertical stroke that presupposes writing made on a sheet in a slantwise or oblique position. This script, preserved in other papyri as well, spread to the eastern reaches of the empire; later it came westward, where it would be known as the half-uncial.

Between the fifth and sixth centuries and extending into the high Middle Ages, a market for manuscripts written in half-uncials developed, perhaps beginning in Africa, as the term *litterae Africanae* is sometimes associated with this font and may suggest its origin. Nevertheless, this writing style never fully supplanted the uncial. One of the oldest witnesses is a copy of the Vulgate Bible from the library at St. Gall, produced as early as the fifth century, perhaps by the circle of Saint Jerome.[44]

Before Jerome's time, however, Christianity had to find its place within the religious spectrum of the Roman world. That independence was a direct inheritance from Judaism, a distinct exclusivity that demarcated Christianity amidst the panorama of cults that proliferated during the Roman imperial age.[45] Its exclusivity thus helped to establish it as a bona fide religion, which certainly had happened by the time of Constantine's conversion and victory at the Milvian Bridge.[46] In spite of Christianity's general resistance to certain practices of Greco-Roman culture,[47] the rapidity of its growth is deducible from the writings of Saint Paul.[48]

A poignant example of the menace that Christianity posed to the pagan en-

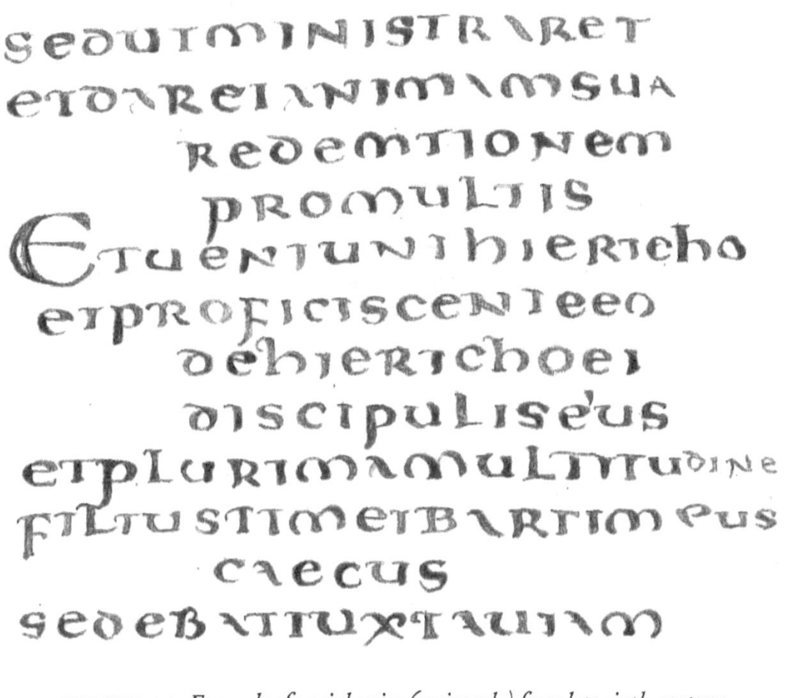

FIGURE 3.2. *Example of uncial script (majuscule) from late sixth-century Rome. Ms. Harley 1775, folio 193 recto (detail). London, British Library. Drawing by Mary Claire Russell MacDonald.*

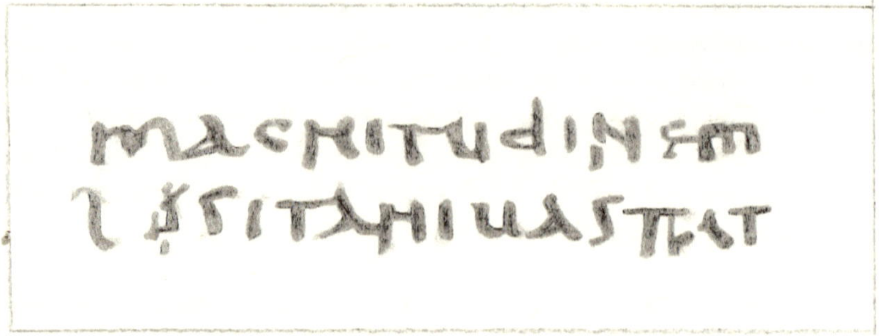

FIGURE 3.3. *Epitome of Livy* fragment (P.Oxy. 668). *Papyrus 1532* recto. London, British Library. Drawing by Mary Claire Russell MacDonald.

vironment in which it was maturing can be seen in the correspondence between Pliny the Younger and the emperor Trajan in AD 112.[49] Yet by the age of Hadrian, certain aspects of the Christian message were being advanced in the Greek philosophical teachings of Aristides and Quadratus, both second-century Athenian philosophers and early apologists. By the middle of the second century, in the school that he founded in Rome, Justin of Nablus, also known as Justin Martyr, bears witness to the affinity between Christianity and Platonic philosophy. Justin, sometimes touted as one "of the two prince apologists,"[50] favored a confrontation between Christian and Greco-Roman cultures. While most of his work is lost, some of the conflict of that century can reliably be deduced from his extant works, which include two apologetic treatises and one dialogue.[51]

Further tension became palpable as religious debates gave way to disagreements that produced an atmosphere conducive to Christian persecution in the second half of the century. Justin himself would be martyred in 165 after the philosopher Crescens the Cynic execrated him and brought him to the attention of political authorities; in his *Address to the Greeks*, Tatian offers a lurid account of Crescens' actions and Justin's fate.[52] Having been tried by Junius Rusticus, the urban prefect from AD 163 to 167, Justin, along with six companions, was beheaded.[53]

Justin's defense of Christianity elicited further strong, calculated, anti-Christian responses from representatives of pagan culture. In AD 180 Celsus' treatise *Alethes logos* was published. The work is no longer extant, but its contents are deducible from Origen's response (*Contra Celsum*), in which he refutes Celsus' three major tenets: that Christ's birth was fictitious, his life ordinary, and his death meaningless.[54] Not surprisingly, Celsus' pamphlet provoked intense reaction in Christian circles, inaugurating a fluctuation between rather more radical and less entrenched positions that would remain in place throughout late antiquity. The attitude of Justin, which some saw as conciliatory, was contested by his student Tatian, also known as "the Assyrian." Other works written in about that same period, such as *The Shepherd of Hermas*, offer an instructional approach that encompasses a blend of messianic references and echoes of the Jewish tradition, giving no quarter to paganism or pluralism.

A similar dichotomy is detectable in some of the oldest extant Latin that was written by the Church Fathers. Tertullian, for example, radically opposes pagan culture, as can be seen in the famous question that he posits in his *De praescriptione haereticorum*, "What has Athens to do with Jerusalem?" (*quid ergo Athenis et Hierosolymis?*).[55] By this pithy query, Tertullian means to call into question how the Church can be directly influenced by any philosophical system.

Contrary to his contemporary, Minucius Felix, who chooses, in his *Octavius*, to expound the Christian message in the form of a Ciceronian dialogue, Tertullian vehemently opposed in its entirety pagan cultural influences on Christian thought. Tertullian's own apologetic work, in which he countered pagan charges of atheism and sedition, is inconceivable without the rhetoric and legal training that he had received in Carthage before his conversion. Thus, even his opposition to pagan culture did not impede him from valuing figures from the Roman tradition, such as a noble Lucretia or Regulus.[56] Tertullian's deep-rooted obduracy, which led him in his old age to adhere to Montanism, was tempered by his sensitivity, not always pervasive in Christian thought, affording him a certain receptivity to elements of Greco-Roman culture that in other quarters might have been considered profane.

The authors of the following centuries who forged a Christian ethos largely based on Greco-Roman culture were careful about how they took up the various literary forms and assimilated the previous tradition with the Christian religion. One might consider, for example, Ambrose's use of Cicero's *De officiis* for his *De officiis ministrorum*, or Augustine's use of Cicero's *De re publica* for his *De civitate Dei*, or the recovery of Virgilian epic by Prudentius. Ambrose and Augustine, in particular, found it expedient to give their own justifications for the reprisal of the pagan material that they had undertaken. Ambrose, in a treatise now lost but quoted by Augustine,[57] apparently suggested that Plato was influenced by the prophet Jeremiah, whom he supposedly met in Egypt. Thus Ambrose could give a historical justification to Platonism, which was popular in Christian circles. At the time of Augustine's composition of *De doctrina christiana* (397), Ambrose's pupil welcomed this explanation, formulating his own program for the selective use of aspects of pagan culture, ever with a view to developing a better knowledge of sacred scripture. Later, in his *De civitate Dei*, Augustine seeks to explain the possible anachronism, stating that Plato had gone to Egypt just after Jeremiah arrived there, but before Ptolemy commissioned the Septuagint translation.[58]

This possibility establishes the metaphorical justification that Augustine advances in *On Christian Doctrine*, based on the account in Exodus of the Jews having left Egypt only after they had essentially plundered the Egyptians. Thus Augustine can write, "In the same way, the teachings of the pagans not only contain discoveries, superstitions, and burdens that are tiring and useless . . . but also contain liberal disciplines suitable to the practice of truth, along with quite useful moral precepts."[59] Presumably, in the case of Plato, such precepts came to the philosopher under the influence of Jeremiah the prophet.

The general concern of Christian authors for legitimizing the use of pagan literature also motivated the composition and promulgation of the spurious correspondence between Seneca and Saint Paul,[60] aimed at promoting the religious contiguity between Senecan philosophy and Christian teaching. A similar motivation lay behind the notion of proffering a biblical interpretation of Virgil's fourth—"Messianic"—eclogue, an interpretation that was destined to enjoy a rich exploration in the course of the Middle Ages.[61] Its reception in the earliest years of that period helped to promote the reading of Virgil in a Christian environment.

Yet, both how Christian writers should appropriate pagan ideas and what the proper limits of allusions to pagan authors should be remained topics of at times heated debate. It was an issue even for Saint Jerome, evidenced by his account of a dream that he relates in a letter to Eustochius in 384. He explains that, inasmuch as he had become quite taken with reading Cicero and Plautus, he fell ill and, in a fever, he fancied himself as being before God, who charged him with being not "a Christian but a Ciceronian."[62] The tale ends with Jerome's solemn promise not to indulge himself in reading any more secular books (*codices saeculares*); after that vow, Jerome states, he miraculously recovered from the illness.

The secular literature of the fourth century was often written within the scope of a pagan revival fomented by the cultural conflict between Christians and pagans. Alan Cameron's important work *The Last Pagans of Rome* presents, however, a serious revision of this thesis. Cameron redefines the supposed revival as a romantic myth fueled by a nostalgic and idealized portrait painted by Macrobius in his *Saturnalia*. Nevertheless, certainly some kind of rebirth is apparent in the culture of the late third century during the reign of Diocletian.

This cultural resurgence is documented by the boost in book production that the widespread use of the codex engendered. Texts from previous centuries preserved only in a slender number of rolls were now put into codex form, which created a demand not only for contemporary books, but also for a fresh dissemination of classic texts. Along with this rebirth in learning driven by the fresh proliferation of texts came the overarching process of the Christianization of the empire, which was greatly hastened by the conversion of Constantine. Though it took time to sink its roots, Christianity enjoyed rather wide acceptance by the mid-fourth century. The revival of book learning played an important role in nascent Christian culture. Such desire for knowledge both vouchsafed Christianity's growth and preserved the Greco-Roman cultural heritage of previous centuries.[63]

Even in the late fourth century, Christian culture sometimes participated in the retrieval and diffusion of classical texts. Cultural homogenization was an aspect of formative education and grammatical and rhetorical studies, which Christian culture tended to complete, not subvert, with doctrinal formation. This ensured, at least until the fifth century, a substantial continuity of educational practices and the preservation of the past at the institutional level. For example, Rufinus of Aquileia, in his polemic against Jerome, reproached his opponent for having had his own students make a copy of the remnants of pagan texts.[64] Important here is the fact that Jerome was obviously not only allowing but even promoting the study of pagan literature.

A more prominent role in the work of transmission, however, fell to aristocratic and cultural elites who, during the course of the fourth century, adopted the Christian faith without denying their own culture. Such conversion, which was widespread, allowed a broad range of possibilities for cultural interaction, fostering a degree of syncretism that might allow one to be a Christian on Neoplatonic terms while also preserving aspects of separatism that might involve personal asceticism and could even lead to monastic life.[65]

A case of particular interest is that of Marius Victorinus, rhetorician and eminent grammarian, of whom a statue in the Forum of Trajan was erected in 353. His conversion to Christianity came late in his life, initiated by an interest in Neoplatonism, as Victorinus had been a translator and exegete of the third-century Neoplatonist Porphyry. His conversion does not appear to have involved the recantation of his scholarly work. Victorinus broke off his teaching of grammar only after the anti-Christian decree of Julian the Apostate in 362. As a solution to the tensions between pagans and Christians, Victorinus proposed a coexistence of secular literature and Christian faith, which was highlighted by another important fourth-century author, Ausonius.

A prominent political figure in Gaul, Decimus Magnus Ausonius was the author of poems that fit within the genre of profane literature, without the intrusion of the Christian faith that he personally professed. This lack of a religious dimension was a deliberate choice, a point underscored by Ausonius' admonition to his student Paulinus, then bishop of Nola, regarding the hazards of radical religious asceticism.[66]

Religious choices intersected with political struggles. A well-documented example is that of Quintus Aurelius Symmachus (340–402), prefect of Rome between 383 and 385. In 382 Gratian, emperor of the West, ordered the removal of the altar of Victory from the senate house.[67] The decision provoked the protest of Symmachus, who went to Milan to make his case, where he was opposed by that city's bishop, Ambrose. The case dragged on for more than a year, until the actual removal of the altar. Symmachus' attitude, insofar as it is detectable in the documents that remain, appears to be mainly defensive: in *Relatio* III, dating to 384, he wrote to Gratian's brother and successor, Valentinian II, offering his support for traditional institutions and calling for the tolerance that had been shown by Christian emperors in previous decades. Proffering a kind of religious pluralism, he writes, "We all contemplate the same stars, we have the sky in common, we are all part of the same universe: what does ideology matter when we all are seeking the truth? One cannot strive after a mystery so great by a single method."[68] Ambrose's response, directed to the emperor, was an affirmation of the exclusivity of the Christian faith: "We know from sure wisdom and divine truth that you are seeking ways that are undependable."[69] To undergird his position, Ambrose threatened the emperor with excommunication if he should withdraw the decree.[70]

The defeat of Symmachus did not seem to have a major impact in terms of cultural politics. Between 395 and 404 the pagan author Claudian served as the court poet for Emperor Honorius in Milan, and between 400 and 402 a bronze statue was erected in his honor in the Forum of Trajan in Rome; that statue's *titulus* (inscription) has survived.[71] Saint Augustine mentions Claudian's devotion to paganism, even calling him a *Christi nomine alienus* ("at a far remove from the name of Christ").[72] Augustine's student, Paulus Orosius, is even more severe, calling him "a most headstrong pagan" (*paganus pervicacissimus*).[73] Claudian's fortune in the Western court would be even more astounding if the hypothesis of Cameron should hold true.[74] He suggests that Claudian left Alexandria, his hometown, to escape the anti-pagan persecution unleashed by Bishop Theophilus in 391.

The case of Claudian confirms the political, more than the religious, value of the story of Symmachus and the altar of Victory. It is significant, moreover, that the outcome of the story does not seem to have had an impact on the editorial activity often tightly bound to the pagan rebirth led by Symmachus, although the full extent of Symmachus' literary contributions in the realm of secular literature is not entirely clear. For example, we do not know whether he ever completed his edition of the Livian corpus.

It is true that families such as Symmachus' own were involved in literary activities, as is documented by numerous *colophons* that record the transmission of a text. The preparation of new editions of pagan authors, however, does not imply that the editor was a pagan. Rather, the transmission of pagan texts, which was unabated in the fifth century, arose in a culturally diverse context, not one that was simply Christian or anti-Christian. Vettius Agorius Basilius Mavortius, for example, who was consul in 527, produced a fresh *recensio* (edition) of Horace alongside that of the Christian writer Prudentius.[75]

Numerous late antique subscriptions (i.e., *colophons*) shed light on how such editing was performed. While the *colophons* discussed in chapter 2 were primarily cited as evidence for monastic conditions, they could also contain vital information about a codex, including documentation for the transmission of a text from late antiquity, and sometimes even the names of the copyist or editor, the date, and the place. The origin of this kind of notation goes back to antiquity. Early examples date to the imperial age, such as the *colophon* Statilius Maximus placed in a copy of the *De lege agraria* of Cicero in the second century AD. Other examples from late antiquity confirm the perception of the codex as a luxury product and status symbol. As such, the possession of codices came to characterize not only the pagan but also the Christian aristocracy.

Perhaps the most famous example of a *colophon* is that of the Codex Mediceus, containing Virgil's works and bearing the name of Turcius Rufius Apronianus Asterius, consul in 494.[76] Beyond his careful attention to the editing of the Codex Mediceus, Asterius also produced an edition of the Christian author Sedulius' *Carmen paschale*. Salustius, who supervised a *recensio* of the works of Apuleius, was a student of Severus Sanctus Endelechius, attested in the writings of Paulinus of Nola,[77] and is probably also the author of a Christian-inspired bucolic composition, the *De mortibus boum*.

The coexistence of secular and Christian culture is also evident in the fifth century in the work of Sidonius Apollinaris, bishop of Auvergne. Sidonius composed eulogies inspired by classical archetypes, as well as a collection of letters modeled on that of a pagan, the younger Pliny. In one epistle Sidonius even describes a library where the works of Augustine are placed next to those of Varro, and the works of Horace next to those of Prudentius, highlighting the integration of the two religious and political traditions.[78]

The delicate balance between Christian and pagan elements evident in Sidonius breaks down in the decades that follow, particularly after the decline of the intellectual class that had supported the rebirth of ancient ideas. Such a cultural trend was replaced in Christian circles by a new one with little interest in (and sometimes even hostile to) the secular tradition.[79]

The monastic movement, which came to have an important influence on the development of early medieval culture and on the preservation of scores of pagan texts, spread from the East to the West between the fourth and fifth centuries. Emblematic of the new climate that this expansion inspired is the story of Caesarius of Arles (470–543), as recorded by Cyprian of Toulon in his biography of Caesarius. Educated in Lérins,[80] one of the most influential monastic centers, Caesarius returned to his hometown of Arles, where a member of the local aristocracy, Firminus (a former correspondent of Sidonius), sought him out, asking whether he might study grammar and rhetoric under Caesarius' tutelage. Caesarius refused to undertake the project, for he had dreamt of a snake devouring a grammar book together with his own arm holding it.[81] The tale shares some similarities with the story narrated by Jerome, but with details resonant with the Christian elite that dominated in the sixth century.[82] Unlike Jerome, whose account of his dream included the detail of him being scourged for loving Cicero as much as Christ, Caesarius neither appears before a judge nor pays a penalty. Rather, he takes note of the portent of the snake, and when he wakes he behaves accordingly.[83]

Several *colophons* are found in medieval manuscripts that were themselves copied from late antique codices. A palimpsest[84] of Fronto bears the name of a certain Caecilius, who would have copied the text around 500. A Veronese tenth-century manuscript containing the first decade of Livy reproduces a number of different *colophons* from an *antigraph* of late antiquity.[85] The principal representative of these purports to have been "emended" (*emendavit*) by Victorianus on behalf of the Symmachus family. Further *colophons* attest other *emendationes* signed by Nicomachus Flavianus and Nicomachus Dexter.

Colophons variously provide dates that can be verified by collating them with historical and literary documents: Symmachus tells us in a letter dated to 401 that he was preparing a revision of the entire corpus of Livy.[86] From an epistle of Sidonius Apollinaris, we hear[87] of the editorial activity of Tascius Victorianus. In that epistle, Sidonius speaks about Nicomachus Flavianus' Latin translation of the *Vita Apollonii* of Philostratus that was corrected by Victorianus, who would seem to have been a collaborator with the efforts of the Symmachi and Nicomachi to preserve pagan literature.[88]

An important Florentine manuscript containing Apuleius' *Metamorphoses*, *Apology*, and *Florida*[89] goes back to a copy dated to AD 395, prepared by Crispus Sallustius, a student of Endelechius' school in the Forum of Augustus in Rome. In an additional *colophon*, dating to 397, Sallustius claims to have been in Constantinople, whither he had evidently brought the manuscript on which he was working. Macrobius' commentary on the *Somnium Scipionis* of Cicero was copied in Ravenna by Symmachus, consul in 485, the father-in-law of Boethius, with the assistance of a certain Macrobius Eudoxius Plotinus, perhaps Macrobius' own grandson. In Ravenna, Rusticius Helpidius Domnulus assembled a manuscript consisting of the *Epitome* of Valerius Maximus made by Julius Paris, the *De chorographia* of Pomponius Mela, and the *De fluminibus* of Vibius Sequester. That manuscript was later copied in the ninth century by Heiric of Auxerre.[90] Other acknowledgments of individual authors by name include Domitius Dracontius and Hierius, who worked in a school located in

Rome in the Forum of Trajan. They edited the transcript of the *Declamationes maiores* attributed to Quintilian; the presence of their two names may suggest that the *emendatio* proper was performed by one of them, while the other's task was primarily proofreading the work.[91]

The term most commonly used in *colophons* to describe work performed on a text is *emendare*, which broadly means "to correct." Given this very general connotation, *emendare* can therefore designate a range of interventions in a text, from correcting trivial or obvious scribal errors to the more intensive task of correcting errors and resolving faults through the consultation of *antigraphs* and the use of Alexandrian philological shorthand techniques.

This former type of emendation appears often in late antique codices of Virgil. In those manuscripts the preservation of incorrect readings often coincides with certain high-value features, such as a monumental (large) arrangement and the inclusion of historiated capitals (decorated initial letters). Such transcriptional errors plagued the Codex Mediceus, whose subscription tells us that a major figure, Apronianus Asterius, was responsible for that manuscript's production. The Codex Mediceus is an example of a luxury codex, not made for study, but chiefly as a display of wealth and refinement; as such, monumental codices were kept primarily for their material, not intellectual, value.

The production of luxury editions was also an aspect of Christian book production, as can be seen when Jerome objects to a noblewoman's practice of adding gems to books prepared with inlaid gold and purple ink.[92] A similar and concurrent objection to ostentation was made in the East by John Chrysostom. He was uneasy not so much with the content but rather the editing of sacred books. He rebuffed the claims of subtlety in the choice of parchment, the practice of writing in calligraphic letters, and the use of gold.[93]

An ancient example of philological emendation may be adduced from a *colophon* found in the manuscript tradition of the *De nuptiis* of Martianus Capella. From it we learn that, in 534, Securus Melior Felix,[94] a rhetorician in the city of Rome, emended the text based on very corrupt copies (*mendosissimis exemplaribus*). The work was composed in Africa in the fifth century, but, because the Byzantines had just overthrown the Vandal kingdom, this *recensio*'s transmission is certainly based on the copy held in Rome.

In Christian circles, where concern for the preservation of textual integrity was paramount, the terminology for such *recensiones* was different. The word most commonly used to designate the editing of a Christian text is not *emendare* ("to correct") but *conferre* ("to compare"), a verb that suggests a less aggressive posture and would even seem to connote a more respectful transmission of a text. Such terminology accords well with the recommendation made by Irenaeus of Lyon, originally written in Greek but translated into Latin by Jerome: "By our Lord Jesus Christ . . . , I beseech you, who transcribe this book, with the greatest diligence to compare and correct what you will have transcribed against the original; and transmit this living prayer, too, just as you found it in the original."[95]

The *colophon* in the Christian text, therefore, assumes a kind of legal value, serving as the proof of authenticity. To grasp fully what it meant to create a *recensio* in its historical context, it is important to keep in mind that the production of a book meant, first, confronting the problems of layout—the order of the document's sections and the entitling of chapters along with any important marginalia—all of which were problems particularly associated with transferring a text from scroll to codex.[96] As the size and capacity of the codex increased, individual manuscripts that had previously taken several scrolls could be assembled into a single volume.

BOOKS AND THEIR CONTENTS IN LATE ANTIQUITY

One of the first expectations for the new book form was that one codex should ideally house the work of a single author. This undertaking presented its own problems, especially since in some cases the attributed author was spurious, and determining authentic authorship was then—and, in some cases, remains—difficult. For example, late antique editions of Virgil privileged his three principal works (*Eclogues, Georgics, Aeneid*). These three works were often compiled into a corpus that previously comprised seventeen rolls, but was later collated into a single codex.[97] These Virgilian codices, however, did not include the *Appendix Vergiliana*, most of which is now considered spurious, although long attributed to Virgil.

Issues in the Virgilian manuscript tradition reflect a consolidated orientation of ancient philology, which is careful to separate the authentic works from those of a dubious origin. It is unclear whether the modern arrangement of the *Appendix Vergiliana* dates back to late antiquity or if it was collated in the Middle Ages based on the list provided by Donatus in his *Vita Virgilii*. In any case, the most ancient testimony that remains is that of a lost codex reported in the ninth century from the catalogue of the monastery in Murbach.

The authenticity of works by several other authors had to be assessed before they could be published. The codex edition of Horace, for example, excluded works that Suetonius deemed spurious. The numerous medieval codices containing Horace's opera that remain all derive from only two late antique manuscripts, designated Psi (Ψ) and Xi (Ξ). A third important witness was the *Blandinius vetustissimus*, a codex of uncertain date, but it was destroyed in a fire that broke out in 1566 in a monastery near Ghent. Beyond the expected textual variants, the two traditions differ in the entitling of the compositions and the ordering of the works. The *Ars poetica* is fourth in the manuscript family descending from Ξ, but second in that of Ψ.

This was not the only way of dealing with spurious works. For other authors, such as Caesar and Tibullus, manuscript copies instead present authentic followed by spurious works. Caesar's *De bello Gallico* and *De bello civili* are included in an edition with the pseudo-Caesarian *Bellum Alexandrinum*, *Bellum Africum*, and *Bellum Hispaniense*, while another edition, which actually includes the names of some ancient correctors, contains only the *De bello Gallico*. In the case of Tibullus, the corpus that has been handed down includes his elegiac works along with other texts related to the circle of Messalla, Tibullus' patron and a prominent supporter of Augustus.

In cases where the manuscript tradition dates back to late antique editions of the same author, we can verify the editors' autonomy in the selection and ordering of the transcribed texts. In addition to the case of Horace mentioned above, we also have two traditions for the comedies of Plautus, as evidenced in some codices of the tenth and eleventh centuries: the Ambrosian Palimpsest and the so-called Palatine tradition. Whereas the palimpsest is in alphabetical order, the Palatine tradition places *Bacchides* after *Epidicus*. The two recensions seem to have been derived from a school text, which was based on the Varronian canon and refined with Alexandrian features, including variant verses, notations, and colometry.

The Codex Bembinus, a late antique manuscript of Terence, contains comedies that are presented in an order different from that in the commentary of Donatus. This is not unlike one aspect of the tradition for Seneca's tragedies. The two recensions (represented by the Codex Etruscus[98] and the so-called A tradition, which includes a number of medieval manuscripts) differ in the order of transcription of the tragedies, and in the A tradition's inclusion of the *Octavia praetexta*. Moreover, these contain textual scraps that, despite their relevance, made it difficult for publishers to establish a single secure text.

In addition to collections of a single author's work, textual editions also gathered works according to their genre or certain shared thematic characteristics. The *De agricultura* of Cato the Censor, for example, has been preserved via the remains of an edition of an agricultural codex that also included the *De re rustica* of Varro, as well as works by Columella and Quintus Gargilius Martialis. Accordingly, that collection linked more ancient works to the contemporaneous work of Gargilius. The same thematic criterion seems to have been paramount for another late antique collection organized around the genre of Latin panegyric. In it we find late antique panegyrics, likely of Gallic origin, together with Pliny the Younger's panegyric to Trajan.

Larger works, often historical, such as the *Naturalis historia* of Pliny the Elder, required the use of multiple codices, which typically gathered groups of five or ten books into one volume. The first five books of Polybius' forty-book *Histories* have been preserved together in their entirety, as well as a late antique volume containing excerpts of eight books. Of Diodorus Siculus' *Bibliotheca* there remain books 1–5 and 11–20 (out of forty), and of Cassius Dio's work only books 36–60 out of a total of eighty survive. In the case of Livy's ponderous *Ab urbe condita*, originally written in 142 books, only thirty-five survive: 1–10 and 21–45, which are preserved in the Codex Vindobonense (now in Vienna, Österreichische Nationalbibliothek, ms. Latinus 15), although this manuscript originally included the entire fifth decade. In some cases, the books that we have received are clustered in groups of five, yet they have suffered from deterioration when the edges of the pages came in contact with destructive external agents such as moisture, abrasion, bookworms, or other misfortunes.

With rolls, by contrast, the most vulnerable section is the beginning of a text, which is the outermost part of the helicoid papyrus. The Ambrosian Palimpsest originally contained the twenty-one comedies of Plautus, arranged alphabetically. However, the first five plays and the last one (the *Vidularia*) are illegible. While the first five recur in the Palatine tradition, there are unfortunately no other witnesses of the *Vidularia*, making this mutilated version the only extant copy of the comedy.

Of the ten books of Curtius Rufus' *Historiae*, the first two in their entirety, the last part of book 5, the beginning of book 6, and the close of book 10 are all lost. The textual degradation goes back to an edition in the form of two manuscripts, each containing five books, both heavily damaged in the initial and final sections before being copied into the archetype of the tradition that has come to us.

Similar manuscript damage occurred in the tradition of the major works of Tacitus, represented by two important manuscripts: the Codex Mediceus *prior*,[99] which contains books 1–6 of the *Annales*, and the Codex Mediceus *alter*,[100] which contains *Annales* 11–16 and *Historiae* 1–5. Books 5 and 6 are mutilated, as is the beginning of book 11 and the end of book 16, which were situated at the beginnings and ends of codices that each contained five books. The combination of the *Annales* with the *Historiae* in the Codex Mediceus *alter* is explained by Jerome in his *Commentarius in Zachariam* 14.1–2, where he mentions a single late antique manuscript comprising thirty books. The number thirty is confirmed in the Codex Mediceus *alter*, where *Historiae* 1 appears as book 17. It can therefore be inferred that the work would have been distributed in three manuscripts (*Annales* 1–10; *Annales* 11–16/*Historiae* 1–5; *Historiae* 6–14), of which only the first two have survived.

These were received, however, through different transmissions: the ninth-century manuscript, Codex Mediceus *prior*, was copied in the imperial abbey of Fulda and then passed to the library of the imperial abbey of Corvey, both in

Germany. It was brought to Rome in 1508 and came into the possession of Pope Leo X, a member of the Florentine Medici family (from whom it was later acquired for what is now the Biblioteca Medicea-Laurenziana).

The Codex Mediceus *alter*, by contrast, was copied at Monte Cassino in the eleventh century and was brought to Florence by Boccaccio or Zanobi da Strada in 1427, coming into the possession of Niccolò Niccoli (from whom it then passed to the Laurentian library). If the total number of Tacitean books was thirty, as Jerome says, the *Annales* then would seem to have originally been comprised of sixteen books and the *Historiae* fourteen. Some doubt about the original structure of the Tacitean corpus is raised, however, by the fact that book 16 of the *Annales* is mutilated, and the damage reveals that a good bit of material is missing. The subject matter of the lost material may possibly be the happenings of AD 66, for the lost section comprised in detail the events subsequent to Nero's death (AD 68) and consequently may have been a topic simply too raw for Roman readers. It has therefore been theorized that the loss of two books conveniently came to pass during the time of transition from roll to the codex. However, the eventual loss of one of the three codices that together had held the entire work was probably not accidental.

Livy's *Ab urbe condita*, conversely, represents more of a random loss, or at least one driven by a lack of interest in the material rather than its content being impolitic. What of Livy's narrative is preserved comes chiefly from the first section of the work, without any trace of books 51–142. It is likely that in late antiquity there was greater interest in the history of ancient Rome than in more recent history, although we know that Symmachus was preparing a fresh edition of the entire Livian corpus.[101] Accordingly, that complete corpus must have then been available in libraries. Yet the odds of loss of any given work in part or as a whole were already in late antiquity rather high, when a general lack of interest in things ancient can reasonably be said to be at least partially responsible for this unfortunate phenomenon.

By the close of the second century AD, smaller epitomies began to supplant the multiple codices of lengthy works. This phenomenon is itself an indication of a cultural shift in that era, as there was a change in the taste of readers and a concomitant decrease in commerce in the book trade. Moreover, manuscript preservation tended to be territorially landlocked, reducing the circulation and the transmission of larger texts. In the third century, epitomization became an issue chiefly for historiographical texts, for the readership of long volumes decreased as the relevance of historical events became temporally distant. In the case of Livy, such problems can be seen already in the Hadrianic period, when Florus appended his epitome to the work, pithily summarizing the lengthy Livian narrative.[102] For Livy, other types of summaries of individual books were produced, known as the *Periochae*.

This phenomenon grew: the second-century Roman historian Justin made an epitome of the much longer *Historiae Philippicae* of the Augustan historian Pompeius Trogus. Some years later, the Emperor Tacitus (275-276) is said to have ordered all libraries containing the work of his own namesake (i.e., the historian Tacitus) to receive imperial subvention. While that story need not be taken at face value,[103] it may nevertheless be regarded as emblematic of the perception that Tacitean works were scarce even in that early period. Similarly, the phenomenon of epitomization occurred for works that were not specifically historiographic: Verrius Flaccus' *De verborum significatu* was summarized by Pompeius Festus, while Pliny's *Naturalis historia* was abridged by Solinus in the *Medicina Plinii*, by Quintus Serenus in the *Liber medicinalis*, and by others.

Despite some lack of interest in certain segments of the reading populace, late antiquity nevertheless played an important role in the transmission of classical culture's textual heritage. The preparation of editions in codex form, arranged either thematically or by author, seems to have been based on a general attempt to preserve what writings of pagan authors had endured. It is difficult, however, to speculate what then survived from such an accumulation of material. Some texts were undoubtedly lost in the transition from roll to codex. Works of previous eras were not all copied and some remained preserved only in papyrus, which decays more easily than vellum.

Overall, however, the most significant works of Latin literature in the early centuries AD were, thanks to the production of codices, likely still in circulation even in the fifth century. The losses that occurred thereafter, between the sixth and seventh centuries, were more significant than those of the previous phase of transmission. In that period, even works housed in libraries were likely to be lost if they were not copied.

The survival of several hundred Virgilian manuscripts is owed not only to general interest but also to the fact that a high number of copies were produced between the fourth and sixth centuries. Unfortunately, some surviving copies were likely not sufficiently circulated in that epoch, and were consequently lost. Despite losses caused by the instability of late antique culture, texts with a slender tradition were valued for their rarity, which in some cases actually fostered their circulation and survival. The most significant cases of the latter are the *Satires* of Juvenal, a work inserted among school readings at the time of Servius, and the *Rhetorica ad Herennium*, a manual by an unknown author that only came into vogue in late antiquity when it was believed to have been penned by Cicero. Others include many works cited by authors in that period, but which are no longer extant. In his *De compendiosa doctrina*, the fourth-century grammarian Nonius mentions that he had read the *Saturae Menippae* of Varro and some tragedies of Ennius; in the fifth century, Orosius cites the *Annales* of Ennius, while Arusianus Messia cites four orations of Cicero of which we have no other records; the *Histories* of Sallust were preserved in over five hundred citations.

In Constantinople the sixth-century antiquarian John the Lydian possessed more complete copies of both Seneca's *Naturales quaestiones* and Suetonius' *Vitae Caesarum* than we currently have today. In the same period in Africa, Fulgentius indicates that he had read parts of the *Satyricon* of Petronius that have not been transmitted to us. Circulation of Seneca's works that are now lost is also well attested in late antiquity: Lactantius knew Seneca's *Exhortationes, De morte et mortalitate,* and *Moralis philosophia*; Jerome knew *De matrimonia*; and Augustine knew *De superstitione*; Cassiodorus possessed a copy of the *De forma mundi*; between 570 and 590, the Spaniard Marcino di Braga availed himself of a lost work of Seneca to compose his *Formula honestae vitae*.

GREEK IN THE WEST

Late antiquity saw the decline of the bilingual Latin and Greek culture that had characterized Rome. Starting in the second century BC, learning Greek was a usual part of the curriculum for the Roman elite. This was accomplished with Greek teachers for school-aged learners followed by an extended visit to the East for immersion in the language. Additionally, Greek culture was continually present in Rome from the second century BC onward by way of representatives such as Polybius and Galen, who lived in the city.

Thus, in antiquity Greek culture was a mark of prestige for Latin-speaking authors who could show their linguistic dexterity by choosing to write in Greek. Conversely, while Greek-speaking authors sometimes might write in Latin, it is significant that this occurred primarily in the first phase of Western literary history, as in the case of Livius Andronicus. This phenomenon does not seem to arise again until late antiquity, as, for example, can be seen in the writings of Ammianus and Claudian. The empire, which was divided geographically between Roman and Greek sectors, saw Latin and Greek prevail as the dominant languages along a boundary that divided the Balkan Peninsula in northern Greece and Egypt from West Africa.

Thus, the division of empire introduced by Diocletian roughly paralleled the partition between the two languages. For a time this arrangement preserved knowledge of Greek in the West, but that knowledge remained important only for cultural elites, resulting in a decreased demand for Greek texts. Even in areas where the cultural transmission had taken place mainly in Greek fields of study, such as philosophy, medicine, and science, a shift to an exclusively Latin mentality slowly began to develop. This transition from Greek to Latin eventually created a demand for the production of adaptations and translations.

Translations in the classical period were occasional and dictated mainly by desires to emulate the original. One can see this not only in the case of early poetic translations, such as the *Odyssey* by Livius Andronicus, but also in the case of various Latin adaptations of Aratus' *Phaenomena* by Atacinus Varro, Cicero, Ovid, and Germanicus. This is also true of the few translations of prose works, including Plato's *Timaeus* translated by Cicero and Aristotle's *De mundo* by Apuleius.

Translations became more frequent in the fourth century. The Christian Chalcidius, who was active in Spain, translated and commented on Plato's *Timaeus*. We also know of lost renderings that Boethius made of works of Plato and Aristotle, as we know, too, of Marius Victorinus' translation of Porphyry's *Isagoge*. There were also numerous renditions of medical works, in particular by African authors: Theodorus Priscianus, a contemporary of Augustine, translated into Latin his own work, *Euporiston*, originally written in Greek. Further, Celius Aurelianus transposed into Latin various works of the second-century writer Soranus of Ephesus. Further translations, including those of the works of Hippocrates, Galen, Dioscorides, Oribasius, and other authors, can be dated between the fifth and the sixth centuries. Much of the intellectual activity of translation was localized in Ravenna, where the Byzantine domination guaranteed direct contacts with the Eastern empire.

This linguistic divide even affected those in Western cultural centers, including Christian culture, whose origins were Greek-speaking. By the latter half of the second century, biblical texts were being translated into Latin, including the *vetus Itala* (also known as the *vetus Latina*), a version of the Bible that found wide circulation. Its popularity was striking, even though it comprised a far from perfect rendering of the Greek and Hebrew scriptures. Because of its inaccuracies, in 382 Pope Damasus I called upon Jerome to establish an office for making a new translation of the entire text. After a sojourn in Palestine, which he dedicated to learning Hebrew, Jerome completed the text of the *Vulgate* in 404. It has remained since then the canonical Latin translation of the Holy Scriptures. Jerome describes, in *Epistle* 57, the criteria that he followed in the translation: under the inspiration of Cicero, he envisioned a version capable of restoring faithfully the meaning of the rendered text, without tracing it out word by word.

As early as the third century one finds within Christian circles many translations of epistles, apocryphal texts, and other works. The *Adversus Haereses*, written by the Greek Irenaeus, has been passed down only in Latin translation. In the fourth century, Tyrannius Rufinus rendered texts of the Fathers of the Eastern Church (Origen and Gregory of Nazianzus) into Latin, along with the *Ecclesiastical History* of Eusebius. Another work of Eusebius, the *Chronicon*, was the basis for the eponymous work of Jerome. Boethius (ca. 480–523) and Cassiodorus (ca. 490–583) were leading figures in the sixth century, which was the last great age of translation into Latin. Boethius planned a translation of all the works of Aristotle and Plato, but only completed a small portion of his grand project.

These translations enjoyed significant circulation during the Middle Ages. Boethius rendered Aristotle's *Analytica, Topica, Categoriae,* and *De interpretatione*. He also arranged for a revision of the translation of Porphyry's *Isagoge* made earlier by Marius Victorinus and then added his own translation and commentary on the work. Cassiodorus (ca. 490–583), to be discussed further below, brought numerous Greek books collected in Constantinople and in other centers to the Vivarium monastery,[104] where he engaged three translators, Epiphanius, Mutianus, and Bellator. The first of these rendered the works of three Greek historians, Sozamenes, Socrates, and Theodorus, from which Cassiodorus compiled his tripartite *Ecclesiastical History* comprising historical events from the time of Constantine to AD 429. Other translations of Cassiodorus included works of Clement of Alexandria and Flavius Josephus (also known as Josephus Latinus). Dionysius Exiguus ("the Little"), one of the last cultural Greek personalities active in the West, was a friend of Cassiodorus. Beyond his translations of hagiographic works, he is remembered as the individual who calculated the number of years since the birth of Christ, which was later adopted as a historically accurate chronology.

After Cassiodorus, knowledge of literary Greek substantially abated, and its remaining linguistic usage was limited to diplomatic and ecclesiastical relations. Some circulation of Greek codices continued in Byzantine areas of southern Italy and Sicily until the Arab conquest of 827, as well as areas where Greek-speaking settlements obtained throughout the Middle Ages. In the rest of western Europe, Greek was incomprehensible to most readers, as is evidenced by the maxim that circulated a century later: *Graecum est, non legitur* ("it is Greek, it is not read").[105] In the Eastern empire, the teaching of Latin was still practiced in the sixth century, coinciding with the attempt of Justinian to restore the unity of the empire. As early as 372 Valens assigned both Latin and Greek copyists to the library at Constantinople.

During the reign of Emperor Anastasius (491–518), Priscian of Mauritanian Caesarea taught grammar in Constantinople and wrote the *Institutio de arte grammatica*, a monumental grammatical treatise that became widespread in the West during the late Middle Ages. In that same period, the aforementioned John the Lydian, the author of Greek works dedicated to ancient Western studies, was active in Constantinople. That his interest lay not only in Roman but also Etruscan culture (*De magistratibus rei publicae Romanae, De mensibus*) shows that Roman studies were strong for many centuries and that interest in ancient Rome persisted in the East as well as the West.

The most significant contribution of that age, however, was made by Justinian (527-565), who published the *Corpus juris civilis* and other legal texts in the Latin language. Moreover, Latin had great flexibility, as even the coronation of Justin II (565-578) was celebrated in Latin by Corippus. Thus, Latin held its own through the sixth century, until, in 610, Greek was finally proclaimed the official language of the Byzantine Empire by the emperor Heraclius.

The shift in linguistic supremacy brought with it significant losses of ancient texts. Among these the most substantial occurred between roughly AD 500 and 650, in the context of a cultural depression that affected important areas of the former Roman Empire (Africa, Italy, and France). Fortunately, the Iberian Peninsula and the British Isles, which conversely enjoyed at this time remarkable intellectual blossoming, were spared such setbacks. In Africa, Vandal culture ended with the close of the fifth century, after which came the Byzantine conquest in 534, which would itself be followed in the next century by the Arab invasion (697). In Italy the intellectual society that had accompanied the reign of Theodoric the Ostragoth declined with the Gothic Wars (535-553), a conflict that preceded the Lombard invasion of 568 within less than a quarter-century.

This transition between late antiquity and the early Middle Ages naturally was characterized by some turmoil. Nonetheless, during that period a new culture was taking shape in Gaul that would offer an intellectual continuity between the fifth and the sixth centuries. That nascent French culture produced authors such as Venantius Fortunatus and Gregory of Tours. Although in the seventh century Gaul underwent marked decline under the Merovingian kings, on the Iberian Peninsula in the same period the Visigoth culture culminated with the work of Isidorus of Seville (570-636) and in the following decades with Braulius and Eugene of Toledo, coming to an end only after the Arab invasion of the eighth century.

In Britain the Anglo-Saxon incursions of the fifth century overturned the Latinized British culture, and the evangelization of Ireland laid the groundwork for the cultural flowering of England in the sixth and seventh centuries, which would come to boast the likes of Aldhelm of Malmesbury (ca. 640-709) and the Venerable Bede (ca. 672-735). In the British Isles, where the vernacular languages were Celtic and Anglo-Saxon, commissioned missionaries learned the languages of the cultures with which they came into contact. The schools they founded, however, focused on Latin grammar, anticipating an educational framework that would only later characterize the whole of Europe.

In continental Europe economic and social transformations that led to the dissolution of the aristocratic classes that had sustained cultural life in late antiquity created a watershed for the preservation and dissemination of texts. The societal downturn brought with it the disruption of traditional school education, the widespread destruction of public libraries, and the crisis of urban centers, which included the decentralization of centers of commerce. The wards of prosperity were no longer cities or towns but rather large villas. Regarding the value of literature and learning, one statistic is particularly telling: at the beginning of the fourth century, twenty-eight public libraries were active in Rome, but, by the end of the century, Ammianus says (if with some exaggeration) that in the city "the libraries were forever sealed, like tombs."[106]

In the fifth and sixth centuries, culture on the European continent was on the whole subdued. Yet relief came from missionaries who were sent from Britain to the continent between the seventh and eighth centuries. The driving force behind these missions arose from the cultural and intellectual maturation that followed the introduction of Christianity to the British Isles. Those who came to the continent were to have a profound influence on the survival of many ancient texts. The Irish Columbán (Saint Columbanus, ca. 542–615), for example, founded monasteries in France (Luxeuil), in the area of Lake Constance (St. Gall), and in Italy (Bobbio, at Piacenza). In these locations the influence of the British insular culture abounded, continuing well into later centuries and playing a major role in the preservation and transmission of proper Latin grammar.[107]

Book production, meanwhile, was commonplace in Rome in the first decades of the sixth century, as is well documented by subscriptions. During that time Boethius assembled a large library for the composition of his works. A change had begun, however, by the middle of that same century. Private libraries collected by members of the old aristocracy fell into disuse, only to be sold off by their heirs. The legacy of these private collections was partly recovered in the monastic educational hubs that began to arise with the arrival of the missionaries from the British Isles.

The cultural activity of reading and writing, fundamental for the transmission of texts, gradually shifted to monasteries and ecclesiastical institutions, but even in those settings, interest in the classic texts began to decline at this time. The attitude of the Christian culture between the sixth and seventh centuries became more selective in preserving pagan classics. Monastic culture was more interested in technical literature (pagan grammatical, medical, and scholarly works), which had no implications for Christian doctrine. Conversely, interest in classical literature waned, except perhaps for certain authors, such as Cicero, Virgil, and a few others.

Cassiodorus is emblematic of the denigration of the value of belletristic writing. He had retired from political life after the Gothic Wars, having served as the authoritative collaborator to the king of the Ostrogoths. In 555 he founded the library in the monastic community known as the Vivarium. The testimonies that remain of this library highlight the privilege given to works of religious interest. Though alongside them are profane texts, they are clearly selected, chiefly with a utilitarian approach. These include the rhetorical works of Cicero, Quintilian, and Atilius Fortunatianus; the agricultural treatise of Columella; the now-lost *De forma mundi* of Seneca; and the *De interpretatione* of Apuleius. A similarly selective approach is detectable in the next century in the work of Isidorus of Seville, the author of a successful encyclopedia, *Origines*, that collected key contributions of grammar, rhetoric, natural sciences, and more. In the compilation of the work, Isidorus, without hiding his impatience with the pagan tradition, drew on a variety of ancient sources, some of which are now lost.

That, generally speaking, Christian doctrine had little interest in pagan culture and could even be openly hostile to it is noticeable in the works of the two most important proponents of Christian culture of the sixth century. The first of these is Saint Benedict of Nursia (ca. 480–547), who in 529 founded the monastery of Monte Cassino and compiled his rules for monastic life, his *Regula*, which would influence monasteries for centuries to come. The second key figure is Gregory the Great (ca. 540–604), who became pope in 590. Gregory is a key figure in the propagation of the Christian faith. His very significant work, *Regula pastoralis* (591), redefined the role of the Church in a new historical and political context.[108] While he had great reverence for carefully crafted Latin prose, he did not value the pagan classics.

Under such a cultural framework, interest in the transmission of a large number of classical authors waned between the sixth and seventh centuries, resulting in the loss of a number of works. Codices dated between 550 and 750 are suggestive of this shift;[109] of the 264 extant codices and codex fragments that were copied in this time, the vast majority—some 238 of them—contain Christian texts, including patristic, liturgical, and exegetical texts. Twenty-four are of a technical nature (eight on law, eight on medicine, six on grammar, one on surveying techniques, and one on military art); the remaining two are represented by a single codex containing two texts, the *Breviarium* of Rufius Festus and the *Itinerarium Antonini*, and a fragment of Lucan. Remarkably, the last entry on this considerable list is the single classical author known to have been copied in that period.

While the disappearance of authors whose manuscripts were found centuries earlier is striking, the fact that to some extent they remained fundamental in education is equally so. For example, although there are many codices and manuscript fragments of Virgil from the fifth and sixth centuries, there is only a single manuscript from the seventh century and no trace of the prodigious author turns up again until the end of the eighth century. Of Cicero, similarly, various codices or fragments dating to the fourth through sixth centuries can be found, but then a chronological gap intervenes until the end of the eighth century. Further, the seventh and eighth centuries show evidence of numerous palimpsests of a religious nature, confirming the lack of interest in classical writings in that period.

In the monastery of Bobbio, the letters of Fronto and the *De re publica* of Cicero were used merely for parchment for, respectively, the Acts of the Council Chalcedon and the commentary of Augustine on the Psalms. Another late antique codex of Fronto was used in the same period in Corbie for making a copy of Jerome's *De viris illustribus*; a biblical text was transposed onto Ambrose's copy of Plautus in the late sixth century; other palimpsests come from the monasteries of Luxeuil and Fleury. In the latter case, the manuscript was dismembered "by a certain deranged monk"—so wrote the most authoritative publisher of the fragments of Sallust.[110] A copy of Sallust's *Historiae* was repurposed to contain a biblical commentary of Jerome; some shreds of the original that were rewritten or used for text bindings remain.

The production of palimpsests highlights the role in preservation that the early medieval monasteries played in the transmission of texts. It is likely that most of the texts that have survived were in fact transcribed in monastic libraries, where the texts left over from the dissolution of the libraries of late antiquity were deposited. Having been stored in monasteries, they resurfaced between the ninth and fifteenth centuries. In general, most of the classical works that disappeared were likely lost between the sixth and seventh centuries. Those works that had a reduced circulation in late antiquity were lost following the destruction of the ancient libraries. Some other texts survived in the early Middle Ages but then went missing or were destroyed, often by accident, without being copied. The classical works that we now have are those that managed to survive the difficult transition from late antiquity and the early Middle Ages to the Renaissance.

SCRIPTS

A consequence of the disintegration of the Western empire was the fragmentation of forms of writing, which afterward took on specific traits in various areas of the remnants of the empire. In late antiquity the dominant hand for book writing and publishing was the uncial, but the rustic capital was also present in many places, while the half-uncial spread in a later phase. In particular, cursive script, which was used for documents and administrative measures, assumed a variety of forms across different regions. Differentiation in scripts was a result of the reduced opportunities for cultural exchange that followed the shift in the political landscape.

A trend similar to the development of various cursive scripts can be seen in scripts adapted to book writing. A common trait was the tendency to use tiny

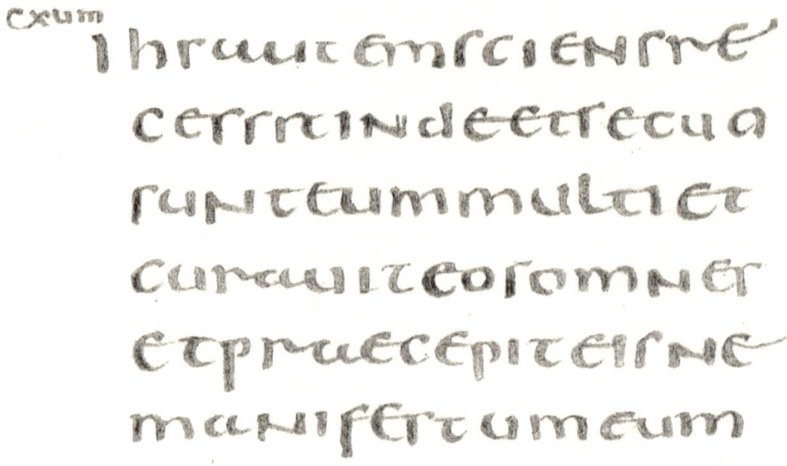

> cxum
> Ihruuicemrcienrre
> cerrrindeerrecua
> runccummulciec
> curauiceoromner
> ecpruecepicerrne
> manifercumeum

FIGURE 3.4. *Example of half-uncial script. Ms. 1395, p. 25 (detail). St. Gall, Stiftsbibliothek. Drawing by Mary Claire Russell MacDonald.*

letterforms, detectable already in the half-uncial (fig. 3.4). Of the scripts that developed in the British Isles after the evangelization of Ireland, the so-called insular script featured both capital and lowercase forms; it likely was an evolution of the half-uncial. Capital letters of insular script are characterized by a marked calligraphic character, round but with slightly elongated and light to dark vertical strokes. In addition to the local British and Irish centers of writing, insular script spread to the continent through missionaries sent by the Catholic Church to found monasteries in Italy, most prominently Bobbio.

On the Iberian Peninsula the Visigothic style developed, a tiny script of uncertain origin with strong similarities to that used in the ninth and tenth centuries in the monastery of St. Catherine at Mount Sinai, in Egypt. Characterized by the particular design of certain letters (fig. 3.5),[111] the Visigothic script continued to be used in the Iberian Peninsula after the Arab conquest. Its continuity in the administration of the Gallic kingdom founded by the Franks created the conditions for the development of a tiny cursive, the so-called Merovingian, characterized by lateral compression of the letters (tight leading). Its strokes are elongated and wavy, with notably ornate features (fig. 3.6).

In the northern regions of Italy during the Lombard period, writing forms were not universally homogeneous. In southern Italy, conversely, the Beneventan script, developed in the monastery of Monte Cassino, became widespread and its use soon extended across the Adriatic to Dalmatia. This style of calligraphy features small letterforms that anticipate cursive (fig. 3.7). It is formed by thinner strokes alternating with thicker and with unique letterforms and its

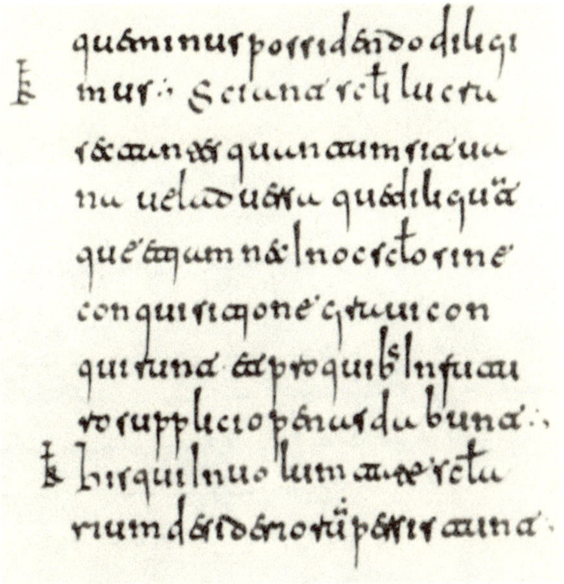

FIGURE 3.5. *Example of Visigothic script. Ms. 10067, folio 83 recto (Isidorus of Seville,* Sententiae*) (detail). Madrid, Biblioteca Nacional.*

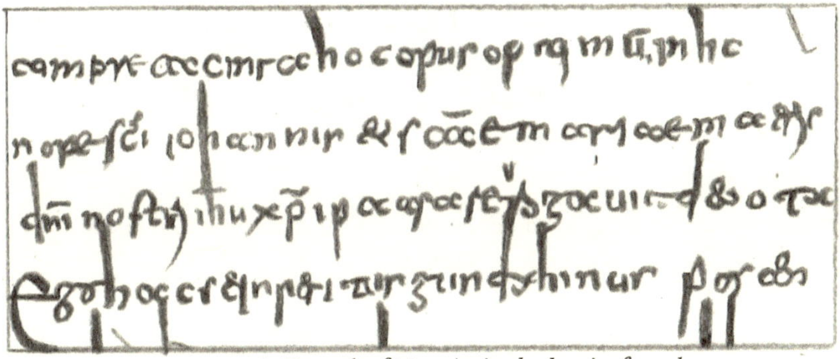

FIGURE 3.6. *Example of Merovingian book script from the eighth century. Drawing by Mary Claire Russell MacDonald.*

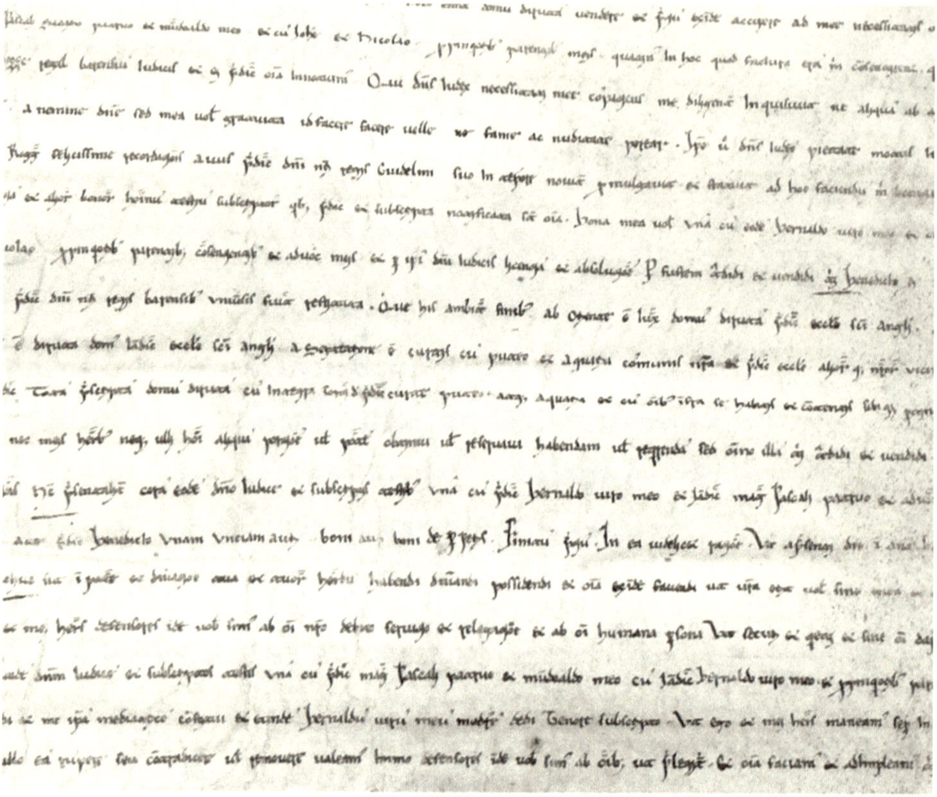

FIGURE 3.7. *Example of Beneventan script. Ms. 444. Monte Cassino, Biblioteca dell'Abbazia. 1075–1090.*

own set of abbreviations. The Beneventan would have some influence years later on the fonts used in book publishing.

THE CAROLINGIAN RENAISSANCE

Cultural revitalization is rarely so profound that it should be described as rebirth. Yet that is the metaphor that Jean-Jacques Ampère used to describe the Carolingian period in his *Histoire littéraire de la France avant le douzième siècle* (1839). That period, which in some ways culturally anticipates the Renaissance, also encompassed a rebirth in interest in the transmission of classical texts, particularly those in Latin. While the seventh century was the low ebb in the processes of dispersion but the high-water mark for the loss of texts, Carolingian culture in the eighth and ninth centuries offers a sharp contrast. Classical texts were once again regularly recovered and copied during this period, dramatically slowing the loss of important authors. This trend continued and stabilized in the tenth and eleventh centuries with the rise of the Ottoman dynasty and the apogee of the Holy Roman Empire.

In just twenty years, Charlemagne, the heir of the Frankish realm, united the remnants of the Western Roman Empire (Spain, the Pyrenees, France, and north-central Italy) with the more distant regions of Germany and the North Sea. His unifying leadership culminated with his coronation by Pope Leo III in Rome on Christmas Eve in AD 800, the occasion that marked the birth of the Holy Roman Empire. Yet even in the years prior to this signal event, Charlemagne had already taken imperial Rome as the model for the construction of his new palace of Aachen, whose foundations were laid in 796.

Charlemagne sought to establish Aachen as a new Rome, and to do so he employed a sundry group of advisers, clerics, and intellectuals for the organization and management of his empire. Charlemagne recognized that it would be to his own advantage to learn Latin, for as a child he had spoken only German. Thus, probably at some point near his late thirties, he commissioned as his instructor Peter of Pisa, formerly master of Latin grammar in the Lombard court of Pavia. Since the late sixth century, that city had been the center of Lombard power, holding sway over Italy until Charlemagne conquered it in 774.

Other scholars attached themselves to the robust intellectual atmosphere surrounding the Carolingian court. Paul the Deacon (ca. 725–799), an increasingly conspicuous exponent of Lombard Italy, became prominent in the Carolinian entourage. He was the author of a grammar, histories,[112] a collection of 248 sermons ordered according to the liturgical year, and a compendium of the lexicographical work of Pompeius Festus presented to Charlemagne to familiarize him with Roman topography and customs. Other members of the diverse group included the Irish scholar Dungal (ca. 811–827), the English Alcuin of York (ca. 732–804), then abbot of Tours, the Visigoth Theodulf (ca. 760–821), who became bishop of Orléans, and the German writer Einhard (ca. 770–840), who studied in Fulda.

As Alcuin returned to England from Rome in 781, he met Charlemagne at Parma and became his close associate and the head of the Palatine school. None had greater influence on Charlemagne's cultural policy than Alcuin. Through his influence the foundation and development of English schools prevailed in Europe, based on the teaching of Latin grammar. In his *Vita Karoli Magni*, Einhard highlights Charlemagne's interest in education: "He decreed that children be educated in such a way that both males and females learn first liberal arts, in which studies he himself engaged."[113]

One of Charlemagne's primary goals was to form a clerical class able to conduct the affairs of the Church and manage the empire itself. The *Admonitio gene-*

ralis of 789 sought to regulate liturgical practices, to ensure uniform application of the Benedictine rule, and to train priests to master Latin, the liturgical language. Thus, an objective for the schools set up in monasteries and cathedrals was to develop a standard form of the Latin language, purified of vulgarisms and sentences lacking in proper coordination, which had in previous centuries become all too common. Prominent among the proponents of this appeal for regularization was the Anglo-Saxon Alcuin, who had learned Latin only as a written language.

In continental Europe during this same period, a movement for purer Latin took root alongside the normalization of Latin phonetics and orthography, so that the scholarly language might be clearly distinct from the vulgar tongues. In 813 the council of bishops meeting at Tours established that sermons were to be read in Latin but also translated *"in rusticam Romanam linguam aut Thioriscam"* —that is, rendered into the vernacular French or German. Nonetheless, Latin, now more coherent because of the regularization that the Carolingian educational model afforded, became the standard language for centuries to follow. Latin's place as the primary scholarly language was vouchsafed by the teaching of grammar, the study of stylistic models, and contemporary literary developments.

This standardization of Latin, therefore, was closely connected to religious policy and the program that encompassed correction and emendation of the Bible, aimed at achieving a uniform text for liturgical use. In the last years of his life, Alcuin labored successfully to establish his edition in Tours. That text is preserved in dozens of specimens, and its distribution was sponsored by imperial authority. Alcuin also displayed an attitude borrowed from the patristic tradition toward classical authors, postulating their appropriate use in respect to the values of the Christian tradition.

The theologian Theodulf of Orléans had some of the same goals as Alcuin. He, too, produced an edition of the Bible, but this was less widely disseminated. Nevertheless, modern scholars often detect greater philological care in Theodulf's version, for it was produced by the scrupulous collation of a significant number of manuscripts.

Among his other literary productions, Alcuin presents in *Epistle* 280 the objective of "perfect knowledge" (*perfecta scientia*), which one can access "with the help of seven steps." Alcuin's seven steps offer a clear allusion to the seven *artes liberales* as arranged by the authors of late antiquity (Martianus Capella, Boethius, Cassiodorus, and Isidorus).[114] In his description, which itself will later develop into seven disciplines, Alcuin divides these studies into two groups, the trivium (grammar, rhetoric, and dialectic) and the quadrivium (arithmetic, geometry, astronomy, and music). Alcuin focused his attention primarily on the trivium, in particular in his work entitled *De grammatica*, which was transmitted from England to Europe as a significant work of grammar.

Carolingian culture recognized the works of Augustine, Boethius, and Cassiodorus as foundational texts that formed a cultural bridge between the classical past and the rebirth of culture in the Carolingian present. Classical authors and their commentators, too, were recovered, even if chiefly for their functionality for improving reading in schools. Nevertheless, the number of classical texts found or copied at this time confirms that the practice was sweeping. Library collections during Charlemagne's reign, preserved in Berlin in a partial catalogue,[115] included authors such as Lucan, Statius, Terence, Juvenal, Tibullus, Martial, Horace, Cicero, and Sallust. Knowledge of other authors can be deduced from allusions to them in poems composed by representatives of the Carolingian court.

The reception of classical antiquity can be inferred also from the convivial environment of the imperial court, in which the names of classical authors, biblical characters, or historical figures were revived. Though Charlemagne himself might well have chosen the name Augustus, he did not, preferring to be called David or Solomon, Einhard tells us.[116] Other members of the court, however, took classical names. Alcuin was named Flaccus (i.e., Horace); Angilbert of Saint-Riquier, who was the ambassador of Charlemagne to the pope, was known as Homer; the poet Moduin of Autun adopted the name Naso (i.e., Ovid).

Another important result of the intellectually fecund atmosphere of the Carolingian court and the cultural revival that it fostered was the development of a new style of writing, the so-called Carolingian minuscule. Amid the robust renewal of book production, the adoption of this calligraphic form reconstituted a communications network on a continental scale and facilitated the circulation of books and, with them, a sense of literary and cultural renewal.

Even in its nascent stages, the Carolingian script was already developing in so-called pre-Caroline letterforms. These are represented by small characters used in some French libraries and centers of writing, such as Luxeuil (fig. 3.8) and Corbie (fig. 3.9), and in the abbey of St. Gall in Switzerland (fig. 3.10). The Carolingian minuscule presented rounded letters spaced apart from each other, lacking features more typical of cursive writing (fig. 3.11). In fact, their shape is, in large part, still in use. Carolingian script rapidly imposed itself as the preferred form of writing not only for book production and publishing but also in the courts for diplomatic and ambassadorial documentation.

The large number of texts in the Carolingian library demonstrates that the Carolingians had collected a vast number of late antique manuscripts from much of the former Western Roman Empire. This is further confirmed by the

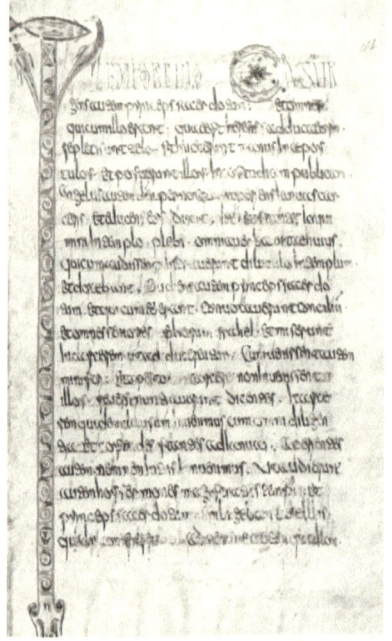 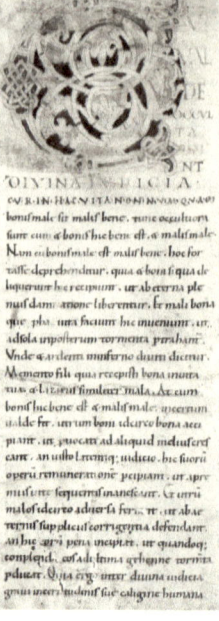

FIGURE 3.8. (left) *Example of pre-Caroline letterforms. The* Lectionary of Luxeuil *(ms. Latinus 9427, folio 144). Paris, Bibliothèque Nationale.*

FIGURE 3.9. (right) *Example of pre-Caroline letterforms. Page from a theological treatise. German School. Eleventh century. Credit: Germanisches Nationalmuseum, Nuremberg (Nürnberg), Germany/Bridgeman Images.*

substantial list of texts of ancient authors whose manuscripts can be dated to the ninth century. Even if the extent of the dissemination of copies of classical texts made in this period cannot be precisely established, a number of additional authors were certainly on the list: those whose works are contained in the codices of the following centuries were most likely copied from examples now lost.

Roughly seven thousand manuscripts survive from the ninth century alone, a number that evidences the great influx of texts from this period, for that is a figure far greater than the total number of books copied between the end of antiquity and the ninth century.[117] The list of included authors favors those of the Republican era, prominent among whom are Terence, who was preserved in a tradition independent from the late antique Bembinus, and Cicero. The majority of Cicero's works passed on by the Carolingians were oratorical: *De inventione, De oratore, Orator,* and *Rhetorica ad Herennium,* the last of which was at the time attributed to him. Also included were his philosophical works (apart from the *Academica posteriora* and *De finibus*) and his *Epistulae ad familiares*. Lucretius was also preserved, as well as the corpora of Caesar and Sallust (including pseudo-Sallustian *Epistulae*).

Augustan and early imperial authors were likewise frequently edited and disseminated in the ninth century; chief among these was Virgil, along with the *Appendix Vergiliana*. Works of Horace as well as the first and third decades of Livy were also Carolingian favorites. Ovid's *Amores*, *Ars amatoria*, *Remedia amoris*, *Heroides*, *Metamorphoses*, and *Ex Ponto* were prominent. Minor works also fared well: the *Cynegetica* of Grattius, Germanicus' *Aratea*, the *De medicina* of Celsus, the *De re rustica* of Columella, and the *De re coquinaria* of Apicius, as well as some of Seneca's works.[118] Petronius survived, too, but chiefly in excerpts. Curtius Rufus, Phaedrus, Pomponius Mela, Valerius Maximus, Seneca the Elder, Lucan, and Persius all found favor.[119]

Among the more popular second-century authors were Juvenal, Pliny the

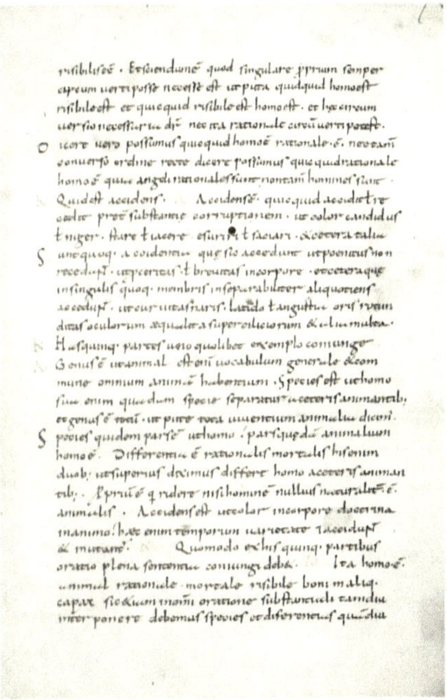

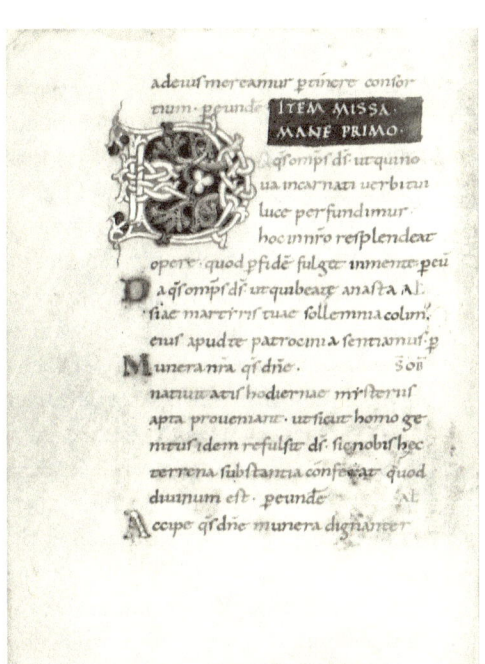

FIGURE 3.10. (left) *Example of pre-Caroline letterforms. Page from ms. 270. Parchment. Ninth century. St. Gall, Stiftsbibliothek. Courtesy of e-codices.*

FIGURE 3.11. (right) *Example of Carolingian minuscule. "Decorated Initial D." Ms. Ludwig V 1, folio 8 verso. First quarter of the eleventh century, France. Tempera colors, gold, silver, and ink on parchment. J. Paul Getty Museum. Digital image courtesy of the Getty's Open Content Program.*

Younger, Suetonius, Tacitus (*Annales* 1–6 and *Agricola*), Apuleius (philosophical works and *Florida*), and Aulus Gellius (*Noctes Atticae* 9–20). Later imperial authors frequently preserved include Censorinus, Justin, and Quintus Serenus. The historical works of Ammianus, Eutropius, and the *Historia Augusta* also found favor at the time, as did authors such as Macrobius and Martianus Capella.[120]

About two-thirds of the manuscripts of the Carolingian period were copied in France, chiefly at intellectual hubs such as Corbie, Reims, Tours, Fleury, and Auxerre. German monasteries also were important, particularly those at Fulda, Lorsch, and Lake Constance. A smaller contribution was derived from the British Isles, whence came a handful of important classical and late antique manuscripts, including the *Ars amatoria* of Ovid and the *De nuptiis* of Martianus Capella.

The location where a manuscript is copied does not necessarily match the origin of the *antigraph*. In many cases it is likely that the recopying was based on imported late antique codices. Thus, one might well assume that there was a significant market for finding and transporting manuscripts from different geographical areas. The generally westward flow of such codices would already have been in place, at least by the seventh century and certainly by the eighth.

From Bede we learn of a number of manuscripts that arrived in England from Italy. Benedict Biscop (ca. 628–690) spent several years in Rome trying to find books. He made his final trip to England in 679, bringing "a huge amount of books of all kinds" with him on the voyage.[121] It is hypothesized that codices were also carried on the reverse route, from the islands to the continent, by Irish missionaries or by those whom Saint Boniface assigned to monasteries founded in Germany. Other codices arrived in Germany through the channels of communication established between the Benedictine abbeys. The origin of many of these manuscripts can be traced back to the library of Monte Cassino.

As early as the seventh century, the quick circulation of books across long distances can be inferred by the astonishing spread of the *Origines* of Isidorus from Spain to England.[122] Also informative is a poem by Alcuin,[123] in which he quotes texts that he had acquired in the library of York: beyond theological writers, he quotes Virgil, Statius, Lucan, Cicero, Pliny, and Pompeius Trogus (i.e., Justin). Two of these authors are found in manuscripts copied in England in the eighth century.[124] Meanwhile, an important manuscript copied in England and dating to that same period is the so-called Spangerberg fragment, a commentary on the Aeneid very similar to the Servius Danielis.[125]

Through Alcuin's intervention, some manuscripts were discovered in England. Others were taken to France by exiled Visigoths who had left the Iberian Peninsula and had found refuge under the aegis of Theodulf of Orléans, in Lyon.[126] Visigothic culture had also recovered texts from Vandal Africa, such as that of the fifth-century Christian poet Dracontius, edited by Eugenius II of Toledo. Other works of Vandal culture arrived in Italy after the fall of the Vandal kingdom to the Byzantines, and were later recovered by this conduit. This was certainly the case, as we have seen, with Dracontius, Corippus, Martianus Capella, and the collection of the Codex Salmasianus.[127]

Though some late antique manuscripts were retrieved by the Carolingians in France and England, the greater part of the manuscripts copied in this period most likely came from Italy, having survived among the book collections belonging to rich libraries existing in late antiquity. The surviving late antique manuscripts were transferred from Italy, where most of them were made, to France between the eighth and ninth centuries: for example, four of the seven late antique codices of Virgil found their way there, as did the Viennese witness to the fifth decade of Livy, which, by the end of the eighth century, was in the possession of the bishop of Utrecht.

The copy of Festus used by Paul the Deacon for his epitome came from southern Italy. During this same period (early ninth century), Ravenna is also likely to have housed an important codex, now lost, containing the corpus comprising Julius Paris, Pomponius Mela, and Vibius Sequester that was annotated by the ninth-century French Benedictine Heiric of Auxerre.[128] Amid such centralization of manuscript collections, the origins of what we now call classical philology took root, evolving in diverse places at various intervals. One vital contributor to Carolingian scholarship is the aforementioned Theodulf of Orléans, who served as bishop from 798 to 818. Theodulf showed a soberly philological approach in his editing and systemization of the Latin Bible. A further early example of philological editorial work can be seen in the production of the codex of Lucretius known as the *Oblongus*, copied in the Carolingian court:[129] the penmanship is identified as belonging to the *corrector Saxonicus*, believed to be the Irishman Dungal, a counselor and member of the court of Charlemagne.

Another Irishman, Sedulius Scotus, distinguished for his erudition and knowledge of the classical texts, was active in Liege in the years 848-858: he wrote a commentary on Priscian's *Institutiones grammaticae* and assembled a sizeable amount of the writings of Priscian and other grammatical writers. This compilation, known as the *Collectaneum*, reveals his prodigious knowledge, particularly of the works of Cicero and other classical authors.

Perhaps the most significant figure of the ninth century, however, with regard to editing classical texts, is Lupus of Ferrières (ca. 802-862). Educated in Fulda under the tutelage of Rabanus Maurus, from 836 on, Lupus served as abbot of Ferrières. He had contact with Einhard and with the Holy See, as well as with cultural centers both in England and on the continent. Lupus availed himself of these contacts particularly to advance his own personal quest for ancient manuscripts. He amassed an important collection, whose codices preserve variants, indicate corruptions, and sometimes retain lacunae found in the transmitted texts. From Fulda, Einhard tells us, Lupus obtained a copy of Cicero's *De oratore* that he copied into what is now known as the Harleian Manuscript.[130]

Einhard offers further details, describing how Lupus acquired a copy of Aulus Gellius' *Noctes Atticae*. He certainly knew the old manuscript containing Valerius Maximus and the epitome of Julius Paris.[131] This is clearly evident from the fact that he could fill gaps in the text.[132] Lupus's student, Heiric of Auxerre, excerpted works by Valerius Maximus and Suetonius; copies of the latter were located in this period in Tours and Fulda. Interestingly, Suetonius' writings provided an important basis for the work of Einhard, who overlooked numerous more recent biographers and instead adopted Suetonius as the primary model for his biography of Charlemagne.

The abbey of Saint-Germain d'Auxerre, whose crypt had been expanded as early as 841 to house Saint Germain's remains, was one of the most important centers of learning for Carolingian culture.[133] There, the Irishman Murethach, author of a commentary on Donatus' *Ars maior*, introduced the teaching of grammar, which emphasis was maintained for half a century by his most famous student, Haimo of Auxerre. Upon the death of Heiric, management of the school was assumed by Remigius of Auxerre (841–908). Remigius studied Platonism under the Irishman Johannes Scotus Eriugena (ca. 815–877), who had enjoyed an active role in the Palatine school during the reign of Charles the Bald. At the dawn of the tenth century, Remigius moved on to teach in the Reims cathedral.[134] Eriugena's particular knowledge of Plato and ability to instruct someone like Remigius of Auxerre in Platonic thought was owed chiefly to the fact that he was indeed one of the very few Western cultural leaders at this time with a knowledge of Greek.

This expertise allowed Eriugena to garner a favorable reputation. He translated works of Origen and texts of other Greek Church Fathers. When in 827 the Byzantine emperor Michael II ("the Stammerer") sent Louis the Pious an important manuscript[135] containing the work of pseudo-Dionysius the Areopagite, Charles the Bald gave a commission to Eriugena to render it into Latin.[136] That translation, accompanied by a commentary, was widely circulated in the following centuries. His influence, however, was limited, as he failed to establish a school that would continue to produce translations.

In that same century, the greatest Western scholar of Greek literature was Anastasius Bibliothecarius ("the Librarian," ca. 810–879), abbot of the Roman Basilica of Santa Maria in Trastevere. Appointed librarian of the Roman Church by Pope Hadrian, he attended the Council of Constantinople in 869–870, overseeing the translation of various Church documents into Latin on behalf of the pontiff.

Many of the texts copied in the pre-Carolingian age or even during the reign of Charlemagne did not enjoy wide distribution. In some cases they had better fortune in the late Middle Ages; in others they would be rediscovered only years later by Renaissance humanists. This fact highlights the scope of the copying launched by the Carolingians, which in part was directed toward the reproduction of texts identified for their antiquity in particular. The value of these texts was, therefore, recognized already in the Carolingian age as lying in something beyond their practical application or daily use.

The reasons for the limited reception of certain texts were various. In the case of the poem of Lucretius there were probably religious motives, which surely led Church leaders to advise caution in the use of the work, as it obviously espouses Epicureanism. Nevertheless, the *De rerum natura* was in fact copied as part of the court, in the aforementioned manuscript known as the *Oblongus*. The *antigraph* of that manuscript served also as the model for another codex, from which derived the so-called *Quadratus*,[137] copied at the end of the ninth century in northeastern France, as well as an additional codex of which a few sheets, produced in the same epoch in southwest Germany, remain. In later centuries only a few isolated traces of knowledge of the poem emerged until, in 1417, Poggio Bracciolini discovered a descendant of the *Oblongus*.

Other ancient works that were copied but had limited circulation during the Middle Ages include the minor works of Tacitus and the poetry of Valerius Flaccus and Silius Italicus. Of the last of these, Silius' *Punica* was probably preserved in a single copy that Poggio tracked down in the same year that he found the Lucretius manuscript mentioned above. There are some clues, however, suggesting that others had knowledge of Silius, mostly in connection with the region of Germany in which that particular codex had survived.

Though during the Middle Ages there were other copies, one important manuscript containing Valerius Flaccus' *Argonautica* was copied in the ninth century in Fulda.[138] Another mutilated copy was discovered by Poggio in St. Gall in 1416; that partial text provided Niccolò Niccoli with what was then the most complete version of the poem; further, that selfsame manuscript was also used in the sixteenth century by the Belgian philologist Luis Carrion. Beyond these copies, the receipt of the poem in medieval culture appears almost nonexistent.

The minor works of Tacitus (*Germania, Agricola*, and the *Dialogus*) along with Suetonius' *De grammaticis et rhetoribus* were copied in the ninth century at Fulda, in a codex of the monastery of Hersfeld, and brought to Italy in the fifteenth century. Of that manuscript, only a section containing part of the *Agricola* has survived; even in this case, the copied text did not enjoy any traceable circulation, although the *Germania* was used in the ninth century by Rudolf of Fulda. The texts most circulated during the Middle Ages are those that Birger Munk Olsen regards as belonging to the early medieval school canon consisting of twenty-five ancient works, of which fifty or more codices remain from the time period between the ninth and twelfth centuries.[139]

There are twelve canonical authors (including some false attributions). Of these, eight are poets: Terence, Virgil,[140] Horace,[141] Ovid,[142] Lucan, Persius, Statius,[143] and Juvenal. Among prose writers, only four made the list: Cicero,[144] Sallust,[145] Seneca,[146] and Solinus.[147] Overall, the canon reveals a remarkable continuity with late antique culture; in late antiquity, well before the Carolingian list, Cassiodorus had defined the "chariot" of scholastic authors as consisting of Terence, Virgil, Cicero, and Sallust.[148]

In the late medieval period, Dante, in his *De vulgari eloquentia* (2.6.7), proclaimed Virgil, Statius, Lucan, and Ovid (i.e., his *Metamorphoses*) as *poetae regulati*, the writers on whom the rule of poetic composition should be based. The canon found in the *De vulgari eloquentia* was the harbinger of *la bella scuola* of *Inferno* 4.88–94, where Dante replaces Statius with Horace and adds Homer, even though he was known only in Greek, and was thus inaccessible to Dante.[149] The fortune of the authors of the canon ensured the survival of a number of commentators, whose interpretive works were assembled in late antiquity, such as Servius' commentary on Virgil, which itself had expanded and superseded that of Aelius Claudius Donatus.[150]

Fortunately, in spite of the odds against their survival, complete corpora of Virgil and Horace, and, in Ovid's case, the *Metamorphoses*, came through to the ninth century in relatively pristine shape. The survival of the last of these was no doubt connected with its unique and important summation of mythological knowledge; even Dante, as we have just seen, specifically touted this work. Ovid's elegiac corpus, too, enjoyed markedly good fortune in the late Middle Ages—the twelfth century is often dubbed the *aetas Ovidiana*—although the transmission of these works had begun earlier. An important codex dating to the Carolingian period preserves the *Amores*, *Remedia amoris*, *Ars amatoria*, and *Heroides*. In the same period, the *Fasti* and *Tristia* were also recopied; the oldest witnesses that survive are datable to the tenth century. Earlier copies have all been lost, as also have been the Carolingian codices of the *Metamorphoses*, although some fragments do remain, providing important testimonies to Ovid's text. The earliest intact manuscript of the *Metamorphoses*, however, dates only to the eleventh century.[151]

In the late Middle Ages the Ovidian corpus included the poet's minor works (*Medicamina faciei femineae*, *Ibis*), but these minor works normally were bound separate from the major works, often compiled with pseudo-Ovidian texts such as the *Nux* and the *Consolatio ad Liviam*. With regard to prose authors, two of Cicero's canonical treatises appear, *De inventione* and *Rhetorica ad Herennium*, the latter of which, though spurious, was the basis for the teaching of rhetoric in the Middle Ages. The *De oratore* and *Orator* circulated in the Middle Ages in a fragmentary tradition of insular origin. One copy, which also included the *Brutus*, existed in Italy but did not form the basis for other copies. Even the *Institutio oratoria* of Quintilian was known in the Carolingian culture only in a fragmentary state, which explains the greater fortune of the *De inventione* and *Rhetorica ad Herennium*.

During the medieval period, so-called *Rhetores latini minores* were fairly widespread. Among them were Fortunatus and Julius Victor, who had in late antiquity supplanted the work of Quintilian. Other works included in the canon were Cicero's *De officiis*, *De amicitia*, and *Somnium Scipionis*. The latter was saved thanks to a commentary written by Macrobius, which granted the work a tradition independent of the *De re publica*, of which it was but the final section. When the larger work was lost, the *Somnium* was nonetheless preserved. Further, Cicero's *De finibus* and *Academica posteriora*, which was transmitted as book 6 of *De finibus*, came back into circulation only at the end of the eleventh or beginning of the twelfth century. Fortunately, thanks to a palimpsest, we do have samples of Cicero's *De re publica*. Other complete texts were preserved in a sufficient number of copies to enjoy relatively robust circulation in the Carolingian period.[152]

As we have seen, the circulation of Seneca in the medieval period was bolstered by interest in the letters supposedly written by him to Saint Paul as well as by the popularity of his *Epistulae ad Lucilium*. The value of the latter work in the medieval period is confirmed in writers such as Louis the Pious. Seneca's *Dialogues* were long unknown as they were virtually sequestered in Monte Cassino. The tradition of *De beneficiis* and *De clementia*, preserved in a manuscript probably copied in Milan around AD 800,[153] began in Italy. That codex found its way to Lorsch by about 850. Neither of these works, however, would enjoy wide circulation until the twelfth century.

A similar trend obtains for other works. In the late medieval period, Seneca's *Apocolocyntosis* underwent a comparable fate, though it fortunately is also preserved in ninth-century French and German codices. Seneca's tragedies, which had been disseminated sluggishly until the fourteenth century, survived in France and Italy.[154] Seneca's *Naturales quaestiones*, too, are mentioned in the ninth century in the catalogue of the Reichenau monastery, but the manuscripts that remain date only from the twelfth century. The last author in prose in this canon is Solinus, whose third-century work, *Collectanea*, is a naturalistic/scientific epitome based mainly on the works of Pliny the Elder and Pomponius Mela. Even Pliny's *Naturalis historia*, while known throughout the Middle Ages, had but limited circulation, owing to the ponderous size of the original work.

The fortune of the twenty-five texts included in the canon was not uniform in any given period within the Middle Ages. Munk Olsen[155] has provided dates, century to century, marking the developments in the dissemination of each work. An important date emerging from his studies is the confirmation of the proposed periodization by Ludwig Traube, who distinguished in the Middle Ages three periods based on the preference for classical authors:[156] the Virgilian age (in the eighth and ninth centuries), the Horatian age (over the tenth and eleventh centuries), and the Ovidian age (in the twelfth and thirteenth centuries). In the ninth century, as the research of Munk Olsen has shown, the works of Virgil (thirty copies of the *Aeneid*, twenty-seven of the *Georgics*, and twenty-two of *Eclogues*) dominate the literary landscape, followed by Lucan, with ten codices of the *Pharsalia*; Terence and Solinus survive in nine manuscripts each; Juvenal's *Saturae* weighs in at eight. In the tenth century the primacy of the *Aeneid* persists in twenty-five manuscripts, but by then Juvenal follows closely with twenty-four codices, and then Terence with nineteen. A preference for Horace would not emerge until the eleventh century.

The *haute culture* of the Carolingian renaissance did not last into the post-Carolingian period. Even if it did not disappear completely, it is certainly true that the central role that Charlemagne held in cultural development was not imitated by his successor, Louis the Pious (814–840). It is striking, too, that during Louis' reign the imperial library was lost and consequently the fate of its books lay entirely with the different monasteries to which they had been distributed.

While some of the reasons for the decline and division of the empire (843) following Louis' reign lay among internal political squabbling, the intensification of external threats, such as Viking raids on the empire's northern coast, as well as the ninth- and tenth-century incursions by Hungarians on the eastern borders, exerted further pressure on the empire. Nevertheless, the educational system introduced by Charlemagne remained active and grew stronger as its institutionalization became more established. The number of monastic centers, too, continued to grow, both in France and in the outlying regions. Scribal activity, while it shows some decline, nevertheless remained important in French cultural centers such as Fleury, which witnessed a remarkable flowering of studies under the direction of Abbo (ca. 945–1004). Meanwhile, in Germany, the Carolingian abbey of Fulda assumed a prominent role in literary culture.

A particularly significant figure of the tenth century was Ratherius (ca. 887–974), bishop of Liège and Verona. In the latter city he studied Catullus, whose work, owing to its erotic content, did not find significant circulation until the fourteenth century, though some allusions to the Catullan corpus are detectable in the *Florilegium Thuaneum*,[157] a compendium that was prepared in France in the eleventh century.

Under the direction of Ratherius in Verona, the manuscript that we know as the *Mediceus*, which offers an important witness to the first decade of Livy, was copied.[158] That tenth-century work testifies for the first time to the important codex located in Germany. In the monastery of Fulda, the only witness of *Carmina Einsidlensia*, which contains two pastoral poems from the age of Nero, was copied in a manuscript containing works of Rabanus Maurus. Germany, too, transmitted important collections of the comedies of Plautus.[159]

At the end of the tenth century in the western regions of Germany, the Codex Monacensis[160] was prepared; now housed in Munich, it contained previously unknown orations of Cicero.[161] From southern Germany comes a codex containing *Invectivae* attributed to Sallust and Cicero.[162] In Italy, the *Partitiones oratoriae* of Cicero resurfaced, as reported in the catalogue of Bobbio, where it was copied, along with Fortunatianus and other rhetorical writers; it is now housed in the Bodmer collection in Geneva.[163] Meanwhile, the *Declamationes maiores*, attributed to Quintilian, were copied in the Codex Bambergensis;[164] the *Declamationes minores* were also included in the ninth-century manuscript of Montpellier, which volume, in addition, preserves one of the two traditions of the work of Seneca the Elder.[165]

THE OTTONIAN RENAISSANCE OF THE ELEVENTH CENTURY

The rise in cultural activity in Germany culminated in the so-called rebirth of the Ottonian age, the dynasty that assumed imperial power in 962. The most important scholastic figure of this period was Gerbert of Aurillac (ca. 950–1003), the tutor of Otto III (983–1002). After studying mathematics and astronomy in Spain, where he came into contact with Arab culture, Gerbert became active in the scholarly centers of Reims and Bobbio. After his educational service to Otto III early in that potentate's reign, Gerbert became bishop of Ravenna and later pope (999), taking the pontifical name Sylvester II.[166] The Ottonian renaissance is exemplified by Otto III's commissioning and creation of the library of Bamberg in Germany, a library similar and, by its particular collection of texts, connected to the earlier Palatine library of Charlemagne.

As a result of the increased emphasis on learning during the eleventh century, texts hitherto lost came back into circulation. In Germany, Cicero's *De optimo genere oratorum* and *De finibus* were discovered, and in Liège the *Pro Archia* turned up. A fragment of a manuscript containing Cicero's *Epistulae ad Atticum* was found at Würzburg. The *Epistulae ad familiares* were transmitted in France in a ninth-century manuscript.[167] In the Loire valley the oldest extant text of Donatus' commentary on Terence was copied.[168] At Gembloux, the panegyrics and invectives of Claudian, known as the *Claudianus maior*, were transcribed. From the same cultural center comes the only manuscript containing the commentary of Favonius Eulogius on the *Somnium Scipionis* of

Cicero.[169] In eleventh-century Italy, too, important ancient texts of which there had previously been no trace were being copied. Of particular importance was an increase in the scribal activity in the monastery of Monte Cassino under the direction of the abbot Desiderius (1058-1085).

Some codices, however, fell out of circulation until their rediscovery in the fourteenth century. The Laurentian codex of the *Pro Cluentio* of Cicero[170] gives rise to an important textual tradition beyond Cicero, as it contains what remains of Varro's *De lingua Latina*. The dialogues of Seneca are preserved in a codex in Milan,[171] while another manuscript combines two originally distinct codices that had separately contained Apuleius[172] and Tacitus.[173] Meanwhile in Rome the Codex Farnesianus[174] was copied; though lacunose, it is an important witness to Festus' *De verborum significatu*.[175]

In the monastery of Pomposa, too, renewed activity arose under the influence of the abbot Jerome (1079-1100). That monastery's catalogue of the year 1093 includes a codex of Seneca's tragedies that can be identified as the *Etruscus*, perhaps copied from a specimen held in Monte Cassino. At the end of the tenth century, a codex featuring fifth-century uncials and containing the fourth decade of Livy was probably still preserved in the monastery of Nonantola. Just a few years before the turn of the millennium, John Philagathus, originally abbot of Nonantola and bishop of Piacenza before being elected Pope John XVI (997-998), gave that codex to Otto III. That same codex would later be donated by Henry II to the cathedral of Bamberg. Today, only a fragment of it remains, but its descendants provide us with information about the intellectual activity of that decade.

TEXTS AT THE END OF THE MIDDLE AGES

The portion of the Middle Ages in the years following the Ottonian flourishing exhibited scholastic features as well, suggesting that the cultural resurgence lasted well into the twelfth century. In his 1927 landmark study, Charles Haskins went so far as to suggest that the late Middle Ages had more features of intellectual rebirth than the Carolingian period and even anticipated modern thought.[176] Certainly, in the view of many of those living at the time, the late Middle Ages inspired cultural fruitfulness. For example, the Englishman Walter Map, in his *De nugis curialium* (1180-1192), speaks of that period's *modernitas* and puts it in opposition with the past, as gold might be compared to old copper, inverting the traditional chronological hierarchy, which always puts the past age on a pedestal.[177]

The twelfth century's sustained cultural expansion is evidenced by a significant increase in book production. This evolution is owed primarily to the development of capitular schools—that is, schools that arose according to the precepts of the imperial court and were organized according to that court's regulatory practices. Universities then took on the task of book production and thus had a role of greater importance than in previous centuries, when manuscripts were copied almost exclusively by monks.

In the twelfth century, manuscript reproduction was advanced by the adoption of a new organizational system for copying, the so-called *pecia* (piece) system. Though this system was a quick and convenient means of producing texts needed by university students and faculty, it was not invented by the universities themselves. Dividing a book to be copied by several scribes at once was a practice that can be traced back to the ninth century.[178] Under the *pecia* system, the scribe's copy was not bound, but left as loose quires known as *peciae* (meaning "pieces" in Latin). These loose quires, usually consisting of four to six leaves, although there could be as many as ten,[179] were then used to copy a book in stages. Stationers (*stationarii*), booksellers who owned the *peciae* but were regulated by the university to ensure quality,[180] would rent them out to be copied. The system's chief innovation was the parceling out of a text specifically for mass production.[181] Texts can usually be recognized as a *pecia* copy by the presence of Roman numerals in the margin, which indicate where a new *pecia* begins.[182]

Each double page of text written by a master was entrusted to a single copyist, resulting in a decrease in the time needed to create a codex, and thus offering the possibility of producing numerous series in a larger number of copies, each of which was called an *apopecia*. The system also guaranteed greater control over the quality of the copies, due to the close regulation of the copyists by the masters.[183] Students, either for their own purposes or because they were a source of cheap labor, would typically serve as scribes. Each student would have a limited amount of time to copy the *pecia*, perhaps a week, before he had to return it.[184] When one *pecia* was returned, the next could be borrowed, until the entire book had been copied. Since the *peciae* were distributed separately, there could be as many copies of a book in production as there were quires.

This system had started to develop by the mid-thirteenth century[185] but began to die out in northern Europe already by the middle of the fourteenth.[186] The system lasted longer in Italy, at least until shortly after the printing press had made its way to the University of Bologna in 1471. The system was not in place at all universities—evidence of it can be found for only seven[187]—and the texts available for copying were limited. Pollard unravels the reasons for the restrictions by noting that *peciae* were generally available only for works on theology and law, since these disciplines required students to have their own texts; in arts courses the instructor read and commented on a text, so no other books were needed.[188]

Beginning in the thirteenth century, the material used to construct books underwent some changes as book size was reduced to accommodate library needs. Paper, first imported from Damascus, began to replace the more expensive parchment. This transition in writing material was, however, not entirely smooth, as the spread of the new copying system raised concerns about the lack of durability of paper in comparison with parchment. Thus it was that in 1226, Frederick II, the Holy Roman emperor and king of Naples, forbade its use for state documents.[189]

The development of universities led to a new demographic for European culture. In Paris the College of the Sorbonne, founded in 1253, quickly established itself as the center for the most important studies in philosophy and dialectic. Meanwhile, the cathedral school of Chartres became the driving force of Platonism; in medicine, Montpellier was held in high regard. Not surprisingly, the main cultural centers were also university seats, such as in northern France, where the University of Paris was founded in 1150. In Italy the University of Bologna, the oldest of those founded in Europe (1088), became known as a principal center for legal studies and rhetoric.

The progressive advancement of study at the university level vigorously renewed education, in both breadth and depth, expanded instruction based on the quadrivium,[190] and introduced a strong interest not only in natural sciences but also in technical studies.

Though considered banal in the ancient tradition, technical studies became revalued, particularly by Hugo of St. Victor (1096–1141), who in his *Didascalicon* connects the study of nature with knowledge of God. The role of Christian thought in the development of medieval technology, as has recently been emphasized by Rodney Stark,[191] provided the primary matrix of the European culture to come. Even the traditional disciplines of the trivium saw notable developments.

The composition of epistles was codified as *ars dictandi*, giving them an established educational role. Dialectic was revamped by the rediscovery of Aristotle, and new developments in the study of logic came at the hands of Peter Abelard of Bath (ca. 1080–1150), an influential figure of the twelfth century. Famous for his teaching in Paris, Abelard fell victim to the judgment of Church authorities—in 1121 his *Theologia* was condemned to the flames by the Synod of Soissons. Abelard's career was largely rendered impotent by his relationship with the scholar Heloise, on which epistolary testimony remains.[192] Nevertheless, Abelard's contributions abide, prominent among which is the rational argumentation of his *Dialogus inter philosophum, Iudaeum et Christianum*, where he creates an open discourse of philosophy with the two major religions of the Western tradition, Christianity and Judaism.

Another vital contribution of late medieval culture is its provision of access to the remains of scientific and philosophical Greek texts, partly through Arabic, into which some lost Greek works were translated. The disciplines that were made available by the contributions of these texts include medicine, astronomy, and algebra. Indeed, our modern numbering system, a system from India known as "Arabic," was also introduced in this era in 1202 in the *Liber abaci* by Leonardo Fibonacci. Such translations were made in the geographic regions in which the West was in direct contact with those Eastern cultures. In Spain many Greek authors that in previous centuries had only been translated into Arabic were now rendered into Latin. These included Aristotle, Ptolemy, and others.

Additionally, authors writing original works in Arabic, including Avicenna and Averroes, were now translated into Latin. Among the promoters of this activity, Peter the Venerable (1056-1092), abbot of the monastery of Cluny in Paris, made a particularly important contribution. Peter traveled in areas of Spain and sponsored the translation of the Qur'an into Latin. Among the many translators recorded from Arabic is Gerard of Cremona, who worked in Toledo in 1134-1178 and undertook numerous Latin renditions, among which his edition of Ptolemy's *Almagest* was perhaps the most important.

In southern Italy, where Greek was still spoken, translations from the original Greek were made of Plato, Euclid, Proclus, and other authors, under the auspices of Henry Aristippus (ca. 1105-1162), archbishop of Catania. Aristippus himself rendered the *Phaedo* and the *Meno* of Plato, both of which translations were widely circulated in the thirteenth and fourteenth centuries. In 1075 Constantine the African, who was born in Carthage, brought to Salerno Arabic texts of Hippocrates, Galen, and other authors. This development laid the foundations for the birth of the medical school in Salerno. Constantine was later welcomed in the monastery of Monte Cassino, where he died in 1087.[193]

In the first half of the fourteenth century, another prolific translator of Greek medical texts, particularly Galen, was Nicolaus of Reggio, active at the Angevin court of Naples.[194] Many other translations were the work of those who learned Greek in the Byzantine Empire and then Latinized philosophical works of Greek science.

In the twelfth century, Burgundio of Pisa, who served as an interpreter in Constantinople from 1135 through 1138, put Hippocrates, Galen, and, at the request of Pope Eugenius III, Saint John of Damascus into good Latin. Another key figure who played a role in the rendering of Greek texts is James of Venice.[195] James is reported to have been in Constantinople at the council in 1136. His translation of Aristotle's *Analytica posteriora* was used in 1159 by John of Salisbury, a student of Abelard in Paris and a member of the school of Chartres. The role of James of Venice as the redactor of Aristotle was emphasized just a few years ago by Sylvain Goughuenheim, who disputes the importance of mediation in the Arab culture of the late Middle Ages.[196]

In the thirteenth century, the most important translator was William of Moerbeke (ca. 1215-1286), who traveled throughout the Byzantine Empire, finally becoming, in 1278, archbishop of Corinth. On the encouragement of Thomas Aquinas (1225-1274), William of Moerbeke completed the translation of numerous works of Aristotle, including the *Organon, Physics, Politics*, and *Metaphysics*.[197] Other authors rendered by William include Archimedes, Galen, Produs, and Alexander of Aphrodisias. The interest in these authors afforded by Latin translations brought fresh attention to the Greek language more broadly. Robert Grosseteste (ca. 1214-1294), bishop of Lincoln and proponent of scientific studies in England, worked on translations and commentaries of Aristotle and pseudo-Dionysius the Areopagite. His pupil Roger Bacon (ca. 1214-1294) wrote an important Greek grammar.

In terms of scholarship, the renewed circulation of the works of Aristotle was perhaps the most vital event in late medieval culture, since previously only a few Aristotelian works, those translated by Boethius, had been known. The growing body of knowledge naturally provoked fresh, if sometimes bitter, debates between Aristotelians, Averroists, and Platonic-Augustinians, as well as between realists and nominalists.

Incumbent upon these developments, there arose a number of instances of censorship. In 1210 the University of Paris forbade the teaching of Aristotle's *Physics* and *Metaphysics*. Nevertheless, Aristotle's influence obtained, for Thomas Aquinas would arrange his theological and philosophical work on Christian thought and doctrine based on Aristotelian principles; that work would become, of course, an important source of doctrine for the Catholic Church.

The teaching practice established in the Carolingian period continued to be the dominant methodological basis for education. With the founding of universities throughout Europe, the *trivium/quadrivium* form was adapted also to higher learning. In the academic programs attested in this period, however, the exclusive study of Latin classics was reevaluated, and a new emphasis on Aristotelian texts emerged. This development produced a shift in prominence that influenced theology and philosophy, as well as the study of logic and law. A further consequence of this scholastic broadening was that the circulation of classical authors increased markedly, generally owing to the escalation in production and circulation of books.

That circulation predictably included authors deemed most vital, such as Virgil and Horace, copies of whose works in previous centuries had been more numerous. In the twelfth century, a period that bears witness to roughly a hundred of their manuscripts, Virgil and Horace maintained their dominant position in the curriculum. Two prose works, the *De inventione* of Cicero and the pseudo-Ciceronian *Rhetorica ad Herennium*, enjoyed a similar rise in popularity, as there was a growing demand for teaching texts.

Literary culture in the late medieval period was also characterized by a desire for the works of Ovid (as we saw earlier, that era was dubbed by Traube the *aetas Ovidiana*).[198] Not only was the demand for Ovid's elegiac works high, but there was also a good deal of imitation in the so-called elegiac comedy, which branched out from the Loire valley into all of Europe. The reception of Ovid's erotic works in the late medieval period is characterized by a difficult balance between a certain ethical relativism and the perceived need to bring Ovid under the sway of Christian morality. The latter program is highlighted in *Ovide moralisé*, a Christianized rendering of the *Metamorphoses* into vernacular French, along with the pseudo-Ovidian *De vetula*, a poem in hexameters edited around the middle of the thirteenth century.

In this period, anthologies represented another means of preserving classical poets. Unfortunately, these kinds of collections often present difficulties in reconstructing the text. The *Florilegium Gallicum*, assembled in northern France, includes selections of authors almost unknown in this period, such as Tibullus, Petronius, and Valerius Flaccus. Richard of Fournival cites rare books, such as the elegies of Tibullus and Propertius, and tragedies of Seneca, in his mid-thirteenth-century work *Biblionomia* (particularly the tragedies of Seneca, for Richard seems to have known the contents of the Parisian manuscript).[199] The Vossianus manuscript of Propertius, preserved in Leiden, was also copied for Richard.[200] Another copy of Propertius, the so-called *Neapolitanus*, dating to about fifty years earlier (ca. 1200), came from northern France, as well.[201] The oldest traces of knowledge of Propertius come from the Loire valley, where, in the second half of the twelfth century, Propertius influenced John of Salisbury and the anonymous author of one of the first comedic elegies, the *Pamphilus*. Richard also perhaps possessed a codex of Tibullus that had been copied in the Carolingian court; in 1272 that manuscript, along with others belonging to him, went to the Sorbonne library,[202] but has since been lost.

Beyond works of Propertius, other classical texts, of which there are no reports in the intervening centuries, reemerge in the twelfth century. These include three orations of Cicero concerning the agrarian law (*De lege agraria*) collected by Wibald of Corvey and Seneca's *Naturales quaestiones*, the dissemination of which was effected by interest in scientific themes, and for which we have no earlier manuscripts. The *Eclogues* of Calpurnius Siculus, an author known even to the poets of the Carolingian period, also turn up in the Loire valley. The incomplete biographical collection of Cornelius Nepos depends on a codex of the twelfth century, which was itself perhaps copied in the Rhineland, before it was lost in the sixteenth century. Also in France the manuscript tradition of Ovid's *Ibis* begins, along with books 1–7 of Aulus Gellius' *Noctes Atticae*, a work that was probably already divided into two codices in late antiquity. The tradition of the two separate parts resulted in the loss of nearly all of book 8; ninth-century copies of books 9–20 survive.

Other texts previously unknown emerge in Italy. Toward 1132 Peter the Deacon at Monte Cassino copied Frontinus' *De aquis*,[203] present also in a ninth-century codex housed in Hersfeld. There is also the Italian tradition of Claudian's *De raptu Proserpinae*, of which there are two twelfth-century codices. The agricultural treatises of Cato and Varro are copied in the twelfth century in what is now the Parisian manuscript Latinus 6842, a codex that in 1426 was in Pavia in the library of the Visconti.

In the thirteenth century, texts that had been hitherto less well distributed saw greater movement. The tragedies of Seneca, already known to Richard de Fournival, were circulated in northern Europe and, at the beginning of the fourteenth century, the learned Oxonian Nicholas Trever wrote a commentary on them. Seneca's *Dialogues*, having emerged in the eleventh century at Monte Cassino, were distributed fairly widely in the thirteenth century and, around 1250, Oxford's Roger Bacon announced their "discovery." Interest in Seneca is thoroughly palpable in Vincent of Beauvais' *Speculum maius* (1190-1264), a monumental encyclopedia that highlights the trend toward systemization of knowledge.

A lesser contribution of the thirteenth century is the emergence of texts previously unknown. The only witness that presents this feature is the codex of the New York Academy of Medicine that contains Sorano's treatise on gynecology, *Gynaecia*, in the Latin translation of Caelius Aurelianus.[204] It is a work that was reported in the library catalogue of the monastery of St. Amand but known only in an edition published in 1533 until a few years ago. This codex of New York came to light in 1948 when it was listed for sale by a Swiss antique dealer.

The fresh cultural climate of the thirteenth century was also marked by the advent of a new form of writing, directly descended from the Carolingian minuscule, which had hitherto been the dominant form of writing. This so-called Gothic script spread from northeastern France in the twelfth and thirteenth centuries and would eventually wend its way across the whole of Europe. The name that the humanists gave it is owed in part to the undoubtedly obvious parallels between the form of Gothic letters and Gothic architecture, which was also widespread in that period throughout northern Europe.

As far as architecture is concerned, the Gothic style is characterized by a tightening of the *ductus* at the top of the doorframe, while in the other parts of the building round shapes contrast with slender elements. In the script, the slender strokes that rise tend to shrink as they go higher in the line, while ele-

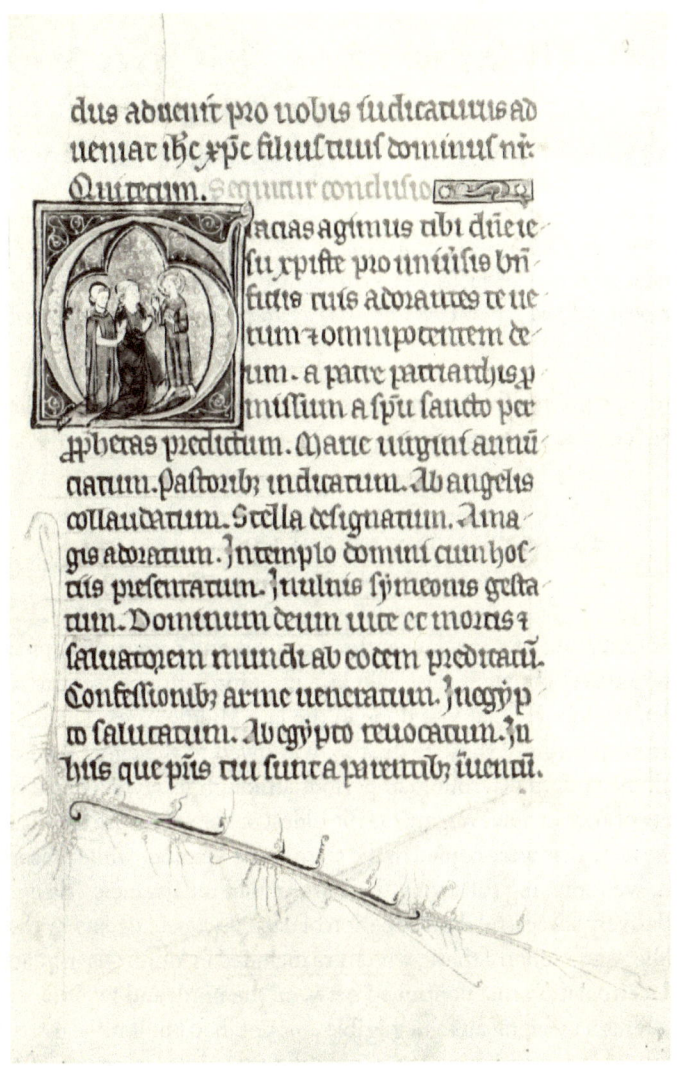

FIGURE 3.12. *Example of Gothic script from the twelfth century. Ms. W.15, folio 5 verso. Baltimore, Walters Art Museum.*

ments that go below the line sometimes disappear altogether, effecting a compression of the words, giving the impression of a greater fullness to the individual letters (fig. 3.12).

The dissemination of Gothic script is connected to the development of universities and the consequent renewal of advanced studies. Parallel to the script's evolution is the development of the cursive form for use in chancelleries. This style incorporates features typical of Gothic and takes on, in France and north-

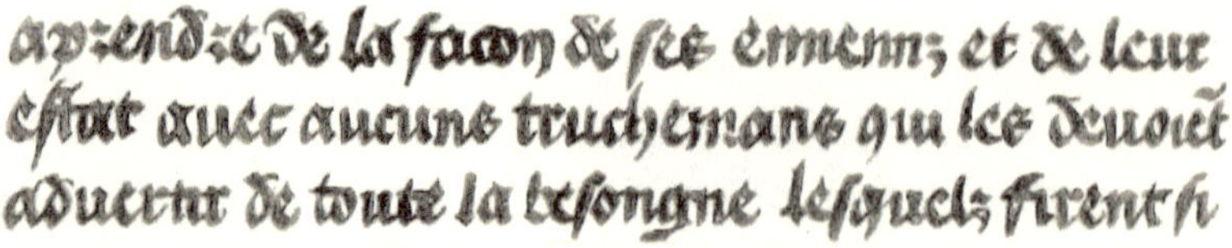

apzendze de la façon de ses ennemis; et de leur estat avec aucuns truchemans qui les devoient aduertir de toute la besongne lesquels firent si

FIGURE 3.13. *Bâtarde script (minuscule). Section from a mid-fifteenth-century manuscript of the* Conquêtes de Charlemagne *(Brussels, Bibliothèque Royale, ms. 9066-68, vol. 1, folio 138b). Drawing by Mary Claire Russell MacDonald.*

ern Europe, a unique form sometimes known by its French name "lettre bâtarde" or "chancery minuscule" (fig. 3.13).

TRANSMISSION AND LOSS OF CLASSICAL TEXTS IN THE MIDDLE AGES

With the exception of a few late antique codices and papyrus fragments, virtually the entirety of extant classical Latin literature presupposes the recovery work, detailed above, that was done in the late Middle Ages. The survival of a significant portion of these texts, too, was owed to the aggressive copying that took place in the Carolingian period, although in several cases only descendants of those copies remain, as the oldest specimens were lost. It is likely that some texts that were copied in the Carolingian period, while preserved for a season, went missing following the Carolingian renaissance. The end result was a relatively successful dissemination of various classical texts by the end of the Middle Ages. Some of these, which are included in Munk Olsen's catalogue, enjoyed a circulation that continued between the ninth and twelfth centuries. Others, though by no means a negligible amount, had but limited distribution. Among these is a manuscript containing the first eight comedies of Plautus; the recipe book by Apicius; the philosophical works of Apuleius; treatises of Celsus, Censorinus, and Columella; the historical works of Curtius Rufus and Florus; books 9–20 of Aulus Gellius; the *Epitome* of Justin; the first and third decade of Livy; the *Epigrams* of Martial; and the *Fables* of Phaedrus. Of this group, very few copies were produced and thus they ran a high risk of permanent loss, but some were rediscovered by humanists.

Many have survived in only a single copy. Among these are the literary works of Apuleius, another group of Plautine comedies, and the poems of Tibullus, Catullus, and Propertius; in prose, Tacitus and Suetonius are noteworthy, although numerous other authors could be mentioned. Some trace evidence about texts that still existed at the time but are now lost can be gleaned from medieval library catalogues in the late medieval period. An important ninth-century booklist, coming from the abbey of Bobbio, mentions the *Opuscula ruralia* of the now only fragmentary poet Septimius Serenus (likely of the third century AD). Despite this reference, however, the work itself later went missing, and the fact that he survived as long as he did suggests Serenus' popularity.[205]

Another catalogue, also of the ninth century, details the works in the monastery of Lorsch and includes mention of the *Medicinales responsiones* of Caelius Aurelianus, a work that survived even at that time only in fragmentary form. The ninth-century catalogue of the Murbach cloister mentions a certain *Bucolicon Olibrii*; that author may be the Olybrius who was consul in 395. Further, in the Carolingian period, the *De verborum significatu* of Pompeius Festus was intact; Paul the Deacon used that work for his *Compendium*, as did the compilers of some glossaries. The eleventh-century Farnese codex of the *De verborum significatu* survived in a fragmentary form. Another text that persisted during the Middle Ages, if only in pieces, was the *Satyricon* of Petronius, which in the ninth century was preserved in Fleury in a copy that included large sections of the work.

The conservation of Greek literature began over a millennium before the medieval period, culminating in the cultural centers established with Alexander the Great's conquests in the fourth century BC. The foundation of the library in Alexandria, in particular, played a major role in the recovery and preservation of Greek literary heritage. The Roman conquest of Greece that followed saw the transfer of entire libraries westward, thus promoting the circulation of Greek texts.

For example, as noted in chapter 2, Strabo (13.608) describes the fate of the works of Aristotle, whose reputation was ensured by "exoteric" dialogues (perhaps those intended for a broad readership, but later lost). Aristotle's extant treatises are known as "esoteric" (intended for members of the Lyceum), to which Theophrastus and, after him, one of his students, a certain Neleus of Scepsis, had access. Neleus took these texts to his hometown of Scepsis on the slopes of Mount Ida in the Troad, where they remained until the early first century BC, when Apellicon of Teos discovered them and brought them back to Athens. From there, according to Plutarch, Sulla carried them off to Rome in 86 BC.[206] In Rome, Andronicus of Rhodes made a new edition at some point between 40 and 20 BC.

Though many texts were thus preserved, some loss probably had already occurred in the Hellenistic period. In the first century AD, Diodorus states that he is not able to track down five of the fifty-eight books that comprised the historical work of Theopompus.[207] In the ninth century, Photius also had access to only fifty-three of Theopompus' books, and everything points to them being the same as those missing eight centuries earlier.

Most book production, buttressed by an increase in literacy, occurred during the long period of relative peace on the Roman frontiers in the Antonine era. Thus the Italian scholar Cavallo has justly referred to that period of the second century as a "renaissance of Greek culture."[208] Even the revival of Atticism, which was, in the West, viewed as an archaizing style of oratory, probably made a positive contribution to the restoration and preservation of ancient literature (particularly in the realm of Attic oratory, but also in respect to Xenophon and Thucydides).[209]

The various processes of preservation and transmission of Greek works during the Roman imperial age are similar to those recorded for Latin texts. As regards the systems of their conservation, the transition from roll to codex is decidedly important, as it affects both literatures and the development of book hands. A similar phenomenon can be seen in the movement from the half-uncial to the uncial, as lowercase lettering continued to evolve.

Analogues between Greek and Latin classics are also evident in the criteria that determined the survival of the texts in late antiquity. It was in this period that the individual canons of Aristophanes and the three great tragedians (Aeschylus, Sophocles, and Euripides) were formed. Of each of these authors, whose ancient tradition enjoyed large-scale production, seven comedies or tragedies were chosen, which are those that have come down to us (for Aristophanes and Euripides, as we shall see, a few more plays were added to the seven canonical works).

Some playwrights were not included in the canon, such as Menander, whose comedies were also well appreciated, even in late antiquity (that Jerome knows his work can be seen in his *Epistle* 58). A similar selection process arose for oratory, historiography, and other genres. In the field of Greek historiography, as for Latin, voluminous books were replaced by compendia and epitomes. Of the 856 historical works for which Jacoby collected fragments and testimonia, only about a fortieth of the total production survived.[210] The number of Greek historians preserved remains higher than that of their Latin counterparts.

In addition to the oldest authors, who were read in schools (Herodotus, Thucydides, and Xenophon), many authors were preserved and kept for centuries in the imperial library of Constantinople. The foundation of this library, inaugurated in 387 by Constantius II, represented a significant step in preserving Greek literary heritage. In his fourth oration (§§59d–60c), Themistius discusses the library's inauguration:

Now I suppose the soul of a wise man is wisdom, intellect, and reason, and
the graves of such souls are books and literature, in which their remains are
interred as if in tombs. These memorials, wasting away like buildings in the
treasury of memory, from a little neglect run the risk of being utterly destroyed, lost—extinguishing with them the souls they contain. Yet these
does he [Constantius] order rekindled, both appointing a magistrate for the
task and contributing funds for the project.... Soon Plato, great in wisdom,
will return to life, as will Aristotle and the orator Paeanieus, and the son of
Theodorus and the son of Olorus.... No matter the number of those who
follow them and their works, it is never enough to guarantee their permanence; but these great men, though mortal, the foresight of our king persists
in making immortal.[211]

Though he provides a somewhat pessimistic picture of the retention of classical Greek texts in the late imperial period, as he fancies that arcs of wisdom
can, due to neglect, crumble "like buildings in the treasury of memory" and are
thus at risk of disappearing altogether, Themistius views the library's installa-
tion as a vital moment worth celebrating, and touts Constantius' key role in
keeping the tradition of classical learning alive.[212]

Yet even Themistius' praise of Constantius II's initiative illustrates the uncertainties inherent in preservation: Themistius goes on to mention the Stoic
philosophers Chrysippus and Cleanthes in this speech; unfortunately, these authors' works went missing, possibly even as early as late antiquity, along with
writers in other philosophical schools that lagged behind those of the more
dominant Platonic-Aristotelian tradition. These included Democritus, Epicu-
rus, and the Skeptics. This loss was only partly offset, in the Byzantine period,
by the recovery of Diogenes Laertius and the survival of various Stoics of the
imperial age, such as Epictetus and Marcus Aurelius.

Early on there were major losses in the ancient scientific tradition. The
medical strand of the Hippocratic corpus and writers of pharmaceutical works,
however, were kept by the schools of medicine, while in late antiquity mathematical treatises and those dealing with astronomy were recovered from Neoplatonist sources. To Neoplatonists we owe the preservation of Archimedes,
our knowledge of whom has been greatly enriched by a Byzantine palimpsest
that came to light in 1998.[213] The work of Euclid was saved by the preparation
of an edition by Theon of Alexandria (fourth century AD).

Theodosius II crowned his predecessor's foundation of the imperial library
by establishing a center of higher learning in 425 at Constantinople. Even after
its founding, other schools remained active in the fifth century, particularly in
the main centers of the Eastern empire: Alexandria, Athens, Antioch, Beirut,

Gaza, and others. Unlike what happened in the West, where in late antiquity the dispersion of aristocrats caused a dissolution of civic infrastructure, in the East educational institutions and libraries continued in urban centers, ensuring the transmission of classical texts.

Eastern religious culture would seem to have played an important role in that transmission. In the first few centuries AD, Christian culture was on the whole receptive to pagan texts: in the fourth century, Basil the Great, in his oration *Ad adulescentes*, praised the educational function of classical love and invited his audience to read Plato in particular. Origen, in the previous century, had invited his students to read the pagan philosophers, though he methodically avoided those who denied the existence of God.[214]

As Christianity became a dominant political force, there was, generally speaking, less than smooth sailing for pagan literature. The period of time under Julian the Apostate in which pagan culture was privileged was brief. While Gratian, who had Ausonius as his tutor, had maintained an ambivalent attitude on the issue of whether the altar of Victory should be removed from the senate house, his uncle Valens had sentenced the philosopher Maximus of Ephesus and other supporters of Julian the Apostate to death.

Anti-pagan repression continued under Theodosius: in 391 the library of the Serapeum at Alexandria was destroyed; in 395 the computation of years by four-year intervals from the first Olympic games was abolished. In 415, also at Alexandria, Hypatia, daughter of Theon of Alexandria, was killed. She had studied Neoplatonic philosophy with Synesius of Cyrene, who converted to Christianity and was then appointed bishop of Ptolemais. In the sixth century, the anti-pagan repression was revived by Justinian, who outlawed pagan teachings and ordered the closure of the philosophical school of Athens in 529. The last heads of the Athenian Academy, Damascius and Simplicius, took refuge with the king of Persia.[215]

Though such episodes of repression and burnings of pagan (and heretical) books are reported to have continued in subsequent years, in general such events had but a limited role in disrupting the transmission of classical texts. Overall, ancient texts were recognized as having value—even the writings of Julian—with the exception only of those that were explicitly anti-Christian. Overall, the intervention of the Christian authorities seems to have mainly affected anti-Christian and heretical literature (two areas in which admittedly we do have considerable loss).[216] The notion, as Paul Maas once noted, that significant losses of classical texts can be attributed to the Orthodox Church's hostility to hellenism is therefore unfounded.[217] Rather it is the case that clerics in later centuries had the greatest role in ensuring the continuity of the classical heritage, as part of a substantial continuity of education, based on grammar, rhetoric, literature, and science—all of which remained central to the education of the ruling classes of the empire.

Nevertheless, as regards the transmission of texts, the Christian reception of the classics was in fact largely passive. The texts preserved do not bear witness to the initiatives of Christians to preserve the material of classical texts or to demonstrate philological correctness.[218] Rather, although philological activity is detectable, it was practiced especially rigorously for biblical texts: the most significant example is that of Origen's multilingual edition of the Old Testament, for which the monasteries took advantage of the system of critical *sigla* that originated with Alexandrian scholars.[219] For the majority of the texts, Byzantine culture chiefly just "transcribed, collected, and punctuated" them.[220]

Yet as late antiquity gave way to the early Middle Ages, learning and textual transmission did not advance. The expression "Dark Ages" is often used for the period of the early medieval papacy, which reflected situations similar to the Byzantine culture of seventh and eighth centuries, in both of which one can see a significant depression in the reproduction and reception of classical texts. In that period Christian culture was torn by the iconoclastic crisis that ran from 726 to 843.[221] Cultural depression was affected by the new unintellectual cultural norm that to some extent resulted from the political and military crisis.

In the seventh century, the Roman Empire was severed by the Sassanids, who between 611 and 620 occupied the area between Antioch and Egypt. The reconquest achieved by Heraclius I (610–641) was short-lived, and from 634 Syria, Palestine, Egypt, and the Byzantine territories of North Africa were occupied by Arabs, who in 674 and again in 717 went on to besiege Constantinople. Cultural activity in Alexandria and Palestine maintained some continuity after the Arab occupation.

From Alexandria, in the years between 850 and 880, a philosophical collection of Neoplatonic texts came to Constantinople.[222] The occupation had other significant effects on the transfer of Greek philosophical and scientific culture into the Arab world. A comprehensive program of translations of Greek texts was promoted at Baghdad by the Abbasids.

As noted above, texts lost in the original Greek are sometimes represented by Arabic translations (or to a lesser extent Syriac and Armenian). Among the Arabic translators active in Baghdad, the multilingual Hunain ibn Ishaq (809–873) stands out. Hunain mastered Arabic, Greek, Syriac, and Persian. An Arab scholar of the tenth century, Ibn al-Nadīm, who was in Baghdad compiling a catalogue of philosophical works, translated some works into scientific language. His writings also speak about the abandonment of these disciplines by the Byzantines, outlining the increase in closed schools and libraries whose books were burned or abandoned to insects.[223]

This picture, which is not reflected in the pockets of cultural renaissance that grew up quixotically elsewhere in the empire, admittedly may be owed to our own cultural bias. Nonetheless, it is sufficient to say that the "Dark Ages," whether dubbed so aptly or not, were certainly characterized by a decrease in book production and by a heightened interest in religious particulars. Though Byzantine culture did not see the massive losses of books and libraries that occurred in the West, traumatic episodes of destruction and devastation occurred from time to time, culminating in the sack of Constantinople in 1204.

A few years after the Carolingian renaissance, Byzantine culture also saw a resurgence under the Macedonian dynasty (867–1056), sometimes known as the Macedonian renaissance, insofar as its flourishing coincides roughly with the dynasty founded by Basil I. As in the West, script hands developed. The uncial letterform, characterized by a much smaller script, was established also for Greek book production. As in the case of Latin, the capital letterform was preserved for initial letters, titles, and other paratextual notation. The Greek minuscule was the reason that breathing marks, accents, and division of words were cultivated. The scholarly community of Constantinople became the driving force of this new style of writing.

The change from uppercase to lowercase later extended to regions on the fringe of the empire where the use of Greek persisted in active monasteries, such as Palestine, the Sinai Peninsula, and southern Italy. One such monastery, Grottaferrata, was located near Rome. Italy generally also played a key role in the preservation of Greek texts that were lost elsewhere: the *Posthomerica* of Quintus Smyrnaeus was handed down in a codex that Cardinal Bessarion discovered in the fifteenth century near Otranto.

The shift to minuscules stimulated the production of books, which not only facilitated the recovery of classical texts based on codices copied centuries before but also ensured their survival. For example, we have the testimony of an important Macedonian renaissance figure, Arethas of Caesarea, who in a letter to Demetrius, written before 907, describes a codex of the *Meditations* of Marcus Aurelius as "old and entirely in tatters."[224]

The Macedonian renaissance was accompanied by a revitalization of the educational system. In 863 the emperor Michael III, under the influence of his uncle and advisor, Bardas, wrote numerous literary works that were, unfortunately, subsequently lost. Yet the most significant figure in the revival of the ninth century was not that emperor but Photius (ca. 810–891), the Eastern patriarch during the period leading up to the schism (formalized in 1054) between Constantinople's Eastern Church and the Catholic Church of Rome.

Photius' cultural contribution is particularly visible in his *Bibliotheca*, a work that in 279 chapters offers numerous comments on the content of some 366 works, one-third of which consists of pagan authors, including historians, rhetoricians, writers of fable, and lexicographers, but notably not poets. The *Bibliotheca* contains the writings of Christian authors, including works of apologetics, theology, and exegesis. These texts include those used for summaries or simply intended to be consulted in libraries in Constantinople or elsewhere. Unfortunately, Photius' own library was lost after his deposition and exile, which he suffered in 867 and then again in 886. His frequent citation of numerous historical works in the *Bibliotheca* coincides with the information we have about the imperial library in that period.[225]

Another hypothesis is that Photius recovered texts from Alexandria, which, over a century before (725), had been reestablished under the sway of the patriarch. Surprisingly, there were in the Alexandrian library, Photius says, many late Hellenistic erotic authors rife with "legendary fictions and shameful obscenity."[226] Some of these salacious works, such as Heliodorus' *Aethiopica* and Achilles Tatius' *Adventures of Leucippe and Clitophon*, did not survive unscathed, while others have come down to us, including the *Metamorphoses* of Lucius Apuleius and Lucian's *Onos*.

The presence of the writers of narrative in the *Bibliotheca* of Photius reveals the ongoing importance of the genre of the novel seen in the Byzantine age. Among the authors who enjoyed the widest distribution are Achilles Tatius, Heliodorus, Longus, Chariton, and Xenophon of Ephesus. An important thirteenth-century manuscript[227] that likely derives from a collection established in the age of the Comneni is an important witness to these authors.[228]

The energetic Photius found an imitator and worthy successor in the previously mentioned Arethas (ca. 860–935), archbishop of Caesarea. His library, while less multifarious than that of Photius, nonetheless stands out, as it included several important non-Christian authors, such as Plato, Aristotle, Epictetus, Euclid, Aelius Aristides, and Lucian. Indeed, it was Arethas who assembled the manuscripts that form the Platonic corpus.[229] This vast undertaking of gathering works was vital for the influential Platonist Gemistus Plethon in his development of his mystery school at Mistra and, later, for the Florentine Neoplatonism of Marsilio Ficino.

In the tenth century, cultural activity was stimulated and supported by the fourth Byzantine king, Constantine VII Porphyrogenitus (913–959). Constantine, who was himself an author of historical and political works, developed and enlarged the imperial library. It is also possible that he wrote an interesting work on warfare entitled *Tactics*, preserved in a Florentine codex; until recently this work had traditionally been attributed to Constantine VII's father, Leo VI.[230]

Near the middle of the tenth century, the Byzantine poet Constantine Cefala assembled the *Anthologia Palatina*, a collection of epigrams and poems from various periods, divided into five books.[231] Shortly after, during the reign of

Byzantine emperor John Tzimiskes (969-976), the *Suda* was compiled. A sizeable, alphabetically arranged encyclopedia, the *Suda* well exemplifies the trend in that period to gather information and to make knowledge more widely accessible. Other classical texts that were recovered in this period are attested in the manuscripts that date to this time: the oldest complete copy of the *Iliad* we have comes from the tenth century.[232]

The eleventh century saw the transcription of another Homeric manuscript, which is of great importance for the scholia that it contains.[233] In roughly that same period, another codex bears witness to four comedies of Aristophanes not included among the seven canonical plays.[234] Another such manuscript, which included these seven, was prepared in late tenth- or early eleventh-century Venice.[235]

In that same period, the polymath Michael Psellus (1018-1096) was a particularly important advocate of a renewed interest in Plato that had begun with Photius about one hundred fifty years before. Meanwhile, at the turn of that century, one of the few female figures of Byzantine literature, Anna Comnena (1081-1115), the daughter of King Alexius I Comnenus, emerges as a scholar of the highest order. Relegated to a convent, she read the ancient historians, wrote a biography of her father (entitled the *Alexiad*), and perhaps commissioned a commentary on Aristotle.

Prominent scholarly figures of the twelfth century include Eustathius (ca. 1110-1198), archbishop of Thessalonica and commentator of the Homeric poems and Pindar; John Tzetzes (ca. 1110-1180), the commentator of Aristophanes, Hesiod, and Homer; and Michael Choniates (1149-1200). From Choniates' correspondence, we learn that he still had access to the *Hecale* of Callimachus, a work that has subsequently been lost.[236]

On April 13, 1204, Constantinople was conquered and sacked by the army of the Fourth Crusade, abetted by Venetian naval support. Constantinople's fall led to the creation of the so-called Latin Empire, of which Baldwin IX, Count of Flanders, became the ruler. Meanwhile, the Byzantine court took refuge in Asia Minor at Nicaea, where a government in exile was formed. Other autonomous governments sprung up in areas not controlled by the new Latin Empire. Such pockets of resistance were located in Trebizond, Epirus, and Bulgaria.

The sack of Constantinople marks the close of the sustained period of cultural growth that the Macedonian dynasty had promoted and the Comnenian rulers (1081-1185) sustained. The loss of texts that the fall of the city engendered is inestimable, likely even worse than that caused in 1453 by the capture of the city by the Turks. Book production and publishing were no longer performed in Constantinople as anti-humanist potentates arose. Nevertheless, Byzantine culture was preserved, to some extent, in nearby Nicaea, where a library was set up. There, in 1160, William of Moerbeke completed his Latin translation of Aristotle.

In the following year, Michael Palaeologus recaptured Constantinople, restoring the unity of the empire. The subsequent Palaeologan renaissance took its name from that of the last Byzantine dynasty, which vigorously promoted fresh cultural activity in Constantinople and other cities (Thessalonica, Mistra, Ephesus, Trebizond). That revival occurred, however, during a time marked by a sustained threat of war with the Ottomans as well as several unsuccessful attempts by the Orthodox Church to reconcile the schism with the Church of Rome.

Among the most prominent figures of this cultural revival was Maximus Planudes (ca. 1255-1330). Skilled in Latin and culturally adroit, Planudes served as a diplomat to Venice. He rendered into Greek several authors of Latin works (Cicero, Ovid, Augustine, Macrobius, and Boethius), thus making them accessible to an audience lacking Latin.

His most remarkable contribution, however, is perhaps the role he played in the transmission of Greek classics. Planudes was responsible for the discovery of the geographical work of Ptolemy, and assembled the corpus of Plutarch's *Moralia* for proper publication as an edition,[237] as well as an important collection of mathematical writings.[238] He played a key role in the preservation of other manuscripts, including an anthology of the writings of Hesiod, Apollonius Rhodius, Nicander, Oppian, Triphiodorus, and Nonnus.[239] Furthermore, he expanded the Palatine Anthology so much that the new version bears his name, the *Anthologia Planudea*, which encompasses an authentic autograph.[240] That edition omits some of the previous epigrams, but adds nearly four hundred more that are otherwise unknown.

Another figure of great cultural importance was the scholar Demetrius Triclinius, active in Thessalonica from 1305 to 1320, during the reign of the Palaeologan dynasty. Demetrius recovered the treatise on ancient meter by Hephaestion and used it to correct and annotate texts of Greek playwrights. He expanded by nine tragedies the canon of plays by Euripides (hitherto but seven), which he included in his new edition in about 1320.[241] It seems that Demetrius drew these additional plays from an ancient volume that included seventy-five Euripidean tragedies, presented in alphabetical order.

Unlike the wide geographic area involved in the preservation of Latin texts, the survival and loss of Greek texts is a tragic story that centers on a single city, Constantinople, a place as responsible for the preservation as it was for the loss of many texts, even after they were successfully transported from the East. As Luciano Canfora has noted about the loss of Greek texts, "the hardest blows were inflicted at the very end."[242]

At or near the zenith of Byzantine culture, the survival of many texts that would later be lost can be documented based on the testimony of Photius and others who substantiate their availability in their lifetimes. Even in the twelfth century, one finds references to the works of Sappho, Bacchylides, Hipponax, Theopompus, and others. The historian John Zonaras, in this same period, knew of books of Dio Cassius that were later lost. Many of these texts were still in circulation when Constantinople fell in 1204. We know that even in the next century Maximus Planudes possessed poetry collections that are no longer extant, and he drew from these when he supplemented the Palatine Anthology. As mentioned above, Demetrius Triclinius, too, would seem to have tracked down at least one volume of an ancient edition of the tragedies of Euripides. From Constantine Lascaris we know that a copy of the *Bibliotheca* of Diodorus was still available in the imperial library of Constantinople, just before the fall of the city, which occurred on May 29, 1453.

The marked decline associated with the final phase of Byzantine history led humanists to work vigorously to preserve the texts that they deemed important, as they undertook the massive task of transferring westward a great portion of classical heritage that the Greek codices vouchsafed. This recovery, we have seen, was neither systematic nor organized, often entrusted, at least at the beginning, to individual humanists and merchants, or to exiles who were moving westward as they fled the perils of political instability. Yet with them they brought ideas more powerful than political sway, for they transported the wisdom of the ancients, wisdom that would inform the art, literature, and culture of the Renaissance.

4 / CLASSICS AND HUMANISTS

THE NINETEENTH-CENTURY SCHOLAR JACOB BURCKHARDT once formulated a concise description of the Renaissance: "The Renaissance is not a mere fragmentary imitation or compilation, but a new birth.... Culture, as soon as it freed itself from the fantastic bonds of the Middle Ages, could not at once and without help find its way to the understanding of the physical and intellectual world. It needed a guide, and found one in the ancient civilization."[1] Along with this view of the Renaissance Burckhardt further asserted that the modern concept of the individual emerged in Italy in the fourteenth and fifteenth centuries. Although Burckhardt's theory has not existed without controversy, it has nevertheless provided a basis for several interpretations of the period.[2] There is great value in his hypothesis, and his formulation undergirds this chapter's range. We thus begin by noting that even a chapter two or three times the size of this one could not do justice to the topics covered herein. With that caveat in mind, after a brief overview we shall launch into an analysis of one of the most delightfully complicated epochs in the history of the Western world.

One major aspect of that complexity is the transmission of classical texts. In a marked departure from the medieval focus on selections, Renaissance scholars began to pay attention to classical works in their entirety. To the Latin texts that had been part of the medieval canon and had enjoyed good repute were added those that had not been in circulation, of which only a few copies — in some cases only a single copy — had survived in some remote monastery. Additionally, Greek texts were now resurfacing as they were carried westward from Constantinople.

By the beginning of the sixteenth century, as the recovery work of the great humanistic researchers began to draw to a close, almost all Greek and Latin authors to whom we now have access became available. The retrieval of those texts came through the activity of humanists, who sought to bring to light all surviving works. Their important search for and appropriation of classical texts, along with the consolidation of them—even if such consolidation was itself too often unsystematic—greatly reduced the risk of the permanent loss of these texts and even led to their widespread reproduction, ensuring the preservation of what was left of classical heritage. The introduction of printing, as we shall see in the next chapter, safeguarded the transmission of these authors to posterity.

The humanistic impulse to gather texts, like the move centuries earlier from roll to codex, was a pivotal moment. Twentieth-century scholarship has sometimes called such transitions "historical or cultural shifts." Some scholars, however, have sought to emphasize cultural continuity in the midst of such moments of stupendous change,[3] and, certainly in the progression from medieval to humanistic thought, there is a good deal of stability. To some extent, then, the continuity view may be justified, for in the case of the Renaissance it can be argued there is a continuum with previous centuries that can be seen in the various "renaissances" that occur throughout the Middle Ages.[4] Those who hold to the notion of cultural continuity tend to present the humanistic age as the final outcome of growing access to classical texts that began in the Carolingian period. From the perspective of continuity, the sense of transition elicited by the humanists' intensive efforts to rediscover ancient texts, which were foundational for the Renaissance, is much diluted.[5] Rather, the process was gradual.

Another category, "pre-humanism," has sometimes been suggested by scholars who believe that some traits of the humanistic period are evident already by the fourteenth century.[6] Those who hold to this view look to Francesco Petrarca (Petrarch, 1304–1374) and a cadre of humanists at Padua, all working between the thirteenth and fourteenth centuries, as prime exponents of this pre-humanism.[7] Chief members of this group include the Paduan notary and judge Lovato Lovati (1240–1309) and his student Albertino Mussato (1261–1329), leading figures in the politics of that city. While humanism can be said to be marked not simply by a search for ancient texts but also by research motivated by a conscious break with the existing medieval cultural structure, the notion of pre-humanism, as seen in the Paduan school, offers an interesting, if not entirely tidy, picture of a movement that would ultimately produce fully fledged humanism.

Whether one subscribes to the notion of pre-humanism or not, the route to humanism seems to have begun, in some senses, with Lovati. Thanks to his admission to Verona's Biblioteca Capitolare, as well as the abbey of Pomposa near Padua, Lovati enjoyed access to authors virtually unknown to other scholars. As an example of this special access, Lovati seems to have studied a manuscript of Catullus in Verona, to whom he alludes in his *Epistulae Metricae*.[8] At Pomposa he copied both the Codex Etruscus, containing the tragedies of Seneca, and another manuscript of the eleventh century, containing an epitome of Justin.[9] Lovati also knew the fourth decade of Livy, which Petrarch recovered only several years later in Chartres. Lovati's interest in antiquity is also evidenced in Padua by a monument he built there. That monument honored the classical Homeric character and legendary founder of Padua, Antenor, whom Lovati claimed to have recognized in a skeleton discovered in Padua in 1283.[10]

As a historian, Lovati's student Mussato was innovative, proposing historiography of the "Livian type." Yet his great contribution was in the realm of drama. He modeled his *Ecerinis* (1315), the first Latin tragedy since antiquity,[11] on the tragedies of Seneca. He constructed the plot of this drama from the conflict between Padua and Verona, and the victorious duke of Ezzelino III from Romano.[12] He could thereby criticize through literature the expansionist ambitions of Cangrande della Scala of Verona.[13] When Padua fell under Cangrande's leadership in 1328, Mussato was forced into exile.[14] Still, before his exile, he was awarded the laurel wreath in 1315 for his work as both poet and historian.[15]

The Paduan school included several other key players, such as the prominent Jeremiah Montagnone, also a lawyer, who compiled a didactic anthology.[16] Thus, in the small city of Padua, a vast stream of research in the humanities was opened through the preservation and study of rare texts, feeding the intellectual impetus that would spur on the first humanists.[17] But as Natalino Sapegno noted many years ago, the Paduans "made inroads in research, but were almost unaware of the novelty of their attitude."[18]

Building on the work of Lovati and Mussato, Petrarch contributed to the cultural change that had begun in Padua and would come to prevail in the fifteenth century in Italy, eventually spreading throughout Europe. Petrarch's view of antiquity often comes to us by way of his own words, especially in letters addressed to ancient authors, such as Cicero, Seneca, and Virgil. These letters are gathered in his epistolary collection, *Familiares*, where Petrarch also inserts letters to friends and contemporaries. In one such letter to Giovanni Colonna dated to 1337, Petrarch describes the ruins of ancient Rome from the texts of classical writers; in another letter the same ruins provide the backdrop for his coronation as poet laureate, which took place on Rome's Capitoline Hill on April 8, 1341.[19] Petrarch's interest in ancient Rome was also laden with political implications, including his support for the republicanism of Cola di Rienzo, a position that put Petrarch into conflict with his major patron, Cardinal Colonna.

Petrarch's penchant for recalling ancient ideas grew side by side with his apparent discomfort with the culture of the late medieval period. That age's debt to ancient culture was imitative, not creative. Before Petrarch's time, Aristotelianism and nominalism dominated the fields of philosophy and theology. The legal culture and formal rhetoric that formed the bases of intellectual education by the twelfth century had begun to fade already in this pre-Renaissance period. Even medieval Latin posed a stylistic contrast with classical Latin. The attention given to ancient thought, foundational for Renaissance humanism, revealed, with every new text discovered, that ancient views about life and culture were thoroughly distant from the medieval mindset. While the Middle Ages merely imitated the ancients, especially in rhetoric and philosophy, the Renaissance would see that the maturation of critical approaches transcended such imitation, producing more original and robust thought.

An early example of the waxing influence of humanistic thought is evident from a meeting that Charles IV of Habsburg called with Petrarch in 1361. On that occasion, Petrarch disapproved of the notion of Austria's independence being attributed to Julius Caesar and Nero.[20] Petrarch made it clear that he derived his own ideas and ideals from ancient texts in order to apply them to a situation current in his day.[21] Generally speaking, in the medieval period, the Catholic Church kept watch over culture, a culture of which Petrarch's various writings reveal that he was personally wary inasmuch as it was often difficult to maintain a balance between the traditional Catholic faith, to which he remained committed, and the ideas that the "new" humanism (even pre-humanism) was advancing.

Nevertheless, Petrarch worked toward such a balance, as is evidenced in an epistle to his brother Gherardo,[22] who in 1343 had become a monk in the charterhouse of Montreux.[23] "Moreover, permit me to use secular attestations with you," Petrarch writes, "which abound not only in Ambrose and our Augustine and Jerome, but even in the Apostle Paul, who does not blush sometimes to use them. Nor prohibit from your home those things which are worthy of my own mouth and not unworthy of your ears." He goes on to express his own regret at having only thoughts for the affairs of this world:

> Thus did my brother sing, quite rightly, singing with his soul inclined toward heaven; for my part, I was pondering earthly things, stooped as I was toward the earth. And perhaps I simply failed to recognize the right hand that could free me; perhaps I based my hope merely on my own strength. Either that, or something else, is the reason why I am not free, even though my chains are gone.[24]

In this context what Petrarch dubs "earthly things" would seem to encompass the sentiments of pagan authors, whom he admires. Such acceptance of profane authors can be seen, too, in Petrarch's acquisition of Cicero's *Somnium Scipionis*, in which Scipio Africanus admonishes Aemilianus to keep his vision earthward and not to fixate upon the celestial world. Later, Petrarch questions his own predilection for Cicero and remembers the story of Jerome's dream.[25]

Denying that he indulges in mere imitation, Petrarch confesses his deep appreciation of Cicero, proclaiming, "But if to admire Cicero is what it means to be Ciceronian, then I am Ciceronian."[26] Nonetheless, the Christian faith is reaffirmed as "the supreme truth, the real happiness, eternal salvation," but with the clarification that Cicero himself "would have been a Christian if he could have seen Christ and known the teachings of Christ."[27] Such mapping of the pagan past onto the spiritual landscape of the Catholic present is clearly a mark of inclusive admiration.

This mindset is detectable in classical flourishes apparent from time to time in Petrarch's poetry, particularly his *Canzoniere*. Yet Petrarch's veneration of antiquity is also evident in the effort that he put forth to acquire ancient books and in his careful research and reading of ancient authors. Among the finest manuscripts that we know Petrarch owned is his codex of Virgil, which by 1335 was in his possession in Avignon. There, Petrarch commissioned the well-known Sienese artist Simone Martini to illuminate it; the manuscript, with its fine illumination, is preserved today in the Biblioteca Ambrosiana of Milan (fig. 4.1).[28]

In this manuscript one can read several of the personal notations that Petrarch made over the course of many years; there is, too, a note to Laura, who figures prominently in the *Canzoniere*. Petrarch also collected the first three decades of Livy, another of his favorite authors, into a single codex.[29] Petrarch had assembled this by collating a manuscript of the first decade in Verona copied by Ratherius and then gaining access to a codex of Livy's fourth decade, previously unrecovered but stored at Chartres thanks to Landolfo Colonna. Collating it with a specimen previously discovered by Lovato Lovati, Petrarch copied this text himself.[30]

From Avignon, the papal seat in which he spent his youth, Petrarch also explored the rich French libraries in search of rare and little-known texts, establishing contacts with scholars of the "whole of Europe, from England to Italy."[31] In his personal quest for codices, Petrarch was a forerunner in the humanist tradition.

In 1333 Petrarch made his own reproduction of a copy of Propertius' *Elegies*, which he found in Paris. His personal copy is no longer extant and may even have fallen out of circulation not too long after his death, but it was the basis for numerous subsequent copies.[32] Where Petrarch discovered the collections of Catullus and Tibullus is unknown, but at some later point he sent copies to Coluccio Salutati (1331–1406). In 1333 in Liège, Petrarch discovered the *Pro Archia* of Cicero in a manuscript also no longer extant that nevertheless likewise provided a basis for numerous reproductions. Propelled by these acquisitions, Petrarch sought to complete his collection of the works of Cicero. In this pursuit, he soon came upon a codex of the *Epistulae ad Atticum* in the Biblioteca Capitolare at Verona. Through his colleague Boccaccio, he discovered the *Pro*

FIGURE 4.1. *Frontispiece to Petrarch's copy of Maurius Servius Honoratus's commentary on Virgil. By Simone Martini (1284–1344). 1340. Milan, Biblioteca Ambrosiana, ms. S.P. 10/27. Italy/Bridgeman Images.*

Cluentio and subsequently acquired the codex now known as Troyes, Bibliothèque Municipale 552, in which there are twenty collected works of Cicero, including a mutilated copy of *Orator*, the *Tusculan Disputations*, and the rare *De legibus*.

Beyond these vital acquisitions, Petrarch recovered other less common authors, including the first-century geographer Pomponius Mela, the late fourth-century geographer Vibius Sequester, and the fourth-century Roman historian Julius Paris, a copy of whose work was at Auxerre in the ninth century. These works came into Petrarch's possession at Avignon in approximately 1335. Significantly, the entire humanistic tradition of these authors stems from his copies. Petrarch soon realized that Verona's Biblioteca Capitolare held a trove of further rare texts, and thus he carefully sifted through that collection. There he found the *Historia Augusta* in the only witness to the work, a ninth-century manuscript copied in northern Italy.[33] He also discovered the work of the fourth-century writer Ausonius, author of *Mosella* and *Ephemeris*.[34]

This was by no means the full extent of Petrarch's interest in manuscript collecting; he forged a network of colleagues who kept him informed of developments and discoveries. Sometime before 1364, Petrarch received news from Verona, brought to him by his friend William of Pastrengo, of the existence of codices of the agricultural treatises of Cato and Varro, as well as of the *Eclogues* of Calpurnius and Nemesianus.[35] By this time, however, Petrarch had already composed his *Bucolicum Carmen*. Even though it was too late for these new discoveries to affect Petrarch's pastoral poem, the influence of these two authors is detectable in the pastoral work of Boccaccio, Petrarch's younger contemporary.

Another author known to Petrarch was Vitruvius, a volume of whose work he found in France in the early 1350s. Fortunately, Petrarch's personal notes were copied into it and are preserved in a later codex.[36] As in the cases of Calpurnius and Nemesianus, Petrarch's acquisition of Vitruvius ensured he would be known not only to Boccaccio but also to subsequent Italian humanists. Petrarch additionally came into possession of two copies of Suetonius' *Vitae Caesarum*, a work little known in the Middle Ages. As in the cases of the other works mentioned here, the *Vitae Caesarum* became available to Boccaccio and later, thanks to Petrarch, was widely circulated.

Giovanni Boccaccio (1313-1375), who relished the role of being a friendly rival to Petrarch in the quest for ancient manuscripts, became also his close friend and confident, each sharing his discoveries with the other. Boccaccio was very interested in manuscripts housed at the monastery of Monte Cassino, a collection that may have derived from a late antique private library.[37] For this particular endeavor, Boccaccio relied on the help of Zanobi da Strada (1310-1361), another member of Petrarch's circle of friends.[38] Zanobi was particularly useful to Boccaccio because he held the position of episcopal vicar at Monte

Cassino from 1355 to 1357. In 1362, just five years after Zanobi's tenure there, Boccaccio would visit Monte Cassino when he learned of the manuscripts it contained.[39]

It seems to have been at about the same time that a duplex volume (two tomes bound together) containing the works of Tacitus and Apuleius was transferred from Monte Cassino to Florence.[40] The Tacitus volume contained *Annales* 11–16 and the *Historiae*, while the other contained Apuleius' *Metamorphoses*, *Apology*, and *Florida*.[41] Although this is the first manuscript of the *Metamorphoses* for which we have evidence, it appears that Boccaccio knew Apuleius from another manuscript as early as 1338–1339, as an *apograph* seems to have been in circulation at some point before the vicarage of Zanobi, allowing Apuleius' work to become one of the models of inspiration for the *Decameron*.[42] By 1427 the duplex codex was in the hands of Niccolò Niccoli (ca. 1365–1437), although how it came to him from Florence is unclear. Niccolò's reluctance to share it is highlighted in a letter from his colleague Poggio Bracciolini: "If you send the Cornelius Tacitus, I shall in fact be delighted; do so, and I'll return your Spartianus; for this I ask you most urgently."[43] At Monte Cassino, Zanobi also recovered an *apograph* of Apuleius, a tome of uncertain origin but datable most likely to the twelfth or thirteenth century.[44] Some obscenity appears in the marginalia of that codex,[45] and Zanobi deemed this a spurious addition, leading him to dub it "spurcum additamentum,"[46] a name for the section that still obtains today.

The role of Boccaccio and the monastery of Monte Cassino in the preservation of a great number of manuscripts, some of them having thoroughly erotic content, cannot be sufficiently stressed. One example is the collection of *Priapea*, an extant codex written in Boccaccio's hand.[47] In his notes Boccaccio attributes it to Virgil, probably on the basis of reading ancient biographies of the poet. Another codex from Monte Cassino that came into Boccaccio's purview contained Varro's *De lingua Latina* along with Cicero's *Pro Cluentio*,[48] a copy of which was sent to Petrarch, who replied with a letter of thanks.[49]

In the course of the Middle Ages, it was rare for any scholar independently to acquire a reading knowledge of Greek. An isolated case is that of William of Moerbeke (1215–1286), of Flemish origin, who traveled widely in Greece and resided briefly in Viterbo and Orvieto. He made a number of personal translations but failed to produce students who might have continued his work. By the fourteenth century, there were a few Greek manuscripts in circulation, but these tended merely to be collectors' editions. At the beginning of the same century, a copy of Plutarch's *Moralia*, owned by Maximus Planudes (1260–1330),[50] was in the possession of Pace da Ferrara in Padua.[51] Not only did Pace not know Greek, but he was so poorly informed that he believed that Sophocles, not Homer, was the author of the *Iliad* and the *Odyssey*.

Ignorance of Greek was not uncommon, even among the learned. Petrarch himself was unable to read Greek, but he had obtained a copy of the *Iliad*[52] along with a voluminous codex of Plato.[53] In January 1354, Petrarch thanked Nicola Sigero for sending him a Homeric codex from Constantinople, marking that occasion with a humorous quip: "Your Homer is here with me; mute or rather in fact deaf am I, in front of him. But I am happy to gaze upon him and often hug him and, sighing, say, 'Great man, how I would love to hear you!'"[54]

A few years earlier Petrarch had tried, with little profit, to learn Greek from Barlaam of Seminara in Calabria (1290–1348), a prelate from Constantinople who had arrived in Avignon in 1341, where he was involved in theological disputes.[55] After he was received into the Latin Church at Avignon in 1342, Barlaam was awarded the bishopric of Gerace in Calabria.[56] In Padua in 1358–1359, Petrarch made the acquaintance of Leontius Pilatus (fl. 1355), another hellenophone from Calabria. Petrarch went so far as to ask him to translate the *Iliad*, but he did not like the result, which he judged unpolished and disappointing. At about that time, Leontius came into contact with Boccaccio, who procured for him a teaching position in the Florentine school known as the Studium. Over the three years he spent in Florence (1360–1362), Leontius managed to complete his own prose translations of the two Homeric epics and to render a number of other texts, including plays of Euripides and a work of pseudo-Aristotle.

Leontius's Florentine teaching post left no significant results, but the fact that he held such a position is emblematic of the fourteenth century's increasing interest in Greek culture. Such interest evidences a significant divergence from the previous centuries, when the primary academic thrust was toward clerical texts or those with philosophical-scientific content. Interest in the Greek classics, primarily Homer and Plutarch, was now beginning to take hold.

Avignon played an important role in the rise in Plutarch's popularity. In 1372 Cardinal Pietro Corsini (1335–1405) commissioned Simon Atumano (ca. 1314–1387), another Greek prelate who had come over to the Roman Church, to translate into Latin Plutarch's treatise on the restraint of anger. The existence of this treatise had been known in the West only because it was mentioned by Aulus Gellius. In 1392 Corsini sent a copy of Atumano's translation to Salutati in Florence. Salutati made a stylistic revision of the Latin text, but he was unable to check it against the original, as he, too, did not know Greek.

Another member of the increasingly learned and culturally diverse curia in Avignon, Juan Fernández de Heredia (1310–1396), discovered a version of Plutarch's *Vitae parallelae* in *koine* and had it translated into Aragonese.[57] In the subsequent years, Salutati began but did not finish a Latin translation of the Aragonese text. Instead it was rendered into Italian in Florence, and this translation enjoyed a rather wide distribution. Salutati, a few years later, was instrumental in the return of Greek works to the West.

The uniquely central role played by Petrarch in the subsequent development of early Renaissance humanistic culture is evidenced in a letter written by Boccaccio in 1372, in which he returns to the image of a path that Petrarch, as his teacher, had opened for him:

> He began to trace the ancient path with so much courage in his heart and with so much ardor in his mind, along with perspicuity of talent, that he could not be stopped by any obstacle nor even by the difficult road. On the contrary, it was overgrown by brambles and bushes with which it was obscured by human negligence; restoring it with solid stones smoothed by water, he paved the way for himself and all those after him who would have the will to ascend.[58]

The metaphorically reopened road brings the present together with ancient culture, which is symbolized by Helicon, the mountain of the Muses. Inspirationally, Florence would become the Helicon of emerging Renaissance culture.

Coluccio Salutati, mentioned above, a younger contemporary of Petrarch and Boccaccio, would play a major role in Florentine culture and politics. By 1375 he was Florence's chancellor, and in the 1390s, much like Petrarch before him, he brought together a group of scholars who would bring about and sustain the humanist movement. The cultural program of the group found a powerful expression in the *Dialogi ad Petrum Histrum* by one of the group's younger contributors, Leonardo Bruni (1370–1444). His use of the dialogue form in this work written in 1401–1403 is clearly inspired by Cicero. The work was dedicated to Pier Paolo Vergerio (1370–1444), who had previously attended the Florentine school of Chrysoloras along with Bruni.[59] Protagonists in that dialogue include Bruni and Salutati as well as two other members of the group, Niccolò Niccoli and Roberto de' Rossi.

In the dialogue, the character representing Niccoli harshly criticizes the "stupid philosophers" of his time, accusing them of knowing Aristotle only through translations. Specifically, such translations are indicted for undermining the philosopher's original thoughts. Disparaging the loss of many ancient authors, Niccoli laments the "unfortunate and impoverished condition" of his age and challenges the prominence of the three great Florentine authors of the fourteenth century—Dante, Petrarch, and Boccaccio—by comparing them to the ancients: "By Hercules, I demand a single letter of Cicero and especially one poem of Virgil from all your booklets." Niccoli also complains about the paltry degree to which writers, even one as accomplished as Dante, held a command of Latin.[60] Niccoli's harsh judgment is moderated in the second day of the dialogue, when Salutati proposes a less radical position.

In spite of the interpretative problems that the dialogue format comprises, the interaction of Salutati and his "brigade of arrogant youth"[61] appears credible. Because there was great concern among ecclesiastical circles in the city about this group's activity, Salutati's moderate position enabled him to act as both teacher and benefactor for his rather radical students. Salutati's other major contributions include both creating humanistic script and, more important, facilitating instruction in Greek in Florence. His presence in the Florentine cultural center gave an impetus to the rediscovery of ancient texts, accomplishing what Petrarch had managed in Avignon.

One important codex possessed by Salutati[62] had probably previously been owned by Petrarch. This manuscript is vital because it contained the little-disseminated *Corpus Tibullianum*.[63] Another important manuscript was Salutati's copy of the *Elegies* of Propertius,[64] which is the oldest witness of the now-lost copy of Propertius that Petrarch himself made in Paris. Furthermore, Salutati owned a copy of Catullus,[65] reproduced in Florence around 1375. Through the agency of the chancellor of Milan, Pasquino di Capelli (fl. 1388), Salutati also came into possession of a ninth-century manuscript[66] containing the *Epistulae ad familiares* of Cicero, almost unknown at the time. This copy may have been made in the court of Louis the Pious in Aachen (778-840); eventually the manuscript would be part of the collection of the library at Lorsch. At the beginning of the eleventh century, the codex was passed on to Leo, bishop of Vercelli (965-1026), where it was identified by Pasquino and acquired by Salutati, whose primary pursuit at the time was the *Epistulae ad Atticum*, known to him through Petrarch. Thanks to this fortunate discovery of the *Ad familiares*, Salutati was the first to possess almost the entire corpus of Cicero's letters. Yet this was only the next step in that age of discovery. Many more discoveries remained to be made in the early days of what was becoming Renaissance culture.

Salutati's finest student, Gian Francesco Poggio Bracciolini (1380-1459), who was born in the Tuscan Terranuova, had access to an even richer trove of manuscripts. By 1403 Poggio Bracciolini had become secretary to Cardinal Landolfo Maramaldo, in which capacity he honed his skill as a theological writer. At the behest of Anti-Pope John XXIII, Poggio moved to Konstanz in Germany in order to participate in the ecumenical council (1414-1417). John had called for this convention in an attempt to heal the schism that had divided the Church and to ascertain who had the legal claim to the papacy: Gregory XII was challenged for the office by both John XXIII and Benedict XI. Poggio's expertise in doctrinal matters was of great value to the Church, and he served as a consultant during that tumultuous period. When Martin V was elected to the papacy in 1417, he appointed Poggio to be his personal secretary (apostolicus secretarius) in which capacity he composed the papal encyclicals dictated and outlined by Martin V.

From Konstanz, Poggio was able to visit numerous Swiss, German, and

French monasteries, where he found a significant number of rare or little-known texts that he put back into circulation. Although these books' passage across the Alps had occurred five centuries earlier when Carolingian scholars had taken numerous codices from Italy to France or to newly founded Swiss and German monasteries, by the fifteenth century a kind of *nostos* of the classics began to take place in Italy. Poggio followed in Petrarch's footsteps by going to great lengths and remote locations to "rescue" ancient codices. Within a few years, Poggio had discovered texts hitherto entirely unknown.

An important period in Poggio's pursuit began in 1415 when he reached the abbey of Cluny in Burgundy. There he discovered a codex that contained two previously unknown orations of Cicero: *Pro Murena* and *Pro Roscio Amerino*. That codex was unfortunately lost. In 1416 Poggio went to the monastery of St. Gall, where he discovered a complete text of Quintilian's *Institutio oratoria*, copied in the tenth century and glossed by Ekkehard IV (980–1056).[67] He also discovered there a codex containing the first four books of the Valerius Flaccus' *Argonautica* and the commentary of Asconius on five orations of Cicero. Of these works, only Poggio's own copies survive.

Two years later, Poggio retrieved codices of Lucretius, Silius Italicus, and Manilius near St. Gall. In the case of Lucretius, the codex discovered by Poggio was a descendant of the *Oblongus*.[68] Although the codices of Silius Italicus and Manilius are no longer extant, valuable transcriptions made by Poggio and his colleagues remain. In the summer of 1417, Poggio found a cluster of Ciceronian orations.[69] He also discovered a copy of Statius' *Silvae* in a codex of the tenth century, preserved in the monastery of Reichenau.[70] In the same year, Poggio acquired from the monastery of Fulda a valuable codex of Ammianus Marcellinus,[71] and he subsequently went as far as England, where he discovered Petronius' *Excerpta vulgaria*, which he promptly sent to Niccoli. In 1423 at Cologne, he tracked down another part of Petronius' *Satyricon*, the *Cena Trimalchionis*,[72] and in 1429 at Monte Cassino recovered a copy of Frontinus' *De aquis* dating to the twelfth century.[73] Through his zealous pursuit of these texts, Poggio gained notoriety, which itself spawned many imitators in the pursuit of manuscripts and the ancient knowledge that they preserved.

Their contributions were no less important and enriched the general library of the texts circulating among the early humanists. In 1421 a ninth-century manuscript containing Cicero's *Brutus*, a work that had not survived elsewhere, was found by a local bishop, Gerardo Landriani, in the library of the cathedral of Lodi near Milan. As in the case of other manuscripts we have seen, the Codex Laudensis itself went missing, but copies survive, one of which wound up in the hands of Flavio Biondo.[74]

In addition, Landriani came upon copies of the *De oratore* and the *Orator*, sending them to Gasparinus de Bergamo (1360–1431), an important humanist of the fifteenth century.[75] By 1429 Niccolò Niccoli had acquired and transcribed them into a manuscript that is still extant.[76] Furthermore, he made a copy of the *Argonautica* of Valerius Flaccus, of which Poggio had discovered a mutilated version only thirteen years before. In the same year, Cardinal Nicholas of Kues (1401-1464) brought from Germany an important eleventh-century manuscript that contained twelve previously unknown comedies of Plautus,[77] which, combined with the eight already known in the Middle Ages, brought the grand total of the corpus to twenty.[78]

Another author virtually unknown during the Middle Ages was Cornelius Nepos, whose surviving *Vitae* (a small part of *De viris illustribus*)[79] had been copied in the twelfth century. A number of copies of the *Vitae* were made in the first decades of the fifteenth century.[80] Similarly, Celsus' *De medicina* was barely circulated in the Middle Ages, but interest in ancient medical treatises resurged with the dissemination of Latin translations of Galen and Hippocrates. Fortunately, Celsus' work suddenly reappeared in Siena in 1426 and was soon copied by Niccoli, who made a number of humanist circles aware of it.

In the same year, John Lamola, a pupil of Guarino Veronese (who will be discussed at length below), tracked down a ninth-century copy of the treatise of Celsus[81] in the basilica of St. Ambrose.[82] Shortly before, Thomas Parentucelli (who would become Pope Nicholas V) had recovered a sixth-century codex of Lactantius in the monastery of Nonantola. In 1433 John Aurispa discovered in Mainz the *Panegyrici Latini*, which had gone missing in late antiquity, along with Aelius Donatus' commentary on the comedies of Terence. As in other cases, although many copies survive, the actual codices that Aurispa discovered are lost.

By positing the bygone era of antiquity as a model for their own, a sizeable group of early humanists appeared to be advocates of tradition, supposing that the past gave birth to enduring values so important that they could rival or surpass those of their own age.[83] Yet, in the same period, another school of thought took root in Renaissance circles that opposed the more traditional group of humanists. This school represented a more "modern" point of view, one that espoused a contemporary culture less indebted to its forbears. These two schools of thought (traditionalists versus modernists) represent the variety in the development of the Renaissance.

Perhaps the most significant figure of the first of these groups was the mid-fifteenth-century anti-modernist figure Lorenzo Valla (1407-1457). In the preface to book 4 of his *Elegantiae Latinae Linguae*, Valla compares the ancients to "bees, flying in remote meadows, which have collected with admirable contrivance sweet honey and wax."[84] He adds that his opponents, the modernists who had little regard for the past, might view him as "similar to ants who, stealing from a neighbor, hide a few grains in their closets. I, as far as I am concerned, would not only prefer to be a bee rather than an ant, but would also prefer to serve in the army as a follower under the queen of the bees rather than to serve as a commander general among the ants."[85]

In their desire to pollinate afresh their own times, the traditionalists could, from time to time, construct an idealistic view of antiquity. The vision of the past that supported the trends of humanistic thought appears to have been characterized, as we shall see below, by a series of particular views that grew out of the humanists' reception of ancient evidence. Such "illusions" have had a significant impact on Western culture, particularly in our attempts to understand the history of the period. But let us back up a bit, to try to understand the process whereby the early Renaissance humanists garnered such a view of the past, both the immediate medieval past and that of deeper antiquity.

The term "Middle Ages," first attributed to the mid-fifteenth-century humanist Giovanni Andrea but not developed conceptually until the sixteenth century,[86] was itself largely a product of the Renaissance. Petrarch had already redacted the Augustinian picture of history, according to which the light of Christ dispelled the darkness that preceded it, into a more complex picture, wherein the light of antiquity was obscured by the darkness of the age that followed.[87] In his *Africa*, Petrarch has one of his characters, Syphax, prophesy to Scipio that his descendants, "having emerged from darkness, will be able to return to their pristine and untainted splendor,"[88] an image that anticipates the notion of "Dark Ages" and the subsequent rebirth of learning in the Renaissance.

The historical organization outlined by Petrarch is taken up by the chancellor and historian of Florence, Bruni, in his *Dialogues for Pier Paolo Vergerio*. His younger contemporary, Valla, would present in his *Elegantiae Linguae Latinae* a framework whereby the corruption of the Latin language and ancient culture is attributed to the barbarian invasions of the Goths and Vandals:

> Pouring into Italy, [they] took Rome, and when [the Romans] had come under their sway, they thus also took the language—[at least] so some say. ... On this account, many codices were written in Gothic characters. If that race was able to corrupt the Roman script, what must one think of the language, especially insofar as they have left a linguistic legacy? Whence, after them, the first dawning of eloquent writers arose, though they were quite a bit inferior to their ancient forebears. Behold how far Roman literature had receded: while ancient Greek had formerly mingled with their language, they now found themselves mixing their language with Gothic.[89]

Valla viewed the barbarian invasions as a corruption that affected both populations in terms of language and writing, causing a kind of cultural eclipse for the Romans. To amend that loss, Valla and his circle offered a comprehensive program for the rediscovery of antiquity, focusing not only on a study of the language but also on the culture and way of life. This "recovery operation" was meant to undo the deterioration that accrued over the roughly seven hundred years since Rome's cultural collapse.

Yet in spite of his harking back to a richer cultural moment, Valla himself was rather imprecise in identifying the characteristics that would mark this transition from antiquity to the Middle Ages. In the preface of the second book of his *Elegantiae*, for example, he suggests that this transition was merely an aspect of late antiquity. He identifies Donatus, Servius, and Priscianus as the last authoritative grammarians:

> There were three men like a triumvirate: Donatus, Servius, and Priscianus, and it is debated among scholars which one is chief of all. I consider them so great that whoever wrote anything about Latinity after them seems to stutter. First among this group is Isodorus, the most arrogant of the ignorant, who, though he knew nothing, taught everything. After him come Papias and other ignorant men, Eberhard, Uguccione, the "Catholicon," Aimo, and others who aren't worth naming, teaching men for a large fee to know nothing at all, or making a student more foolish than he was when first they got hold of him.[90]

Valla cites late medieval works, including Papias's *Elementarium doctrinae rudimentum*, Eberhard of Béthune's *Graecismus*, Uguccione of Pisa's *Derivationes*, and Giovanni Balbi's *Catholicon*. While the identity of Aimo mentioned in the quote above is not entirely certain, he may be Haymo of Auxerre, although that author is not believed to have been a lexicographer. Overall, Valla's list reveals a harsh tone and divulges his unfavorable judgment of late medieval culture, even though he does so by mentioning the principal texts of lexicography that were still in use in the fifteenth century. Paradoxically, these technical treatises helped preserve some of the ancient literary tradition in an era when books were not circulated widely.

The idea of medieval decline was not new. Even before Valla, Flavio Biondo (1392–1463) had marked the end of antiquity at 410, when Rome was sacked by the Visigoths.[91] Indeed, the decline of grammar that occurred between Priscian and Isidorus, beginning in the sixth century, represents a period already roughly outlined by Salutati in an epistle to Juan Fernández de Heredia.[92] In that letter, he complains about a cultural decline spanning the previous six centuries.

Yet Valla, ever hawkish about language and style, places the end of antiquity further back than Salutati. To wit, the body of literature that Valla outlines in his *Elegantiae* is rather thin. After Plautus, he includes Cicero, and then a selection of authors from the Augustan period, and the most recent author whose value he touts is Quintilian. Such a schema suggests a very early literary decline, beginning in the first century. Valla was not alone in his aggressive view of literary decline. Sicco Polenton (1375-1447), in his *Scriptores Illustri Latinae Linguae*, shifts the "beginning of the decline of literature" to the period following the first-century satirist Juvenal, after whom literature "remained silent for many years."[93]

The key to the origins of Renaissance thought, however, is not the precise date at which any one early Renaissance figure placed the beginning of the decline of Latin letters. Rather, it is the fact that they recognized a decline and that, in a variety of ways from the humble and wary to the bold and brash, they positioned themselves and their age to resurrect ancient traditions. They achieved this feat through their grand contribution of what we now call humanistic writing.

Valla, as we have seen, describes the "Goths" putting the final nail in the coffin of Roman culture, as it were, and says that tangible proof of the corruption of Roman rustic capitals into Gothic letterforms can be seen in numerous codices. In point of fact, however, Gothic script has absolutely nothing to do with the Goths: it is merely an evolution of the Carolingian script that had become widespread in the late Middle Ages.[94] The name "Gothic," introduced by humanists such as Valla, is merely a consequence of the chronological distortions that devolved from the inaccurate appraisals rendered by those scholars based on the libraries in their purview.

Late medieval books written in Gothic script were simply considered "medieval" because of the notion of a dark age propounded by the early Renaissance humanists. On the other hand, the script of the Carolingian period was consistently deemed to be ancient, as evidenced by the name that humanists adopted for it: *litterae antiquae*.[95] This misconception arose because in the monastic libraries of the fourteenth century, the oldest books available dated only to the ninth or tenth centuries, as late antique codices were rare. The humanists were not, in fact, "rediscovering" ancient works, but rather works copied in the Carolingian period that had not enjoyed broad circulation in the medieval period. The chronological assessment based on the findings of the early humanists was therefore flawed, and their miscalculation explains the emphasis on the rediscovery of ancient authors.[96]

The lively search for codices in the first decades of the fifteenth century gave rise to the sense that with proper scrutiny such ancient relics could be found. It is true that, while rare, such discoveries did occur. Poggio, for example, boasts in an epistle to Guarino Veronese that he had found a manuscript containing the *Institutio oratoria* of Quintilian (he had discovered it in St. Gall in 1416). The manuscript is likened to a man, once "splendid . . . , neat and elegant," who has become "sad and sordid, as they are accustomed to be when sentenced to death."[97] The description of the monastery in which the codex was kept highlights the attitude of contempt with which the humanists viewed the medieval culture of the monks. This dismissal comes despite the fact that the medieval dedication to the preservation of knowledge had vouchsafed the authors that they were now recovering. Poggio says in that same epistle:

> And there, among many manuscripts, too long to list, I found Quintilian intact, albeit full of mold and dust. Those books were in fact not in the library, as their dignity required but, as it were, in a frightening and dark prison where those condemned to death would have been held in chains.[98]

This tenth-century manuscript, now known as the Ambrosian Quintilian, includes the corrections of Ekkehard IV, an eminent cultural figure of the eleventh century and a teacher in the school attached to the cathedral of Mainz. Poggio seems to have no doubts about the antiquity of the manuscript and, on the basis of this certainty, interprets as ancient what is actually a valuable product of medieval culture.

Such misappraisal of correct dates of manuscripts continued throughout the Renaissance, as can be seen in the case of the Mediceus Virgil, a late antique codex housed in the Bobbio monastery before being brought to Rome in murky circumstances in approximately 1470. This manuscript was incorrectly believed to have been the autograph of Virgil. Indeed, the *incunabula* printed in Venice in 1472 refers to it as derived from the Virgilian *exemplar*.[99]

The identification of the Carolingian script as antique had an important consequence, namely, its adoption as a humanistic script and the ensuing eclipse of Gothic. This situation arose from preference for the Italian tradition, in which the Gothic script was not firmly established. Perhaps a greater factor in this development was a deep-seated and growing sentiment in favor of reviving the script used by the ancient Romans.

Petrarch had already tried to assimilate the style of script, which in his day would have been essentially a semi-Gothic, into the Carolingian. He had himself become familiar with that style from a codex that is still extant, containing the commentary of Augustine on the Psalms. This volume was a gift from Boccaccio in 1355,[100] and in a letter that he wrote to Boccaccio shortly after receiving it, Petrarch praises the dignity, the sobriety, and above all the "solemnity of the older kind of writing" (*vetustioris litterae maiestas*).[101] In another letter addressed to Boccaccio a few years later, Petrarch is critical of the modern style of writing, contrived and lavish, and praises the "sober and clear style, which is easily accessed by the eyes."[102]

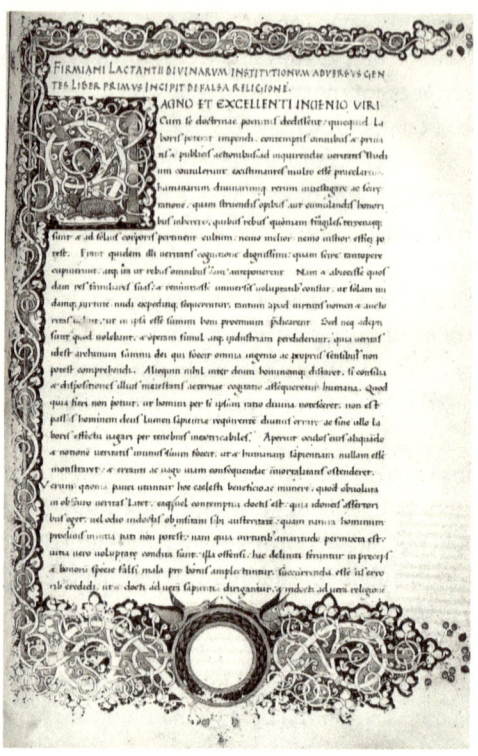

FIGURE 4.2. *Example of humanistic script (minuscule) with decorated initial (ca. 1425–1460). By Guglielmino Tanaglia. Schoyen Collection, ms. 1369 (Lucius Caelius Lactantius*, Divinarum institutionum adversus gentes*). Florence, ca. 1420–1430.*

The new script was developed at Florence in the circle of Salutati, a group that included Poggio and Niccoli. As restored, it took the form of the Carolingian minuscule supplemented with Roman capitals. Niccoli developed also a cursive version. From Florence the use of this new style of writing gradually spread throughout the Italic peninsula. Poggio himself popularized it in the Curia and then, at the Council of Constance (1414–1418), exposed it to a wider and more influential audience. The introduction of humanistic script is the most striking novelty in book production of the fifteenth century. Yet it was not the only important development, for suddenly there appeared a technique known as *mise-en-page*, that is, the layout of a page, which could, for example, feature multiple columns or the inclusion of a decorated initial (fig. 4.2) or a historiated initial[103] (fig. 4.3).[104] Taken as a whole, the new humanistic script appears to be functional in terms of the dominance of visuality that occurs during the Renaissance, a development one finds pronounced in the fluid forms that characterize Renaissance art.[105]

Humanistic script was also influential for many early printed books, known as *incunabula*. Although the Gutenberg Bible, published in Mainz, was printed in Gothic characters (fig. 4.4), in the following decades Italian printing dominated. Particularly influential were the typographical characters developed in Venice by Aldus Manutius (1449–1515), who also introduced the cursive form. Gothic characters were used, but they remained dominant only among the Germans.[106]

Valla's goal was to revive classical Latin, for he believed that it had been corrupted under Gothic influence. To do so, he undertook the ambitious project of composing his *Elegantiae Linguae Latinae*, which, though at the time was among the most significant and influential works of humanistic culture, had very little appeal later. One can find this treatise among his *Opera Omnia* printed in Basel

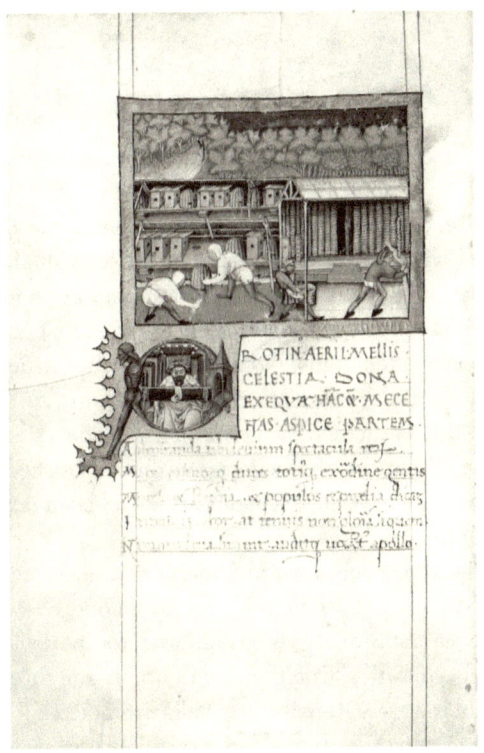

FIGURE 4.3. *Example of historiated initial. By the Master of the Vitae Imperatorum. Ms. Rawlinson G. 98, folio 49 verso (Virgil, Georgics). Milan, mid-fifteenth century. Oxford, Bodleian Libraries.*

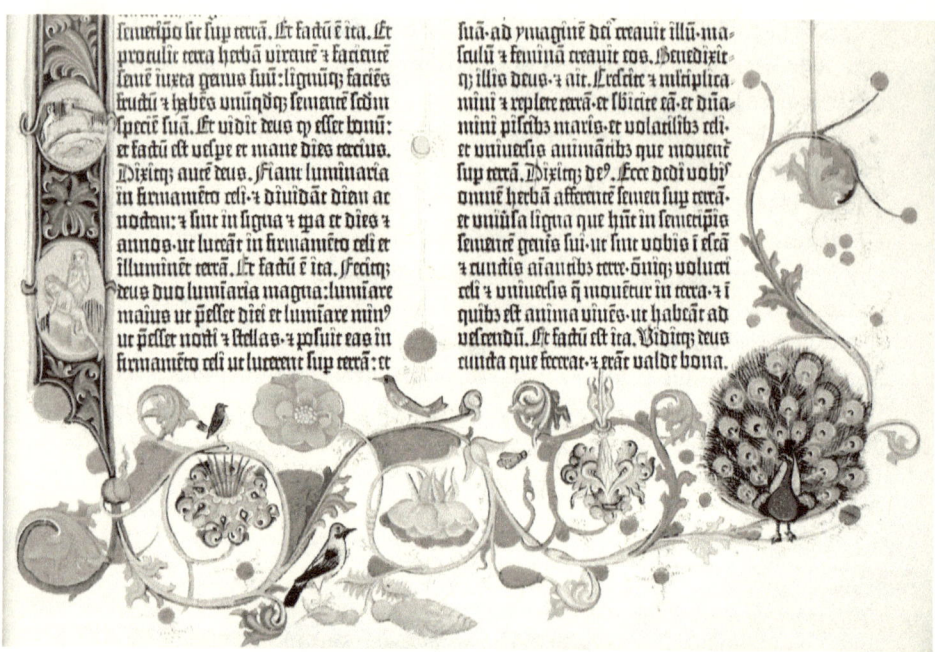

FIGURE 4.4. *Gutenberg Bible* (detail). *1455. London, British Museum.*

in 1540, a work rarely found in its own freestanding edition. That treatise's decline in popularity was perhaps engendered by the early Renaissance scholars' misguided attempt to place Latin language and literature not only in primary position, but in one exclusive of the vernacular. While modern scholars, such as Eduard Norden, have often quipped that Valla's attempt to declare Latin's primacy was mere "fantasy,"[107] the notion of Latin's centrality did not originate with Valla. Indeed, a century earlier Petrarch had made a similar proposition, as he and other early humanists grew impatient and dissatisfied with the clumsy Latin of medieval authors, whose wooden style had been dominant well into the fourteenth century.

Even this earlier movement toward Latin purity, however, was not new. A similar impulse had arisen in the ninth century among Carolingian scholars, when the recovery of traditional grammar allowed for the development of standardized linguistic rules that would govern Latin usage throughout the centuries to follow. In the early Renaissance, Valla promoted a fresh undertaking, which enabled the Latin of the humanistic age to find its inspiration in Ciceronian style. This new style would prove to be useful for writing and for teaching in schools.

The humanists' widespread commitment to the restoration of classical Latin encompassed a vision of a privileged role for this new brand of Latin. Unlike the age in which the Carolingian scholars had operated, the historical and cultural context of the early humanists was one in which the vernacular languages found their own literary traditions. The Tuscan dialect, for example, was well represented by Dante, Petrarch, and Boccaccio. Yet Latin's unique status was seen by these scholars as an important aspect of the continuity between the learning of the Middle Ages and the budding of humanistic thought. The medieval viewpoint is evidenced by Dante in his *De vulgari eloquentia* (1303-1305), in which he assigns to Latin a grammatical characterization superior to other languages. The Latin language was envisioned as a kind of remedy or antidote to the linguistic disorder with which God punished humankind following the events involving the tower of Babel. In the medieval treatises, therefore, Latin is synonymous with grammar, especially in terms of its sociolinguistic character. The mastery of Latin thus became the prerogative of the clergy and *litterati*, while the *idiotae* and laity had their own dialect.[108]

This view of Latin as the quintessential grammatical language was related to the notion in antiquity of the *diglossia* (Latin and the vernacular) that characterizes medieval culture. In the *Divine Comedy*, Dante has Virgil speak in the Lombard vernacular,[109] as Dante believed that the vernacular was already spoken in antiquity and that even in ancient Rome Latin was the language of intellectuals. This medieval conception of Latin was co-opted also by Petrarch and Salutati as a part of their program for the revival of classical Latin. Just over half a century after Petrarch's death, this issue would come to a head in Florence (1435) in a famous debate between the secretaries of Pope Eugenius IV. This event, held at the seat of Santa Maria Novella, was attended by a number of leading intellectuals, including Poggio, Bruni, Antonio Loschi, the historian Flavio Biondo, cleric and humanist Andrea Fiocchi, and papal secretary and architect Leon Battista Alberti.[110]

In the pope's absence, the main event was a debate that would only be settled in the sixteenth century when Tuscan philologists came to consider Latin a "dead" language.[111] Biondo and Bruni published their countervailing positions in two letters written in the weeks that followed the debate.[112] While Bruni advanced a positive understanding of continuing the medieval maintenance of Latin, Biondo proposed the use of vernacular languages as merely deviant forms of the Latin spoken by the ancients.

The debate among the humanists over Latin and the vernacular was far from settled with this signal event. Just fifteen years later, Guarino Veronese would resume the dispute in a letter addressed to Leonello d'Este.[113] While referring to the arguments of Biondo, Guarino draws a sharp contrast between the vernacular and cultured Latin.[114] In his reconstruction, Guarino avails himself of Isidorus, who had drawn attention to a corruption of Latin in the imperial age.[115] Disputing the view of the grammatical character of Greek and Latin, Guarino cites his experience in Constantinople, where Greek remained a language learned naturally, not simply from schoolbooks.

Poggio proffered similar arguments in his *Disceptatio* in 1450, as did Francesco Filelfo in two letters addressed to Francesco Sforza a year later. In these writings a distinction between the literary Latin and the vernacular is delineated, with Filelfo offering a new designation for *Litteratura* and *Latinitas*. Despite the diversity of positions, the intent to revive the language remained widespread. To achieve it, cultured and literary sources were often cited, and these were rife with classical references that would help legitimize the claim for the reestablishment of Latin as the language of culture.

Among those who were in favor of Biondo's argument, Leon Battista Alberti (1404-1472) was the only person who also consistently maintained the literary dignity of the vernacular. In 1441 Alberti, a humanist, mathematician, and architect,[116] promoted a *certame coronario*, a contest that took place in Florence intended to reward a composition in the vernacular on the theme of friendship. No award was assigned because the jury, formed by the apostolic secretaries of Eugenius IV, did not deem any of the compositions presented worthy of the prize. While the unfortunate result of the contest, whose very existence could be seen as a confirmation of humanistic ideas, perhaps presented Alberti with a minor blotch on his escutcheon, he nevertheless went on to prepare a grammar of the Tuscan dialect. This grammar demonstrates the assumption that all languages, not only Latin, are governed by rules of grammar.

Valla's position on the vernacular question, which sparked such contrasting opinions, is less clear. Like Biondo, Valla seems to have attributed a decline in Latin to the Gothic invasions of the fifth century. He thus takes a historical approach to the issue of Latin's authority. Regarding the language's future, however, it is more difficult to establish Valla's viewpoint without distinguishing it from that of Bruni. Though Valla conceives of Latin as the favored language in literature and culture, like Bruni he is nonetheless amenable to the coinage of neologisms aimed at adapting Latin to the demands of contemporary culture. The historical approach, insofar as it pertains to the decline in Latin that the Gothic incursions ostensibly brought about, does not seem to have affected his views about the necessary evolution of Latin into the Romance languages. Rather, he seems to privilege classical over medieval Latin, while keeping an open mind about Latin's future, acknowledging that the medieval ossification of the language is precisely what the humanistic restoration sought to remedy.

Valla also reveals a significant historical-linguistic sensibility. In his *Elegantiae Linguae Latinae* he considers the transmission of ancient authors and carefully marks the stylistic differences between Plautus, Cicero, and Quintilian, of whom he favored the last in terms of literary elegance. Still, he never firmly attaches the idea of a perfect form of Latin to the imitation of a single author. This opinion is reflected in a disagreement that arose between Valla and Poggio, who markedly favored Cicero. Their erudite tête-à-tête evolved into a spirited debate even though they both shared the conviction that Latin had a role fundamental to Renaissance literature. Valla, however, envisioned a language that would draw upon classical forebears not narrowly but liberally. For all Valla's efforts, the position of Poggio and Biondo ultimately prevailed, opening the road to slavish imitation of the Ciceronian model.

This specific compositional practice would grow widely in the second half of the fifteenth and the first half of the sixteenth century. Yet the debate about style did not end here, as Angelo Ambrogini, known as Politian (1454-1494), would counter such Ciceronianism, replying to critics who had turned against him that he was his own person, not Cicero.[117] When a generation later Erasmus returned to Italy from Rotterdam (1528), he would in fact ridicule, in his work entitled *Ciceronianus*, the sort of Ciceronian orthodoxy that was still affirmed in humanistic circles.

The restoration of classical Latin was important for the rediscovery and reassessment of ancient Latin literature. Sicco Polenton, as we have seen, posited Juvenal's work as the stylistic upper limit for classical literature, but he thought also that Latin literature had been reborn in modern times with Dante, Petrarch, Boccaccio, and Mussato, and included these authors in his history. In any case, Sicco Polenton's study looked to historical models to establish literature's foundational aspects.

In the subsequent period, numerous authors would also seek to reconstruct the development of individual genres. In one of his inaugural lectures (published later as *Nutricia* in 1485-1486) in the courses held in the Florentine Studium, Politian rehearses the history of ancient poetry. Giglio Gregorio Giraldi of Ferrara (1479-1552) inserted Greek literature into literary history more systematically through his dialogues entitled *De Historia Poetarum tam Graecorum tam Latinorum*.[118] Giraldi's account of literary production and others like it encompassed the primary genres of antiquity, especially those revived after the medieval period.

Petrarch, for example, had relaunched the ancient epic in his *Africa*. With his *Familiares*, he rekindled the legacy of the epistolary genre that had disappeared since the ancient collections of Cicero and Pliny the Younger. Following Petrarch, collections of epistles would be offered by Salutati, Bruni, and others. Bruni in particular reprises the form of the Ciceronian dialogue in his *Dialogi ad Petrum Histrum*. The fifteenth century would witness the reconstituting of the genres of encomium and epigram. Classical models also provided a basis for the humanistic disciplines of historiography and biography.

The dramatic genre of tragedy, too, was rekindled. As noted above, Albertino Mussato wrote a tragedy in the tradition of Seneca, entitled *Ecerinis*. Meanwhile, the comic genre was reinvigorated through the activity of Pierre Paolo Vergerio, who wrote a play meant only to be read, known as an "armchair comedy," entitled *Paulus*. The humanistic recovery of performed comedy, lost at the close of antiquity, came only later with the first stage production (Plautus' *Menaechmi*) at Ferrara in 1486.

The massive Latin literary production in the course of the fifteenth century obscured the development of literature in the vernacular. Dante, Petrarch, and Boccaccio had written prevalently in Latin, and the importance that these authors attributed to their own production in the vernacular is not readily apparent. Petrarch, in particular, clearly gave greater weight to his Latin poem *Africa* than to his *Rerum Vulgaria Fragmenta*. A great number of humanists wrote almost exclusively in Latin. A few notable exceptions to this rule are Alberti and Politian.

THE RETURN OF GREEK TO THE WEST

Subsequent to these developments in Latin and the vernacular, a desire to know Greek literature arose. The first major proponent for the return of Greek to the West was Salutati. In 1395 he urged his own student, Jacopo d'Angelo da Scarperia (ca. 1360–ca. 1410), who was going to Constantinople, to pore diligently over Greek works, such as the Homeric poems, and to seek out much-needed Greek dictionaries.[119] In Constantinople Jacopo came into contact with Emmanuel Chrysoloras (ca. 1350–1415), inviting him to move to Italy to teach Greek in the Florentine Studium, of which Salutati played the role of sponsor, if only at a distance, owing to his governmental post. This followed upon a meeting some years before when Chrysoloras, as an ambassador, was in Venice on a diplomatic mission to seek aid for an ongoing war with the Ottomans. On that occasion, he taught a few Greek lessons to Roberto de' Rossi, a member of Salutati's Florentine circle. Accordingly, Chrysoloras welcomed the invitation issued officially by the Studium, and by 1397 he was offering Greek courses in Florence. A school of learning soon developed around him.[120]

Where Pilato had failed forty years earlier, within only three years Chrysoloras managed to introduce Greek thought and language to the West. After a short time, Chrysoloras went to Milan, where he began teaching. In 1414, in the capacity of diplomatic representative of the Greek Church, Chrysoloras would set out to the Council of Constance, during which time of service he passed away.[121]

By the time of his death, however, Chrysoloras had produced a number of good Greek students in Florence and elsewhere in Italy. Those students were thus able to replicate the experience for others, spreading the teaching of Greek throughout the peninsula. Chrysoloras' method, which his students adopted, was marked by a simplified approach to grammar and direct reading of texts. From his instructional notes he had compiled a complete Greek grammar, entitled *Erotemata*. That book enjoyed lasting success, and in 1471, roughly half a century after Chrysoloras's death, it was published as the first Greek grammar printed in the West. When in that same century Florence experienced an influx of Greek manuscripts from Constantinople, there were already in place a number of bright scholars capable of reading them.[122]

Niccoli and other representatives of Florence's cultural elite purchased these manuscripts as they arrived. Among those representatives, Chrysoloras's student Bruni was particularly well known, in part because he produced several Latin translations from Greek originals, including Saint Basil's *Ad adulescentes*, the *Hiero* of Xenophon, and Plato's *Phaedo*. Other Florentine translators include Jacopo d'Angelo, who first rendered into Latin the *Geography* of Ptolemy and some works of Plutarch. The Camaldolese monk Ambrogio Traversari (1386–1439) made an important contribution to patristic studies in that period by translating John Chrysostom, Basil the Great, and, in 1433, Diogenes Laertius's *Vitae philosophorum*.

Chrysoloras also trained students between Pavia and Milan, to which metropolitan hub he moved in 1400. There, over the course of the next two years, he collaborated with Uberto Decembrio, secretary to Gian Galeazzo Visconti, duke of Milan (1395–1402), to make a complete translation of Plato's *Republic*.[123] About the same time, he met Guarino Veronese (1374–1460), an important figure in the development of Greek studies in the first half of the fifteenth century. Between 1403 and 1408, Guarino lived in Constantinople in the service of the Venetian ambassador, perfecting his own linguistic competence alongside Chrysoloras's nephew Giovanni, and Chrysoloras himself, who had for a season returned to Constantinople.[124] In 1413 Guarino welcomed the proposal to teach Greek in the Florentine Studium, but soon abandoned the Tuscan city because of strong disagreements with Niccoli. While Guarino was teaching in Venice, Vittorino da Feltre (ca. 1378–1440) became his student. In 1423 Vittorino founded another successful school in Mantua. By 1429 Guarino had moved to Ferrara, where he enjoyed the protection of the Este family and established a school that would play an important cultural role for the next thirty years.

Although Greek schools in northern Italy were now more common, many scholars still saw fit to travel to the East, often as far as Constantinople, to acquire their own firsthand knowledge of Greek. For example, the Sicilian Giovanni Aurispa (1373–1459) was in Chios in 1413. In the course of that eastward journey, he was able to perfect his own linguistic competence in Greek. Significantly, Aurispa brought with him a considerable number of manuscripts from the Byzantine Empire to Italy with the intention of selling them. Aurispa's activity made an important contribution to the dissemination of Greek texts. Returning to Italy, he found himself in Rome by 1420, where he introduced Valla to the Greek language. In 1423 Aurispa returned to Constantinople, becoming secretary of the emperor John VIII Palaeologos, whom he accompanied on his next trip to the West. Soon, however—in 1425—Aurispa was invited to teach Greek at the Florentine Studium. He did not linger there for very long, as in 1427 he turns up in Ferrara, where he lived during the decades that followed, protected by the Este family, the same family that would offer sanctuary to Guarino when he arrived there just two years later.

Among his many contributions to Renaissance learning, Aurispa played a significant role in the search for both Greek and Latin manuscripts. Among the codices discovered by Aurispa, Athenaeus' *Deipnosophistae* stands out, itself a work that had remained virtually unknown in Byzantine culture. The manuscript in which it is found was brought by Aurispa to Constantinople in 1423.[125] It would later come into the hands of Cardinal Bessarion, whose collection would form the basis of Venice's principal library.

Other scholars were connected or indebted to Chrysoloras. Francesco Filelfo (1398–1481), who at the behest of the Venetian government had gone to Constantinople and lived there from 1420 to 1427, actually married into the family of Chrysoloras. In the following years, Filelfo returned to Italy. First he taught in Florence, where he eventually found himself quarreling with the powerful Medici family. As a result, he spent most of his subsequent career in Milan, where he worked for the Visconti and the Sforza families. Like Aurispa, Filelfo also contributed to the recovery of Greek codices, encouraging importation that intensified after both the sack of Thessalonica (1430) and the pressure that was increasingly exerted by the Ottomans over the Byzantine Empire.

Although he had not been a pupil of Chrysoloras, Giovanni Tortelli of Arezzo (ca. 1400–1466) also advanced Greek studies, going so far as to live in Constantinople from 1435 until 1437 in order to learn Greek. In 1449 Tortelli went to Rome, in the service of Pope Nicholas V, to help establish the Vatican library. Tortelli displayed his knowledge of Greek in his learned work *De Orthographia*, a monumental tome dedicated to the pope that dealt with the spelling of Latin words of Greek origin.

The students and heirs of Chrysoloras initiated one of the most striking literary phenomena of the fifteenth century: Latin translations of Greek texts. In the space of little more than a century (1397–1527), 154 Greek authors, with a combined total of some 784 works, were translated by 165 different translators.[126] Many of these, especially in the first phase, were translations of Plutarch's *Vitae parallelae*; individual *Vitae* were translated by Jacopo d'Angelo, Cencio de' Rustici, Bruni, Guarino, and others. As we saw earlier, Bruni also translated the works of Plato, Aristotle, and Demosthenes, while Guarino rendered Strabo and Isocrates.

Bruni was highly intentional in his work, even formulating a theoretical basis for his translations in the treatise *De Interpretatione Recta* (1420), which defines, on the basis of Cicero and Jerome, the criteria for proper translation. One can see in the humanists' translations a tension with their medieval forebears, who had generally rendered a work word for word. Bruni, who also took into account the lessons of Chrysoloras, argued the need to translate faithfully but using the syntactic forms and expressions typical of humanistic Latin.

A significant stimulus for translations from the Greek came from Pope Nicholas V, who commissioned and subsidized the production of numerous translations for the Vatican library (1451). Not only did Nicholas acquire a number of important Greek texts that he wished to keep in what he envisioned would be a public library, but he also wanted them translated into Latin to grant access to those without knowledge of Greek. Among the translations of that period, those of Niccolò Perotti and Theodore Gaza stand out. The former rendered Polybius, Epictetus, and Plutarch, while the latter translated the scientific works of Aristotle and Theophrastus along with the *Tactica* of Aelian.[127]

In the course of the fifteenth century, nearly all known Greek literature came to be rendered into Latin. Some works were translated many times by different authors. Such translated editions, in circulation in the second half of the fifteenth century, were published, and were often reproduced in the next century and appended to the editions of the Greek texts.

Thus, Latin translations played a central role in understanding Greek texts. Linguistic study, which had considerable success in the schools of Mantua, Ferrara, and others, did not permit the direct reading of the Greek manuscripts, except for those who had mastered linguistic competence — and very few had a strong command of the language.[128] Only thanks to these translations did those who had some education have the opportunity to access such authors as Plato, Plutarch, and Lucian. Such translations also played an important role in the first phase of the history of printing. In the sixteenth century, translations of Greek texts were found in vernacular languages, such as Plutarch's *Vitae parallelae* and *Moralia*, translated into French by Jacques Amyot between 1559 and 1572. Many of these translations into the vernacular, it should be noted, were made on the basis of the Latin, and not the original Greek.

A significant milestone in the spread of Greek studies in the West occurred in the 1430s, as Pope Martin V gathered a council to settle the theological disagreements with the Eastern Church. The council first met in Basel (1431), then in Ferrara (1438), and ultimately in Florence (1439). During the reign of the Byzantine emperor John VIII Palaeologos, numerous Orthodox prelates attended that council, which sanctioned a short-lived union between the Catholic and Orthodox churches. That union (known as *Laetentur caeli*) was swiftly dissolved after the Orthodox delegates returned home.[129]

One member of the Florentine enclave was the influential Byzantine Bessarion (1403-1472), an Orthodox archbishop of Nicaea. After the council, he joined the Catholic Church, was appointed cardinal, and played a leading role in Pius II's papacy.[130] As the author of theological writings and as a considerably cultured individual, Bessarion was a patron of Greek exiles and the proponent of a humanistic circle active in Rome between 1465 and 1470. He acquired a large number of codices, which he gave to the Republic of Venice in 1468; that collection became the basis for the Marciana library.

George of Trebizond (1395-1472) also immigrated to Italy, in his case from Crete, and rapidly became so deft a Latinist that he went on to be a renowned translator, particularly of Aristotle, valued by the popes Eugenius IV and Nicolas V. In Rome a dispute with Poggio arose over translating these authors, which in 1452 escalated to fisticuffs between the two men.[131] As an Aristotelian, Trebizond sharply debated Bessarion about Plato's value and wrote the *Comparatio Philosophorum Aristotelis et Platonis* (1458), in which he denounces the anti-Christian character of Platonic philosophy. Bessarion replied to Trebizond with a work pointedly entitled *In Calumniatorem Platonis (Against Plato's Critic)*. After the fall of Constantinople in 1453, Trebizond went on a peace mission to the new monarch of Byzantium, Mehmet II.[132] Upon his return to Rome, Trebizond was imprisoned for a few months in the Castel Sant'Angelo.

Another important fifteenth-century Greek who came to the West was Theodorus Gaza of Thessalonica (1415-1475), who would become a translator of several Greek texts. His activity was primarily centered in Rome, whither he came at the behest of Pope Nicholas V. Gaza undertook for the pontiff the vast work of translating Greek works, focusing on the texts of the Peripatetic school, chiefly Aristotle and Theophrastus. Gaza also had ample contact with Cardinal Bessarion.[133]

Other exiles followed, particularly after Constantinople's fall. Prominent among this group was Giovanni Argyropoulos (ca. 1416-1487), a professor of Greek in Florence in 1456. Demetrius Chalcondyles (1423-1511), the first professor of Greek at the University of Padua in 1463,[134] played an important role in the transfer of much of the Byzantine library to the West after the Ottoman sack of that city.[135] Lorenzo de' Medici, too, found himself in a position to acquire a number of Greek books after Constantinople's fall.

Of those not born in Greece or Greek-speaking places, Politian, some aspects of whose contribution to the intellectual climate of this period will be discussed further, is said to have been the first to gain a complete mastery of classical Greek. So adroit was he that, beyond those that he wrote in Latin and Italian, Politian composed many poems in ancient Greek. In 1485, in the *Oratio in Expositione Homeri*, he proclaimed that Athens was not in barbarian hands but had been transplanted to Italian soil in Florence.

HUMANISTIC CULTURE AND EDUCATION

The growth of humanistic culture was mediated by the success of new ideas encountered in the royal courts and among other positions of leadership. New needs, unmet in late medieval culture, found an effective system of training and tools for communication in the humanistic schools.

In Florence, humanism arose in a municipal context. Among its early chief proponents was Salutati, chancellor of the Florentine republic from 1375 to his death in 1406. Bruni, who had by then enjoyed a long curial career, took up that same chancellery in 1427. Salutati and Bruni, facing the need to defend Florence from the expansion of the duke of Milan, Gian Galeazzo Visconti, and to justify the Florentine constitution, revisited and bolstered the ancient Roman ideals, which provided precedent foundational model for Florence's form of government.

It is not surprising that Plutarch's biographies of Republican Rome's leading figures stand out among the first translations of Greek works created in Florence in the first years of the fifteenth century. While some might deem this propagandistic, the use of these ancient sources seems on the whole to have been chiefly didactic, connected to civic responsibility rather than empire building.[136] An example of this can be seen in 1400 when Salutati responds in his treatise, *De Tyranno*, to a question posed by a Paduan student, Antonio Dell'Acqua. The student's particular concern was the pain Dante assigns in *Inferno* to Caesar's assassins. Salutati takes a generally moderate position, denying the legitimacy of Brutus' and Cassius' actions.

Even the subsequent evolution of the Florentine political situation, in which humanists eventually had to find a way to conciliate and cooperate with the Medici regime, shows the flexibility of humanistic culture. That culture offered, rather than a monolithic ideology, an adaptable model of civic identity, one that could work with the variations in government — even monarchy — that asserted themselves during the fifteenth century.

Central to the humanistic program was the idea of *studia humanitatis*, a term borrowed from Cicero and to some extent also from Gellius.[137] The term is already used with programmatic value by Salutati and soon after by Bruni.[138] Whereas the word "humanist" means, in the fifteenth century, the student or scholar engaged in the *studia humanitatis*, it was only at the beginning of the nineteenth century in Germany that the term *Humanismus* was coined to designate the entire historical and cultural phenomenon.[139]

Cicero's sense of *humanitas* encompassed a strong sociocultural connotation. In *De oratore*, he refers to the education of Rome's "free" men, trained in the *artes liberales*.[140] This last expression is vital to a proper understanding of Cicero's educational vision, especially when considered in light of his comments in the *Pro Archia*, where he advances not so much the idea of education being for freeborn men as the notion that the arts are fundamental to a particular way of thinking. This is fitting because in the *Pro Archia* Cicero makes his appeal on behalf of a Greek poet, thus revealing that Greek poetry and arts were already in Cicero's day representative of the classic ideal. Cicero proffers a eulogy of culture less influenced by the Roman social context than by the notion of *ars gratia artis*.

The importance of that single work of Cicero, certainly more than any other of his speeches, cannot be underscored sufficiently. Petrarch's discovery of the *Pro Archia* in Liège allowed it to become accessible to many of the new humanists and, by 1370, in particular to Salutati, whose understanding of that slender oration was decisive for the development of the humanistic program.[141]

The effectiveness of humanistic training has been well expressed by François Rabelais (ca. 1484-1552), whose third book of *Pantagruel* (1546) contains a letter from Gargantua to his son, Pantagruel, a student in Paris. In that letter Gargantua compares his son's training to that which he himself had experienced. He tells Pantagruel,

> The time was not so favorable to letters and as comfortable as nowadays, and I had no abundance of tutors such as you have had. The time was still dark and the infelicity and calamity of the Goths had destroyed all good literature. But, thanks to divine goodness, light and dignity have been returned to the letters in your age and one can see such progress that today I would hardly be promoted to the first class of schoolchildren.... Now all disciplines are held in honor, languages especially: Greek — it is a shame that without it a person can be called "scholarly" — Hebrew, Chaldean, Latin; and wonderfully elegant and correct printed editions are in use that by divine inspiration were discovered in my day. By contrast, at that time diabolical counsel gave birth to artillery. But today, the world is full of educated people, highly learned tutors, very large libraries, and I think that even in the days of Plato, Cicero, or Papinian, there were as many opportunities to study as there are today.[142]

The new educational system, to which Rabelais calls attention in these lines, had been developed a century before in Florence, where pedagogy was one of the main interests cultivated by the humanists.

Works that appeared in this period reflect just such a renewed educational emphasis. Vergerio wrote *De Moribus et Liberalibus Studiis Adolescentiae* in 1402-1403, dedicated to Ubertino, the twelfth son of the lord of Padua, Francesco Novello da Carrara. Guarino translated the *De liberis educandis* of Plutarch in 1411. At about the same time, Bruni translated Saint Basil's admonition to learn Greek literature, which he entitled *Ad adulescentes*, and, in 1424 he addressed the *De studiis et litteris* to Battista Malatesta, wife of the lord of Pesaro, Galeazzo Malatesta.

As an alternative to traditional education based on grammar, the hitherto prevalent manual of style known as *Ars dictaminis*,[143] and the medieval system of the liberal arts, humanistic education taught Latin through direct reading of the classics, followed by the teaching of poetry, history, and the moral philosophy of Cicero and Seneca. That didactic program was inspired by Cicero's *De oratore* and Quintilian's *Institutio oratoria*, in which the learning of various disciplines aims at the formation of the citizen, who in the context of fifteenth-century Italy would become not only the government official or the diplomat, but also the banker, the man of commerce, even the prince himself. Indeed, many political protagonists of the time had training of this type and served as patrons.

To take but one example, near Mantua, Vittorino da Feltre founded a school for educating the children of famous nobles. Among these was Federico da Montefeltro (1422–1482), who would become the duke of Urbino, perhaps best known today from the portrait of him by Piero della Francesca (fig. 4.5). Protected by the Gonzaga family, Vittorino's Mantuan school was known as La Casa Zojosa, "The Joyous House," sometimes called School of Princes because of the noble sons educated there.[144] Though it emphasized primary and secondary education, the school also encompassed the teaching of Greek in a context that brought together students from different social strata. Participation in the life of the school was central to its educational community. This approach anticipated how humanistic education led the way from knightly culture, founded on honor, to a society in which good manners and behavioral models were central. That development would be codified in the first quarter of the sixteenth century by Baldassare Castiglione, whose dialogue entitled *Cortegiano* (*Book of the Courtier*) would be published by the Aldine Press in 1528. It would be followed, just a few years later, by Giovanni Della Casa's *Il Galateo, overo de' costumi*, a book touching on proper Renaissance comportment that was published in Venice two years after the author's death in 1558.

Guarino funded Ferrara's Studium, which essentially served as its university. The program of study there included Greek, which was taught on the basis of Chrysoloras's grammar. The studies of the Guarinian school included mythology (through close reading of Ovid's *Metamorphoses*), rhetoric, composition, and the philosophical works of Cicero and Plato. The sciences were not neglected, but included readings of Pliny the Elder, Strabo, and other ancient authors who treated scientific subjects.

Previously existing universities, which long remained strongholds of Aristotelian and scholastic thought, were slower to accept new humanistic ideas. In Rome's Studium, and in those newly formed, such as Guarino's at Ferrara,

FIGURE 4.5. *Piero della Francesca, portrait of the duke of Urbino, Federico da Montefeltro (1490–1492). Tempera on panel. Florence, Uffizi Gallery.*

humanism became dominant toward the middle of the fifteenth century. Still, in the older universities, traditional ideas stayed tenaciously in place. Thomas More, in a letter from 1518, laments the prejudices of the Oxford dons and denounces the practice of reading the Fathers of the Church without knowing Greek:

Anyone who boasts that he can understand the works of the Fathers without an uncommon acquaintance with the languages of each and all of them will in his ignorance boast for a long time before the learned trust his judgment.[145]

The modern scholar Étienne Gilson (1884-1978) once wrote that the "difference between the Renaissance and the Middle Ages is not a difference by addition, but by subtraction. The Renaissance ... is not the Middle Ages plus humans, but the Middle Ages minus God."[146] Clever as this *bon mot* sounds, it is not entirely true, as the idea of "Christian humanism" finds its roots in the patristic synthesis of classical culture and Christian doctrine,[147] thus representing an important strand of humanistic culture. Nevertheless, Renaissance humanism, itself not explicitly anti-Christian, provided not only a stimulus for secularization[148] but also an impetus for reformation. Long before the Reformation, Catholic thought, too, had offered ample stimuli for Christian humanism, which the writings of the Church Fathers anticipated.

Indeed, certain debates and ideas associated with various ecclesiastical schools of thought were echoed in the concerns of Salutati's circle, particularly in regard to his interest in ancient pagan poets. One member of the circle provoked the wrath of the Camaldolese monk Giovanni of San Miniato (1360-1428). In 1405, Salutati responded forcefully to Giovanni's criticism, writing in a letter, "if one searches for the truth in the poets or other pagan authors, one does not thereby cease to walk upon the streets of the Lord" (*Epistulae* 12.20).[149] A further attack upon Salutati came at about the same time from the Dominican monk and later cardinal Giovanni Dominici (1356-1419). Dominici responded to Salutati's epistle with a work entitled *Lucula Noctis*, published in 1405, in which the prelate wrote that it is more useful for Christians to plow the ground than to read the books of the pagans.

A more aggressive attitude was assumed by Bruni against Ambrogio Traversari (1386-1439), who represented another strand of Christian humanism. Having acquired a knowledge of Greek from Chrysoloras, Traversari devoted himself to the translation of patristic texts. He also rendered the work of Diogenes Laertius, doing so only at the behest of Cosimo de' Medici, as we learn from an epistle written by Traversari himself. In 1431 Traversari became general of the order of the Camaldolese, and in 1438-1439 was among the members of the Council of Basel-Ferrara.

In response to staunch criticism from Traversari, Bruni promulgated a scathing retort in the form of a pamphlet in 1417, *Ad Hypocritas*, based on Cicero's *Ad Pisonem*. In it Bruni contrasts the pristine origins of Christianity with the hypocrisy of the contemporary friars. The pamphlet highlights the opposition between the active life, represented by Bruni and other civic-minded humanists, and the contemplative life that Traversari embodied.[150] The conflict between these two Renaissance figures reveals a delicate balance, which the first humanists had to navigate. It is no coincidence, in this regard, that one of Bruni's earliest translations was Saint Basil's *Ad adulescentes*, which is one of the Christian texts most open to pagan literature.[151]

In his *De studiis et litteris*, Bruni critiques the view of certain readings as ill-suited for the Christian faith. In this work, he suggests that his interlocutor actually read the ancient poets instead of merely dismissing them. To the objection that "love and lust are discovered in them,"[152] Bruni replies that erotic descriptions in Homer are treated symbolically and that there are numerous examples of virtue in Homer, such as "Penelope's practice of most faithful chastity as she awaited Ulysses."[153] He adds that even the sacred books narrate questionable events, such as the "love bordering on insanity"[154] of Samson, "David's romantic love for Bathsheba, his evil behavior toward Uriah, Solomon's murder of his brother and his greatly numerous flock of concubines."[155] Significant, too, is the response of Bruni to those who object to the pagan faith of the ancient authors:

> But did they not perhaps live according to their own morality? As if honesty and the seriousness of morality were not then the same as they are now![156]

Such a response highlights the distance rather than the opposition that occurs in this period between moments of secularization and the persistent religious impulse.

Some twenty years later, Valla faced more serious issues when he questioned the apostolic origin of the Creed. He was tried for heresy in Naples by Friar Antonio da Bitonto. Some of the Church's hostility against him may have continued from a few years before, when Valla wrote a famous pamphlet in which he demonstrated the falsity of the "Donation of Constantine," a document that traced the temporal power of the popes. Nicholas of Cusa (1401-1464), who would later become a cardinal, had expressed doubts about the authenticity of the "Donation" a few years earlier, at the Council of Basel (1431). In his treatise, Valla expounds in great detail the document's spuriousness, as he would also do a few years later in a pamphlet[157] that contested the authenticity of the correspondence between Seneca and Saint Paul. While Valla never completely lost standing with the Church—indeed, at the behest of Eugenius IV's successor, Nicholas V, he would translate many a document for the newly formed Vatican library—he nevertheless anticipates Luther and the reformers who, less than a century later, would boldly challenge papal authority.

The trial in Naples never reached a conclusion, as it was broken off by the intervention of Alfonso V of Aragon, who in 1444 was king of Naples and patron of Valla. Valla's treatise on the "Donation" was of particular interest to

Alfonso because he found himself in conflict with Pope Eugenius IV, who was hostile to his accession to the throne of Naples. This hostility was reversed by his appointment of Nicholas V as apostolic secretary. Returning to Rome, Valla taught at the Studium while continuing his research on the Greek text of the New Testament. This line of research irritated traditionalists. One aspect of his research, preserved in his *Adnotationes in Novum Testamentum*, reveals an ambitious effort to verify Jerome's Vulgate based on Greek patristic textual citations and documentation. That work would not, however, be published until 1505, nearly five decades after his death. Similarly, the treatise on the "Donation" was not formally published until the Lutheran Reformation.

Both Pope Nicholas V (1447-1455) and Pope Pius II (1458-1464) actively protected the humanists from traditionalist attacks. The papacy of Paul II (1464-1471), however, was not so sympathetic. In February 1468 Paul II incarcerated Pomponio Leto, Bartolomeo Platina, Filippo Buonaccorsi, and other members of the Roman humanistic school on charges of heresy, pederasty, and even conspiracy to assassinate the pontiff. Though details of the historical account of the assassination attempt remain uncertain, the cultural implications are clear enough. In a report given by the Milanese ambassador, the defendants were accused of reading "Juvenal, Terence, Plautus, Ovid and other books," which were deemed morally questionable.[158] They defended themselves by asserting that not only does "Juvenal demonstrate vices and scorn them," but he also "makes the reader aware of them and helps each reader to learn a lesson."[159]

Although some of the defendants were subjected to torture in the Castel Sant'Angelo, they were released in less than two years' time. Buonaccorsi fled to Poland, where he served the classical scholar Gregory of Sanok, bishop of Lwów (now known as Lviv). Leto resumed his activities as a teacher in the Roman Studium after his discharge, while Platina was invited by Pope Sixtus IV to head the Vatican library.

It is not entirely clear which activities of Leto and his followers had aroused the suspicions of the pope. Perhaps it was their sympathy for Platonism, which could have been inferred from Leto calling their scholarly circle the "Academy." In previous decades, accusations of neopaganism had been lodged against Gemistus Plethon.[160] One of Plethon's admirers, Sigismondo Malatesta of Rimini, was attacked by Pope Pius II, despite his humanistic leanings.[161] Suspicions of neopaganism would also dog the Florentine Neoplatonism of Marsilio Ficino. Further anti-humanistic feelings emerged in Florence in 1497, evidenced by Savonarola's infamous book burnings in the Piazza della Signoria.

In general, however, the humanism of the fifteenth century remained opposed to openly anti-religious sentiments. Bruni and Valla never placed themselves outside Christendom, nor did Leto, even after his detention in the Castel Sant'Angelo. The same can be said of such figures of the sixteenth century as Desiderius Erasmus and Thomas More. Epicurean, skeptical, and materialistic positions along with professions of libertinism that would find a certain tolerance in the full Counter-Reformation at Venice only emerged in the late Renaissance.[162]

LATE HUMANISM

In Renaissance culture, new discoveries and renewal of interest in classical learning encountered rivalry, controversy, and competition. As humanism waxed central, the discovery of ancient codices by individual scholars offered them a modicum of fame, the possibility for or the culmination of a career in a noble court, and personal enrichment. The keepers of the monastic libraries realized the value of the collectable manuscripts that they oversaw and thus made indiscriminate searches of their collections more difficult. Accordingly, they tried to keep their collections intact. On those occasions when they were "pillaged," they demanded adequate remuneration. In light of their success, it is rather difficult to reconstruct the history of the recovery of many of the codices.

A case in point is that of the codex that at the beginning of the fifteenth century was kept in the monastery of Hersfeld, Germany, in which four texts were transcribed. Beyond Tacitus' *Agricola*[163] was a particularly important triad of single-copy texts: Tacitus' *Germania*, his *Dialogus de oratoribus*, and Suetonius' *De grammaticis*. Poggio discovered this manuscript in 1425 but did not manage to obtain the codex, although he negotiated with the monastery's abbot. Some thirty years later, that very manuscript turned up in the hands of Enoch of Ascoli, who in about 1430 had been one of Filelfo's students in Florence. In 1451, at the behest of Pope Nicholas V, Enoch had gone to northern Europe in search of manuscripts. He returned in the fall of 1455 with a substantial harvest, among which the most prominent was this copy of Tacitus and Suetonius.

The death of Nicholas V in March of that same year was unfortunate, as his successor, Pope Callistus III, had little interest in acquiring Enoch's trove for the Vatican library. Thus, Enoch privately sold the codices that he had retrieved. An unfortunate consequence was the sale of the Hersfeldense manuscript, which, after being copied, was sold in pieces for higher prices. Of it only one portion, a section of the *Agricola*, survives, itself having been incorporated into another manuscript gathered a few years later by the humanist Stefano Guarnieri.[164]

Obscure circumstances characterize the origins of other texts. One persistent problem is the difficulty in distinguishing between forged and authentic documents, as forgeries were often produced for commercial purposes or even out of professional rivalry. Prominent among such unsolved mysteries is the case of a section of book 8 of Silius Italicus' *Punica*. Lines 144-225 are absent in the known manuscript tradition but nevertheless were published by the humanist Giacomo Costanzi in 1508. These lines, Costanzi averred, were copied by him from a manuscript that has never been identified. Modern editors of Silius have not reached consensus about the authenticity of this section of the text, known as the "Additamentum Aldinum."[165] In the preface of his recent edition (1987), Joseph Delz asserts that the lines are spurious additions made by Costanzi or other Renaissance humanists and that they came to be a part of the accepted textual tradition merely by virtue of their long-standing position in the text.[166]

The origin of a fifteenth-century manuscript containing Phaedrus' *Fabulae* is also unknown, but it is widely accepted as a genuine part of the manuscript tradition. It was first reported in the second half of the fifteenth century by Niccolò Perotti and is thus called the Appendix Perottina. Despite the assumed veracity, both Perotti and Leto in their commentaries cite numerous fragments of otherwise unattested Latin authors, giving the impression that these citations and the manuscript itself are forgeries.

One case of forgery is clearly established: *De Magistratibus et Sacerdotiis Romanorum*, a text on Roman magistrates and priesthoods. Although this text was first falsely attributed to Fenestella, a Roman historian whose floruit was in the imperial period, it is indeed the work of Andrea Fiocchi. It was spuriously published under Fenestella's name in 1510, but by 1563 a fresh edition had appeared, properly published under its true author's name.

Moreover, one finds often unreliable references to Latin works of which few other traces remain. Among these are the grammatical work of Palaemon, the commentary of Probus on Persius, the missing part of the fifth decade of Livy, the commentary on Cicero by the grammarian Grillio, and the *De virtutibus* of Cicero, known also to the Frenchman Antoine de la Sale. Owing to sundry circumstances, it may be the case that these texts survived into the Renaissance and have since been lost, but it is also quite likely that these claims are the vaunting of humanists, borne out of rivalry and the desire to give one another false leads.

The fifteenth century was marked by the growth of personal libraries. This came about both by the acquisition and by the copying of manuscripts in monasteries, as is particularly clear in Florence as a result of Poggio's activities at Monte Cassino, as well as by the further dispersal upon the death of early patrons, such as Petrarch. Beyond the Western monastic collections, a number of Greek texts arrived in Venice from Constantinople, facilitated by Chrysoloras. Two particularly noteworthy examples of this merging and movement occurred in Dominican libraries. In Venice in 1437, Niccoli merged a collection he inherited from his mentor Salutati with the Dominican library of St. Mark's Square. Leonardo Dati (1360-1425), upon his death, bequeathed his books to Santa Maria Novella in Florence, the same city where, a few decades later, Lorenzo de' Medici merged his collection with the Biblioteca Laurenziana.[167] Meanwhile, in Rome, Tortelli became the director of the Vatican library. It was not uncommon for the patronage of individuals to produce remarkable collections.[168]

The first printed book was produced in Mainz, Germany, by Johannes Gutenberg, between 1452 and 1455. It was a two-volume work of the Latin text of the Bible. A mere 180 copies were produced. In 1470 there were only nineteen European cities in which printing presses were in operation, but by 1550 that number had grown to 255. The advent of the printing press led to a decrease in costs of both supplies and paper, resulting in an unprecedented dissemination of books. During the first fifty years of the existence of the printing press, more books were produced than had been copied in the previous millennium.[169] These books — that is, those of the fifteenth century, known as *incunabula* (Latin for "swaddling wraps")[170] — represent printing's first phase.

The first classical work printed was the *De officiis* of Cicero, published in 1465 in two different editions, in Cologne by Ulrich Zell and in Mainz by Johann Fust and Peter Schöffer, two apprentices of Gutenberg. In the years that followed no other classical author was more widely published than Cicero. By the end of the century, 18 percent of the classical works published were those of Cicero, while Virgil held second place at 11 percent.[171] In the first phase, classical works comprised a significant percentage of printed editions and Latin texts.

The new technology quickly spread to Italy. At the behest of Cardinal Nicholas of Cusa in 1465, Gutenberg's former associates Conrad Sweynheym and Arnold Pannartz established a printing press in the monastery of Subiaco, near Rome. Nicholas gave the press a few items for publication: *De oratore* by Cicero, the *De civitate Dei* of Augustine, and a collection of works by Lactantius. In 1467 the two printers moved to Rome proper, where they published a large number of issues, mainly Latin classics. The editorial work was first performed by Giovanni Andrea Bussi, bishop of Aleria, who formerly collaborated with Nicholas. The second phase of that press's production fell to Perotti, who had been Cardinal Bessarion's secretary.

Also in 1467 Ulrich Han began to press books in Rome, and other German printers established book production in various cultural centers. In Venice, Wendelin von Speyer and Nicolas Jenson opened a press. At the end of the fifteenth century, Aldus Manutius (1449-1515) founded the famous Aldine Press, an institution that would make Venice, for a number of decades, the principal center of book production in Europe. In addition to Latin texts, Manutius published numerous Greek editions. Owing to the difficulty of producing freestanding Greek characters, such works had been overlooked by printers in the early decades of book production.

The dissemination of printed texts was rapid and swiftly supplanted the copying of manuscripts. Despite the fact that hand copying was only a marginal aspect of textual transmission by the end of the century, some collectors continued to acquire parchment codices, as Federico da Montefeltro did for the library of Urbino. Another strong supporter of the codex was the French abbot Jean Trithème, who warned of the dangers of printing in his *De Laude Scriptorum Manualium*. Ironically, the diffusion of that book was entrusted to the printing press at Mainz in 1494.

Alongside these technological advances, the quest to rediscover manuscripts continued. The monastery of Bobbio was an important source. There, just one year before his death, Giorgio Merula (1430–1494), a former student of Filelfo, discovered several important codices. Amidst those unpublished works a considerable body of grammatical texts was discovered.[172] In part because of Merula's sudden death after the discovery, not all of these texts were disclosed immediately. Merula had communicated the finding to the duke of Milan, Ludovico il Moro, but he did not make it public. This may have been, in part, because of the virulent controversy that Merula had with Politian, whom he had earlier even accused of plagiarism. Following the death of Merula, Politian tried in vain to recover the codices and, some years later, a portion of them came into the possession of Aulus Janus Parrasio (1470–1522).[173]

Such discoveries continued into the next century. Jacopo Sannazaro (1457–1530) found Ausonius[174] in a ninth-century copy in Lyon. That copy, in script, had been transcribed by exiled Visigoths. Other late discoveries include the opening books of the *Annals* of Tacitus and previously unknown sections of Petronius' *Satyricon*. Continuing his research in Germany, Sannazaro discovered a late eighth-century manuscript containing the pseudo-Ovidian *Halieutica* and the *Cynegetica* of Grattius.[175] In 1508 Sannazaro brought a ninth-century codex[176]—which contained books 1–6 of the *Annales* of Tacitus and was probably copied in the German monastery of Fulda—from the French abbey of Corvey to Rome. Another manuscript, containing the largest part of the *Anthologia Latina*, only reappeared in 1615, in the hands of the French scholar Claude de Saumaise.[177]

The reconstruction of the text of Petronius' *Satyricon*, which had been discovered only bit by bit, was undertaken between the sixteenth and seventeenth centuries. Some sections of that work, three of the sixteen books that had survived the Middle Ages, were reconfigured. Some of these sections had been in France, in Auxerre, Fleury, and Orléans, and had been dispersed in four different collections of *excerpta*, prepared after the ninth century. Some *excerpta* were included in the twelfth-century *Florilegium Gallicum*. In 1420 the *excerpta brevia* were discovered by Poggio, who also discovered the *Cena Trimalchionis*, which was subsequently lost, only to reappear in 1650 in Trogir, Dalmatia, owing to the labor of Marino Statileo. The *excerpta longa*, discovered between 1562 and 1565 by the jurist Jacques Cujas, were published in 1575. The history of Petronian editions reflects the times in which these different acquisitions occurred; we know that the Satyricon was printed as a single work only in 1669. Thus, the loss of original manuscripts was often the result of such assemblage for printing. A further example is the 1508 Aldine edition, the sole witness of the *Liber prodigiorum* of the fourth-century author Julius Obsequens.

In 1515 the German Beatus Rhenanus unearthed the historical work of Velleius Paterculus in a codex probably dating from the Carolingian age. It was owned by the abbey of Murbach, which lost track of it after 1786. Fortunately, that tome had been put to press in 1520. The 1527 Basel edition contains the *Laus Pisonis*, a composition of the early imperial period, perhaps attributable to Calpurnius Siculus.[178] The collection of recipes of Scribonius Largus was put to press in 1529, though the sixteenth-century codex was rediscovered only in 1976. Additionally, two medical treatises of Caelius Aurelianus were found solely in printed editions of 1529 and 1533.[179]

The famous scholar Wilamowitz wrote incisively that "for a long time the humanists were not at all philologists, but rather only writers, journalists, teachers."[180] That assessment reflects clearly the claim for the primacy of philology that was settled and institutionalized after Lachmann, whose "method," observed also in previous centuries, consisted of copying the text furnished from the *antigraph*, trying to correct errors and adjusting for textual corruptions. Occasionally this procedure was carried out by a scholar who would use codices, relying on what he deemed the best one for copying, and turning to others only for the passages that appeared to be problematic.[181] The quality of the output was highly dependent on the overall quality of the copyist, his learning, sensitivity, and intuition. Obviously, results varied.

Exceptional cases aside, the outcomes of the Renaissance practice of philology were largely rather poor, aggravated both by the inadequate care given to the surviving codices and by the losses of numerous medieval manuscripts, whose recovery in this period were short-lived. Fortunately, the texts they contained were preserved in copies made by humanists. These losses preclude the possibility of knowing the date and provenance of the codices that they used and it is also often difficult to reconstruct the text as it was transmitted to the humanists, particularly in cases where their penmanship is illegible or the copy is damaged. In 1429 Poggio wrote that the Cassinese manuscript of Frontinus was written with "gaps and in the worst handwriting [*mendosus et pessimis litteris*], so much so that it is barely readable."[182] The majority of medieval manuscripts copied by Poggio during the period of the Council of Basel-Ferrara

(1430s), however, went missing, with the exception of the codex of Ammianus obtained by Fulda. Copies were rarely carefully wrought. One fortunate exception is the codex from Cluny that Poggio discovered containing the work of Cicero.[183]

Because of the presumption that printed copies supplanted the originals, the history of manuscripts was difficult to reconstruct. This is the case with the codex of *De medicina* of Celsus recopied in Siena by Niccoli. Despite the fact that, in 1426, this text was well regarded by the Panormita, who held it as "marvelous for its age" (*prae vetustate mirabilis*), it was soon lost. Niccoli can also be blamed for the loss of the codex of Petronius' *Cena Trimalchionis*, which he had acquired from Poggio.[184] After being copied, a codex of Cicero from Lodi was deemed unnecessary by Gasparinus de Bergamo, who in 1428 wrote to Landriani that he had no further need of it.[185] In any case, the practice of disposing of the manuscripts after using them became common with the introduction of printing. Printers frequently did away with manuscripts once they had used them to set print typeface, while at other times the manuscripts were destroyed simply to facilitate the work of copying.

The philological contribution of early humanism becomes more recognizable if we consider, as an aspect of philology, the evaluation of the authenticity of the texts and their attributions. Valla stands out in this regard, as we have seen in the case of his work on the spuriousness of the "Donation of Constantine." Additionally, Valla's philological studies on the text of the New Testament and his *Emendationes Sex Librorum Titi Livi* (1446-1447) are remarkable. In the latter work, his annotations and amendments on books 21-26 of Livy were made on the basis of Harley ms. 2493, which Petrarch had also once possessed. In it we find Valla's personal notes alongside those of Petrarch. Valla's work thus paved the way for the practice of textual-critical commentaries, which would become standard in the second half of the fifteenth century. Other documents of philological value that Valla produced remain confined to his notes in the codices of his personal collection.[186]

The generation of humanists active after 1460, stimulated by the advent of printing, performed yeoman's work in textual editing and exegesis. Yet amidst that activity, certain disputes arose. In the time of Bracciolini and Valla, such disagreements had been confined to historical problems and ideals, such as the 1450 controversy between Valla and Trebizond on the superiority of the Roman generals compared to Alexander the Great. But by the 1470s, a new type of dispute arose, based on philological principles: problems of interpretation of individual passages, on which the humanist might exhibit his philological erudition and insight. Among the most clamorous disputes in these years is that of Perotti and Domitius Calderini, both members of the group that gathered around Cardinal Bessarion.

Another aspect of the 1470s was the ever-widening breadth of the centuries in which an author might still be considered classical: while Valla had chronologically marked Quintilian as the end of truly classical Latin, the next generation would embrace the study of Martial, Silius Italicus, Statius, Pliny the Elder, and other imperial authors. This broader philological approach was founded on the importance of the discovery of new witnesses, each to be appraised individually, as well as the ongoing need for the careful practice of emendation. Leto was among the first to tout the value of the Medicean Virgil, for example, a manuscript that had been recovered from Bobbio. Leto put in circulation variants that he found in that codex along with others that he jotted down. Such manuscript annotation was not uncommon at the time. This practice, previously undertaken by Petrarch and Valla, reveals an increasingly more direct and active relationship of the scholar with the codex.[187]

Politian towered over his contemporaries, inaugurating a new form of philological literature that would have enduring success, namely, that of the *Miscellany*, modeled on Gellius' *Noctes Atticae*. Politian's work comprised a collection of one hundred chapters devoted to various linguistic and philological problems. This work became a model for subsequent scholarly literature.[188]

In his harsh judgment of Renaissance philology, Wilamowitz made an exception for Politian as a true philologist.[189] Wilamowitz's opinion was based in part on his high regard for Politian's *Miscellany*.[190] In that tome, we find, for the first time, specific references to codices that allow the identification and detection of certain manuscripts. Dealing with works attested in other codices, Politian introduces the criterion of *sigla* (manuscript acronyms) assigned to different codices: for the edition of Pliny the Elder of 1473 Politian collated five manuscripts, distinguishing each of these with an individual letter. He also uses this demarcation for an edition of Ovid in 1477.[191]

In this way, Politian spoke life into a new system of evaluation of the manuscript witnesses, thereby defining their value to the transmission of the text. He was able to identify agreements with the *antigraph* (i.e., the dependency of one codex on another) and to sketch the principle that will later be called *eliminatio codicum descriptorum*, which, simply put, is the situation that arises when there are two manuscripts and one of those two depends on the other for its existence (sc. it is "descriptus"). That manuscript is, strictly speaking, unnecessary for the formation of the text.

Politian delineated also the principle in which the conjecture—that is, the systemization made by the editor of the text—must take into account the oldest text that can be accessed. Although these advancements, easily discernible in the work of Politian, long remained unknown, they would be recovered by the philologists of the nineteenth century.

5 / CLASSICAL TEXTS IN THE AGE OF PRINTING

FROM THE FIFTEENTH TO THE SIXTEENTH CENTURIES, THE products of antiquity were no longer viewed as merely objects to possess but rather as desirable subjects of study. This assessment took the air out of the humanistic idealism that the transition from the late Renaissance to the Early Modern period had inadvertently produced. "The fifteenth century," as Mark Pattison observed in his biography of Casaubon, first published in 1875, "had rediscovered antiquity, the sixteenth was slowly deciphering it."[1]

This new approach is highlighted in the aphorism "dwarfs resting upon the shoulders of giants," coined by Bernard of Chartres in the twelfth century and made famous by Isaac Newton.[2] With a posture of respect and gratitude, the humanists admired the ancients. They looked to these progenitors as models on the basis of which to excel in the world they inherited. The way in which humanism avowed its abiding connection to its roots stabilized humanistic studies during the progress of the sixteenth and seventeenth centuries.

This transition is perhaps best exemplified by a new interest in exploration, which ultimately led to the discovery of the Americas and found its impetus in various discussions surrounding the Council of Florence (1439-1445) in the 1430s. These deliberations, held on the council's fringes and in its back hallways, were prompted by the recovery of a work of classical learning, Strabo's *Geography*. Leading the conversation were Gemistus Plethon (1355-1452) and Paolo dal Pozzo Toscanelli (1397-1482), whose interest in the world's natural features had been piqued by the rediscovery of Strabo's work, which was one of the most important of the Greek books that had come westward. Strabo's observations about the earth's circumference, based on the Greek mathematician and geographer Eratosthenes, were vital for Plethon and Toscanelli's own calculation of the distance between Europe and the Indies. They would eventually present this calculation to Christopher Columbus in a 1474 letter that influenced Columbus to sail westward some eighteen years later, leading to the discovery of the New World.[3]

Significant scientific innovations in the sixteenth and seventeenth centuries were not perceived as incompatible with the cultural tradition of humanism. Such advances were undertaken in the context of the humanist rediscovery of ancient science. "Endeavouring to see in nature what Greek writers had declared to be there, European scientists slowly came to see what really was there."[4] The detection of the circulation of blood did not immediately tarnish Galen's reputation, nor did the discovery of heliocentrism by Kepler or Galileo negate Ptolemy's importance.[5] As alluded to above, in a letter to Robert Hooke on February 5, 1675, Isaac Newton (1643–1727) presented his findings by slightly recasting the aphorism of Bernard: "If I have seen further it is by standing on the shoulders of giants."[6]

By Newton's time, classical culture was a pervasive and integral part of the ethos of the West, as the formative models of Italian humanism had by now prevailed throughout Europe. Moreover, Latin had become the *lingua franca* of the humanists, particularly employed to write to other intellectuals of various countries. Not only did Newton do so, but Copernicus, Harvey, and many others as well. Although printed volumes were widely disseminated, not all connection to the old manuscripts was lost: the characters in the printed volumes imitated the script of the Carolingian minuscule that Salutati had revived at the dawn of the fifteenth century.[7] More important, thanks to the printing press and the growth of publishing houses throughout northern Europe, classical authors had become more accessible than ever before.[8]

Amidst the tapestry of cultural relationships being woven together in Europe in the twelfth century, the environment that would welcome and foster Renaissance humanism grew. Petrarch was read not only in Tuscany but even in Paris, where Jean de Montreuil (1334–1418), famous debater of Christine de Pizan, was among his admirers. In the years of the Council of Constance (1414–1418), Montrueil worked alongside Poggio in his quest for fresh manuscripts on the continent.[9] In the fifteenth century, Janus Pannonius fostered the humanistic movement in Hungary, while at Oxford, Robert Fleming undertook studies of the classics in Lincoln College. When traveling in Italy, Fleming studied Greek under Guarino at Ferrara, and through this arrangement became the first Englishman to know Greek since Robert Grosseteste and Roger Bacon, more than a century earlier. The German Albrecht von Eyb returned to Germany with manuscripts of Petrarch and Poggio in tow after studying in Italy at Pavia and Bologna. In Spain Iñigo López de Mendoza, marquis of Santillana, was in contact with Bruni and introduced a new level of humanistic eloquence to his Spanish colleagues.[10] Finally, Gregory of Sanok, from northern Poland, studied in Rome in the 1430s; returning to his home region, he would later make Lwów a significant center of classical studies.[11]

Representatives of Italian humanism, in turn, moved to various countries in Europe and played a large role in the advancement of Renaissance culture. For example, Pier Paolo Vergerio went to Hungary, where he enjoyed the patronage of King Sigismund of Luxembourg. Guiniforte Barzizza was in Catalonia, where he served in the court of Alfonso V of Aragon before Alfonso's conquest of the kingdom of Naples in 1442. Filippo Buonaccorsi (also called "Callimachus"), who was involved in a conspiracy against Pope Paul II in February 1468, avoided the pope's subsequent call for his arrest and found shelter in Poland, where he was prominent in cultural activities.[12]

Support for humanistic thought grew in Renaissance chancelleries and courts of power. In the second half of the fifteenth century, King Matthias Corvinus of Hungary became a strong advocate for humanistic principles and had, in particular, a good deal of interest in classical texts. Classical ideas continued to spread further east. After he conquered Constantinople in 1453, Mehmet II (1432-1481), a young and enlightened Ottoman sultan, assembled a large library, in which were many important books of the Western canon.[13] While some scholars have raised questions about who may have tutored Mehmet, the possibility that Cyriacus of Ancona was among them, specifically with regard to Mehmet's acquisition of the classical languages, cannot be ruled out.[14] Mehmet assembled a court in Constantinople, to which he invited not only Greek scholars but also Italian humanists and artists. Notably, the Venetian artist Gentile Bellini was invited to render a portrait of him.[15]

Meanwhile in the West, humanistic ideas grew in the universities, replacing the approaches to learning carried over from late medieval culture. After these new methods of liberal studies had taken root, Politian emerged, taking the position of the last major proponent of humanism in Italy. Before his premature death at age 40 in 1494, he saw the arrival of the French troops of Charles VIII in Italy, an event which marked the political (although not yet cultural) decline of the peninsula.[16] In this period Rome would witness the spectacular careers of Michelangelo and Raphael, while in Florence Niccolò Machiavelli would soon write his controversial masterpieces. Meanwhile, in Venice, Pietro Bembo (1470-1547) not only established the linguistic canon of vernacular Italian but also confirmed Cicero as the gold standard of Latin composition.[17] About the same time, in 1528, Baldassare Castiglione published *The Courtier* in Rome. Yet in spite of this rise in intellectual activity, the political crisis in Italy had now become irreversible, poignantly marked with the sack of Rome in 1527 at the hands of the duke of Bourbon, Charles V, who led a cohort of Spaniards, accompanied by a large number of German mercenaries (Landsknechte).[18]

The displacement of Italy as Europe's political epicenter permitted other hubs of Renaissance culture to emerge in northern Europe. This shift is highlighted by the rise of new centers of book production, which broke the monopoly that the Venetian printers had enjoyed for many years. In Basel the typography of Johann Froben, friend and publisher of Erasmus, attests to the city's waxing influence. The theologian Guillaume Fichet, who served as the librarian of the Sorbonne, founded the first publishing house in Paris in 1470. In 1507 Josse Badius (Ascensius), who had studied in Ferrara with Battista Guarini, the son and successor of Guarino Veronese, founded his own publishing house, known as Prelum Ascensianum.[19] In the middle of the sixteenth century in Antwerp, Christophe Plantin (1520-1589) founded the Plantin Press, one of the most important of these new presses.

Aside from Erasmus, whom we shall consider below, the most prominent European humanists of the first half of the sixteenth century were the Spaniard Juan Luis Vives (1492-1540), the German Johannes Reuchlin (1455-1523), the Frenchman Guillaume Budé (1468-1540), and the Englishman Sir Thomas More (1478-1535). The bold stances taken by these important figures in their individual lives reflect the new habits of independent thought that grew out of Renaissance humanism.

Reuchlin traveled to Italy, where he had contacts with Politian and Pico della Mirandola (1463-1494). He developed a keen interest in Hebrew and the esoteric teachings of the Kabbalah, leading him to advocate for the establishment of professorships in the Hebrew language in Germany. Refusing to join the Lutheran Reformation, he retired to Tübingen, the university founded in 1477 by his patron, Eberhard I of Württemberg.[20]

In France Reuchlin's contemporary Guillaume Budé became well known not only for his published translations of Plutarch but also his important work entitled *De Asse et Partibus*, published in 1514, which dealt with both numismatic studies and carefully reconstructed ancient systems of measurement. Further, he made an important contribution to Greek lexicography with his *Commentarii Linguae Graecae* (1529). He also wrote commentaries on legal texts, and in 1503 founded the Collegium Trilingue, later called the Collège de France.[21]

Budé's friend and correspondent, Thomas More, is best known for his *Utopia* (1516), the story of an imaginary journey to an ideal city inspired by Plato's *Republic* and modeled on some aspects of Lucian's *Menippus*.[22] Because of his personal political convictions, More was condemned to death by Henry VIII for remaining faithful to the Catholic Church after the Act of Supremacy (1534), which established the independence of the Anglican Church.[23]

More's younger contemporary, Juan Luis Vives, who also served in the court of Henry VIII as a tutor, was likewise unable to avoid the consequences of the Anglican rift. Though in 1522 he had dedicated his commentary on Augustine's *De civitate Dei* to Henry, he had to flee London and take refuge in Bruges because he remained unwavering in his support of the queen and his stance against Henry's polygamy. There he was able to write a polemical work against Aristotelian authority and another work demonstrating the spuriousness of the *Letter of Aristeas*, the document famously purporting to witness the transcription of the Septuagint.[24]

Perhaps the most eminent and influential personality in the early decades of the sixteenth century, however, was Erasmus of Rotterdam (1469–1536). While he made a few trips to Italy, he was active primarily in Paris, Leuven, Freiburg, and Basel. Though an admirer of Valla, he was nevertheless an opponent of Italian-style humanism. In his *Ciceronianus*, for example, he takes issue with Bembo.[25] He calls attention to the spread of humanistic ideas in Europe, writing in 1517: "Polite letters, which were almost extinct, are now cultivated and embraced by Scots, by Danes, and by Irishmen."[26] Erasmus advocated Christian humanism, inspired by Cicero but also owing a great debt to the Church Fathers and, of course, thoroughly disassociated from the Lutheran Reformation. In spite of his dedication to patristic thought, he was regarded with suspicion even by the Catholic Church. To wit, his works would be included in the Church's "Index of Prohibited Books."

The most famous work of Erasmus, *Praise of Folly* (1511), was inspired by the satire of Lucian. Not coincidentally, it was composed while Erasmus was a guest of Sir Thomas More, to whom he dedicated the work. As we saw above, More was also an admirer of Lucian. Erasmus's network of relationships also included Vives and other prominent representatives of European humanism. Among the many scholarly contributions of Erasmus was his important reconstruction of the pronunciation of ancient Greek, *Dialogus de Recta Latini Graecique Sermonis Pronunciatione*, published in 1528.[27] In 1516 Erasmus published his edition of the Greek text of the New Testament, a work for which he used the *Adnotationes* of Valla, an edition of which he himself had edited for publication in 1505.[28] This work, perhaps his most important, was published just as the Catholic Church was facing the Protestant Reformation.

In 1517, by nailing his ninety-five theses on the portal of the Wittenberg cathedral,[29] Martin Luther opened a vital chapter in European cultural history, one that featured deep religious division both on the continent and beyond. The fourteenth-century zeitgeist of humanistic revival and rebirth naturally brought with it questions of the Church's authority. This skepticism harked back to the work of the anti-papal Ulrich von Hutten, who had published the pamphlet of Valla entitled "On the Falsely Believed and Deceptive Donation of Constantine," directly challenging papal authority.[30]

Yet the peace between humanism and Protestant thought was an uneasy one. The anti-humanistic strains are clear in Luther's response to Erasmus's

De Libero Arbitrio Diatribe sive Collatio (1524), which claimed free will as a vital theological concept. In *On the Bondage of the Will*, Luther expounds upon the idea of predestination, which runs rather far from the free-thinking individuality that characterized humanistic thought. Philip Melanchthon (1497–1560), the chief confidant of Luther, would resume this debate but never settle it, leaving the question open for others who followed in Luther's wake to consider, such as John Calvin.[31]

In the following decades, both the reformed and the Catholic churches took a selective attitude toward classical culture. Each acknowledged the formative role that it had assumed in humanistic education but looked askance at critical approaches that seemed corrupt, and marginalized certain aspects of antiquity and even certain authors that appeared incompatible with Christian doctrine. One signal example is the philological study of sacred scripture inaugurated by Valla and continued by Erasmus. In the seventeenth century, the French priest Richard Simon (1638–1712) was targeted by Catholic authorities and French Protestant theologians for two critical histories: one of the Old Testament (1678) and the other of the New Testament (1689).[32] The Catholic Church's defense of the Vulgate text against the intrusion of philological criticism long remained a point of dogma, together with the prohibition against new translations of the Holy Scripture. The research conducted by Simon and continued by Jean Leclerc would later be valued in the Enlightenment, leading to the *Dictionnaire philosophique* of Voltaire (1764) and the *Encyclopédie* of Denis Diderot and Jean le Rond d'Alembert (1751–1772).[33]

Despite the fluctuations in the outlook of fifteenth-century popes toward Renaissance ideas, the Council of Trent (1545–1563) allowed the Church to establish a policy with regard to humanistic learning. One particularly alarming development was the Church's "Index of Prohibited Books" (1558; known simply as the "Index"), which established a tight control on book production. Among the classical authors censored by Church authorities, Lucian stands out on the Greek side. On the Latin side, it is perhaps not surprising that Lucretius drew the Church's ire, as his long poem, *De rerum natura*, espouses Epicureanism, which to many seemed synonymous with impiety and atheism. Other poets, such as Horace and Ovid, were admitted into circulation but in an expurgated form, with the elimination of the sections that seemed obscene. The line of the Catholic Church toward classical authors is well summarized by Ignatius of Loyola, who in 1555 advised his followers to make use of pagan writers as Israel had when, as described in the book of Exodus, it famously escaped Pharaoh and took with it the treasures of Egypt. Thus, in the Jesuit schools, the use of such authors as Cicero, Quintilian, Virgil, and Ovid was duly censored.[34]

The theater, which in late antiquity had been opposed by Christian culture, was transformed into an edifying educational tool, not only by Jesuits but in the reformed churches. The production of Latin plays in the sixteenth and seventeenth centuries was widespread. The events staged were often drawn from ancient history, mostly on the model of Seneca.[35]

Nevertheless, severity was reserved for positions that were at odds with Catholic doctrine. In 1556 the *De Immortalitate Animae* of Pietro Pomponazzi, professor at Bologna, was publicly burnt. Pomponazzi had argued that Aristotle viewed the soul as mortal. This thesis was revived in the last decades of the century by Cesare Cremonini, professor at Padua, who enjoyed the protection of the republic of Venice, a city that then found itself in conflict with the Jesuit order. In Germany, Reuchlin was attacked for showing interest in Jewish books and for his *De Arte Cabalistica* (1527), while in France Robert Estienne, a philologist and a leader in the printing industry (on whom see further below), was censured by the theologians of the College of Sorbonne for his translations of sacred scripture. He then fled to Geneva, where he joined the Reformation.[36]

Beyond these important figures and historical events, the age of the Counter-Reformation saw the roots of classical culture deepen through the activity of the Jesuits. They built a strong educational system based on the study of Latin and Greek grammar, literature, and rhetoric, a system whose influence would obtain for centuries. The Jesuits also afforded the expansion of classical culture and humanities beyond the borders of Europe. Their humanistic education, transplanted to America, produced the remarkable figure of Garcilaso de la Vega (1539-1616), the son of a Spanish conquistador and an Incan princess. Garcilaso moved to Spain, where he used Neoplatonism to defend Native Americans and their lotus cults from accusations of barbarism and idolatry. The Jesuit missionary Matteo Ricci introduced Catholicism in China between 1582 and 1610. There he popularized not only Christian texts but also the work of Epictetus.[37]

In spite of such dissemination, Latin's humanistic supremacy rapidly waned in the sixteenth century. This was in part because national literatures adopted their various vernaculars, but also because different genres no longer relied upon Latin for universality of expression. Two of these genres were epic and satire. Epic was revived in Italian by Ludovico Ariosto in *Orlando Furioso* (1516) and Torquato Tasso in *Gerusalemme Liberata* (1575), in Portuguese by Luís Vaz de Camões in *Os Lusíadas* (1572), and in English by John Milton in *Paradise Lost* (1667). The tradition of satire, in its various ancient forms such as are found in works by Horace, is filtered by François Rabelais in *Gargantua et Pantagruel* (1532-1552) and imitated by Nicolas Boileau in *Satires* (1660-1668), by John Dryden in *Discourse concerning the Origins and Progress of Satire* (1692), and by Alexander Pope in *Imitations of Horace* (1733-1738).

The rebirth of theater, propelled by Senecan tragedy, finds in the sixteenth century several literary representatives in Europe. The *exemple célèbre* is William Shakespeare (1564–1616), who drew upon ancient history for numerous plots. The Greek tragedies, combined with the instructive observations of Aristotle's *Poetics*, nourished the French neoclassicism of Pierre Corneille (1606–1684) and Jean Racine (1639–1699). In Germany the recovery of the Aristotelian theory of *catharsis* by Gotthold Ephraim Lessing in his work *Hamburgische Dramaturgie* (1767–1768) influenced the writings of Friedrich Schiller (1759–1805).[38]

REFUGEES AND PHILOLOGISTS

The literary forms revived in the sixteenth century owed a great debt, of course, to the ancient exemplars that had been preserved. Yet not all the texts of antiquity had survived the late Renaissance, as many codices used for the early editions of manuscripts went missing, whether because they were deemed unnecessary after the creation of the print copy or because they were damaged during print production.[39] For example, Giovanni Andrea Bussi produced an edition of Pliny's *Naturalis historia*, printed by Sweynheym and Pannartz in Rome in 1479. He availed himself of a codex copied a few years earlier,[40] but introduced into the printed text some interpolations and a number of changes that the great twentieth-century Italian textual critic Remigio Sabbadini regarded as "superfluous."[41] The lack of attention Bussi gave to any particular edition he produced may be attributed to a demand for quick publication. Thus, his haste to publish a number of remarkable texts in so short a time allowed them to be put into circulation in variously corrupt and unreliable states.

The importance of the first printed edition of a work (*editio princeps*) is determined by the multiplication of copies that followed. When manuscripts were still being copied, a codex would be read by only a single person. With the dawn of the printing press in the mid-fifteenth century, the printed edition became the standard text instead of the earlier codices. Once a large amount of copies had been printed, the other exemplary manuscripts would be overlooked even though they sometimes preserved a more accurate text.

By the last few decades of the fifteenth century, humanists were already becoming aware of problems propagated in the texts of the first printed editions. In the preface to the edition of Plautus' comedies printed in Venice in 1472, Giorgio Merula attributes the corruption of the text "to the arrogance of interpolators, the inattention on the part of the copyists, and the ignorance of the librarians," and even compares the task of the publisher to the labors of Hercules.[42] For the faults and errors in the edition of Pliny, mentioned above, Bussi was heavily criticized by Perotti, who in a 1470 letter proposed the establishment of a kind of "philological censorship" that would oversee the examination of the texts to be given to press. Shortly afterward, Perotti managed to replace

Bussi as curator of printed volumes for Sweynheym and Pannartz. He published a second edition of Pliny (1473), which, despite Perotti's earlier thoroughgoing criticism, presented only minor changes to the text already published by Bussi.[43]

Even in cases where there were ancient and authoritative manuscripts available, the printed editions often resorted to less reliable codices. One example is *De re coquinaria*, attributed to Apicius. This work was preserved in Italy in two medieval codices: one was in the library of the duke of Urbino,[44] and the other was the Codex Fuldense, brought to Italy by Enoch d'Ascoli.[45] The Fuldense manuscript was known to Politian, who also produced part of it in an edited volume printed in Milan in 1498 that was based not on the original manuscript but on an *apograph* copied in 1490.

Nevertheless, Politian had practiced the collation and the comparison of several manuscripts copied from the same text, a scholarly pursuit facilitated by the large availability of various codices in Italy of which he could avail himself. Yet philologists of the following centuries, active mostly in northern Europe, did not always enjoy such opportunities and frequently had to work from fragmentary texts.

In the Early Modern period, most manuscripts were kept in private collections, not always accessible and in some cases even unknown, as catalogues of individual libraries were but rarely available. Indeed, only after the French Revolution did major public libraries develop the policy of acquiring manuscripts. Moreover, long-distance travel often prevented individual scholars from consulting certain codices: before photography was developed in the nineteenth century, any type of collation of these manuscripts had to be done in situ.

Another factor that made it difficult to trace out manuscript traditions was the lack of any codified knowledge of paleography, a field established only in the eighteenth century, which would help in reconstructing the dates of the codices. The first steps in this regard were taken by Politian, who dubbed "Lombardic" the codices that were written in difficult-to-read minuscules.[46]

Until the nineteenth century, publishers tended to work chiefly from the texts of the printed editions. To improve each new edition, two forms of emendation were available. The first was *emendatio ope codicum*, "correction performed by the aid of manuscripts," whereby variants were established merely on the basis of what partial collation evidenced to be preferable. The second was *emendatio ope ingenii*, which is recourse to conjecture meant to improve the text on the basis of the editor's linguistic and stylistic sensitivity.[47]

The humanists and philologists of the sixteenth through the eighteenth centuries did not systematically record their collations of manuscripts—indeed, the critical apparatus did not yet exist—but simply applied individual readings to individual codices. Where they made changes in the text, they did so on their own initiative using criteria that they rarely stated explicitly. In general, they were guided by their own grammatical preferences, their own knowledge of historical antiquity, and sometimes even by their own political motives. Anthony Grafton has gone so far as to suggest that a textual intervention made by Joseph Scaliger in Seneca's *Apocolocyntosis* was motivated by political concerns. Instead of referring to the population of British brigands and their shields as *scuta Brigantas*,[48] Scaliger "corrected" the text to read *Scotobrigantas*. This innovation allowed the name of the Scots to be historically associated with brigandage, a declaration that would become a propagandistic tool in the conflict between Scotland and England.[49] The fateful result, though no less interesting for it, of an *emendatio* produced under these circumstances is that it would become a permanent part of the history of the text.

Accordingly, the process of discerning what the *textus receptus* originally must have been was by no means systematic. Even the most excellent later editions continued to propagate the banalizations transmitted in the first editions. Proper assessment of the value of manuscripts and the editor's personal interest in the particular codex that he worked from were often muddled, and therefore he would sometimes transmit errors that would recur in subsequent studies. Recourse to *emendatio ope codicum* could be rather random, often entrusted merely to the sensitivity of the editor, who frequently lacked either sufficient paleographic knowledge or an awareness of the historical value of an individual textual witness that would be necessary for him to make a well-informed decision.

Finally, the use of conjecture itself was quite various, and editors showed greater or lesser propensity to indulge in it. This feature has been and still is a constant in the history of philology, a profession divided between textual conservatives and those willing to make conjectures. In the sixteenth century, the conservative view was most prevalent, as evidenced by the writings of Justus Lipsius (1547–1606). "There are two paths for the corrector," he writes in 1581, "those of codices and conjecture; the first path is sure and safe, the other path, conjecture, is slippery, especially when bold and brash young people set upon it, or old men who act like youths."[50] In the following centuries the two approaches remain variously present in philological circles, with the majority of textual *emendatoria* performed by the most eminent philologist of the eighteenth century, Richard Bentley, who proffered a famous dictum in his 1711 edition of Horace: "as far as I am concerned, the use of reason and the matter in question itself are of greater value than a hundred manuscripts."[51]

By the end of the fifteenth century, philological activity had already achieved excellence in France. Most significant was the work of Janus Lascaris (1445–1535), who arrived in Paris in 1495. He served as the editor of a number of first editions, including Callimachus, Apollonius, and, near the end of his career, the Homeric and Sophoclean scholia.[52] Another critical contributor to Greek philology was Girolamo Aleandro (1480-1542), best known as the chief prosecutor of Luther at the Diet of Worms. He had produced his major lexicographical contribution, a *Greek to Latin Dictionary*, in 1512. In Paris, a decade later, Aleandro served as an ambassador from the Vatican to King Francis I's court, and in Rome, a further decade later, he was appointed cardinal.[53]

Among printers, Robert Estienne (better known by his Latinized name, Stephanus) had developed exceptionally high-quality typography at the beginning of the century. He passed the firm to his son (1503-1559), of the same name as his father, who in 1532 published his famous *Thesaurus Linguae Latinae*. The work was continued by the younger Robert's son, Henri (1528-1598), who printed several editions of Latin texts and, in 1572, published the *Thesaurus Linguae Graecae*.[54]

Beyond Estienne and Budé, sixteenth-century France saw several other important contributors to the field of philology. Chief among these is the medical doctor Julius Caesar Scaliger (1484-1558), originally from Riva del Garda near Trento. In 1525 he became the personal physician of the bishop of Agen, France. His education in classics led him to write commentaries on the botanical works of Aristotle and Theophrastus, as well as his own work on the features of the Latin language, *De Causis Linguae Latinae* (1540). He also published his own *Poetics* in 1561, in which he developed the ideas of Aristotle and Horace, offering an overview of poets from antiquity to his contemporaries.

Another major figure of this era was Adrien Turnebus (1512-1565), a professor of Greek in nearby Toulouse, a city roughly one hundred kilometers from Agen. He was then director of the Royal Press of Paris and editor of a new edition of Cicero's *De legibus*.[55] Denis Lambin (1520-1572) must also be added to this list. A careful researcher of codices, he published editions of Horace and Cicero, as well as an important edition of Lucretius. Pierre Daniel (1530-1603) also deserves mention. He had access to the codices of Fleury and published an *editio princeps* of the late antique Latin comedy *Querolus* (or *Aulularia*), the only Latin comedy other than those of Plautus and Terence to survive; Daniel, drawing from several medieval manuscripts, also published an expanded version of the commentary of Servius on Virgil. The importance of this work is reflected in the name by which it is typically known, the "Servius Danielis."[56]

The most important figure of French philology in the sixteenth century, however, was undoubtedly Joseph Justus Scaliger (1540-1609), the son of Julius Caesar Scaliger. Joseph dedicated himself to difficult-to-read texts, such as Festus, Manilius, and Catullus, of which he published important editions. He was also interested in ancient history and wrote a book, the *Thesaurus Temporum* (1606), that was fundamentally important at that time for research in ancient history. The great season of French philology came to an end with Robert Estienne, whose bold venture in printing has already been noted, and Isaac

Casaubon (1559–1614), the scholar and commentator on Greek authors such as Diogenes Laertius and Strabo, only sparsely studied at the time.[57]

After these figures, however, philology in France began to decline. Its loss of influence can be connected with the religious conflicts that shook France in the seventeenth century. The son of an outspoken Calvinist in Geneva, Casaubon first went to Paris, where he became a supporter of the Huguenots. He was then forced to leave in 1610 and found refuge and protection in London. Joseph Scaliger, also a Calvinist, found refuge in the Netherlands. That country played a leading role in philological studies, which were undertaken both in the reformed university in Leiden and in the Catholic University of Leuven.[58] In publishing, the typographer Plantin of Antwerp produced a famous polyglot Bible (1568–1573). At Leiden, Lodewijk Elsevier was also active from 1580 and his press produced several classical editions.

Scaliger's chair at Leiden had previously been held by Justus Lipsius, notable scholar of Seneca and Tacitus. Scaliger's successor was the Frenchman Claudius Salmasius (Claude de Saumaise, 1588–1653), who discovered not only the *Latin Anthology* in the manuscript that bears his name but also the *Anthologia Palatina*[59] in the codex of Constantine Cefala, a manuscript that came into the possession of the Palatine library of Heidelberg. Johann Gerhard Voss (1577–1649) was trained at Leiden and then became a professor at Amsterdam. After establishing himself as a scholar of rhetoric and grammar, he gained fame as the author of both *De Historicis Graecis* and *De Historicis Latinis*, which were studies of historians from the earliest historical writings to the contemporary age.

In the decades that followed, Johann Friedrich Gronovius (1611–1671), a German active in Leiden, worked on the text of the main prose writers of the Roman imperial age, including Livy, Tacitus, and Gellius. He also demonstrated the high value of the Codex Etruscus, which preserves the tragedies of Seneca.[60] Nicolas Heinsius (1620–1681) concentrated on poetry, publishing editions of Ovid, Virgil, and Valerius Flaccus. He also collaborated with Queen Christina of Sweden in search of manuscripts, which were known at the time as the *Reginenses*.[61] While visiting many libraries, he discovered important manuscripts of Ovid, which were vital for his own and future editions of that poet.

Between the seventeenth and eighteenth centuries, Richard Bentley (1662–1742), rector of Trinity College, Cambridge, was the single most dominant figure in English philology. His propensity for *emendatio ope ingenii* takes center stage in his edition of Horace, where there are about seven hundred examples of his emendations.[62] Bentley was interested in Greek as well as Latin texts. His contributions to Homeric studies are particularly important. Chief among these is his documentation of the dropping of the Greek *digamma*, a loss that permanently affected the Greek language and produced strange metrical patterns. Bentley's recognition of its disappearance solved the mystery of Homer's unscannable lines. He also demonstrated the spuriousness of the epistles attributed to the tyrant Phalaris. Competent also in theology, he brought forth a new edition of the New Testament.

Beyond the edition of Horace, Bentley is recognized for his editorial work on Latin texts of authors such as Terence, Phaedrus, and Manilius. His countless conjectures are based not only on his great sensitivity to those authors' styles but also on his unique brand of linguistic rationalism that can, at times, come off as excessive. A poignant example comes from a Horatian emendation, wherein Bentley changed *Epistle* 1.7's "greedy fox" (*volpecula*), who enters a barn to devour grain but cannot escape because it is overstuffed, to a "mouse" (*nitedula*), because foxes do not eat wheat. Despite Bentley's evidenced close attention to detail, his proposal was later dropped by the publishers, who no doubt regarded it as simply too clever.

Amidst the flurry of activities that the humanists undertook, the pursuit and collection of codices was soon augmented by the quest for other articles of material culture. In particular, inscriptions, vital for our understanding of history, were sought and gathered into collections. Historical and antiquarian interests were also growing, propagated in Rome particularly by Flavio Biondo, who had opposed Bruni in the debate about the vernacular. Under the patronage of Pope Eugene IV, Biondo published the first topographical survey of ancient Rome, *Roma Instaurata*, in three volumes (1444-1448). By the 1470s we find Leto collecting memorabilia and inscriptions. Along with the followers of his academy, Leto explored the Roman catacombs.

Numerous Greek inscriptions had been collected in the first half of the fifteenth century by Cyriacus of Ancona (1391-1452), a humanist and businessman dedicated to commercial trade. As mentioned briefly above, Cyriacus made many trips to the eastern Mediterranean and had much contact with Byzantine culture and the Muslim world. In the course of his travels, he transcribed a large number of Greek inscriptions, providing sketches and giving information about their provenance. He was among the first to draw the site of the Parthenon in Athens, a sketch fortunately made before the Parthenon sustained heavy damage in warfare. Unfortunately, his *Commentarii* were destroyed in the fire that damaged the Sforza palace in Pesaro in 1514, although various partial transcripts survived.[63]

In the sixteenth century, collecting inscriptions became widespread, with the creation of *lapidaria* in which various archaeological troves were assembled. These finds included ancient inscriptions, coins, and works of art. Among the last of these, the turn of the sixteenth century saw the recovery of masterpieces such as the *Apollo Belvedere* (1490) and the *Laocoön* (1506).[64]

The results of this research, however haphazard it may have been, began to be dispersed in the second half of the sixteenth century. In 1555 Carlo Sigonio published the *Fasti Consulares*, a collection of lists of Roman consuls discovered in the Forum Romanum and elsewhere in Rome. In 1565 Onofrio Panvinio drew a fresh archaeological map of ancient Rome. Meanwhile, Fulvio Orsini, in 1570, published a collection of portraits and ancient depictions of characters, the *Imagines Virorum Illustrium*, followed, in 1577, by his *Familiae Romanae*, a detailed work on Roman coinage.[65]

The first systematic collection of Latin inscriptions was published by Martin de Smet in Leiden in 1588. A few years later a wider collection, to which Joseph Scaliger also contributed, was published by Janus Gruter in Heidelberg, entitled *Inscriptiones Antiquae Totius Orbis Romani* (1603). In the eighteenth century, the shipping industry transported numerous inscriptions, particularly from Greece and destinations farther east, turning epigraphic acquisition into a trade that had been, after the fall of Constantinople in 1453, difficult and quixotic.[66]

PALEOGRAPHY

One of the biggest challenges in assessing the importance of a codex is presented by the difficulty of establishing a date for that manuscript. As we have seen, various misunderstandings grew out of the chronological difficulty, leading not only to discrepancies but even to quarrels among humanists. This situation did not improve between the sixteenth and seventeenth centuries, in spite of the efforts of philologists such as Scaliger and Heinsius, who developed criteria for the empirical evaluation of codices.

In the course of their own research agendas, two French Benedictine monks offered insights that helped address the issue of chronology, and thus they are not inappropriately credited as the founders of paleography.[67] The older of the two, Jean Mabillon, in 1681 published the *De Re Diplomatica*, a work in which he discusses the care with which he goes about weighing one manuscript against another. Indeed, Mabillon systematically catalogued the chirography preserved in various codices, outlining their historical succession and in this way laying the foundation for more accurate dating of those manuscripts based on the style of the scribe's handwriting. The younger of the two monks, Bernard de Montfaucon (1655–1741), published his *Palaeographia Graeca* in 1708, in which he expanded upon Mabillon's principles to create the first paleographic handbook. In it, Montfaucon first used the term "paleography."[68]

A significant boost to Latin paleography was given in the eighteenth century by Francisco Scipione Maffei (1675–1755), who was able to study the manuscripts of the Biblioteca Capitolare of Verona. As that library's collection had not declined significantly during the Middle Ages, it provided to Maffei enough codices for him to study penmanship and letterforms over many centuries. This broad distribution was key for his research. Maffei, who was also a playwright and an antiquarian, founded the lapidary museum of Verona, one of the first public museums dedicated to inscriptions.

Excavations carried out between 1738 and 1748 at Herculaneum and Pompeii were popular throughout Europe. The rediscovered cities allowed a close comparison with the views of antiquity hitherto derived only from texts and a few scattered monuments. Using Vitruvius' *De architectura* as a guide, Karl Weber, a Swiss scholar known for his reconstruction of ancient environments, directed the first excavations.[69] The frescoes and paintings discovered offered scholars the opportunity to confirm in person the descriptions of paintings provided by Pliny the Elder in his *Naturalis historia*. In the decades following the excavations, individuals of culture from the whole of Europe visited these ancient cities, giving a significant boost to the study of Roman antiquities and material culture. The structure at Herculaneum that housed many examples of charred papyrus was dubbed the "Villa of the Papyri," although at that time it was not yet possible to read these ancient rolls.

The papyri that had begun to circulate in the course of the eighteenth century came from Egypt, often purchased by travelers. The first papyrus published was the so-called Charta Borgiana, a documentary papyrus of the second century AD, recovered in 1788 in the Egyptian resort town of Giza. Cardinal Borgia entrusted the study of this papyrus to the Danish theologian Niels Iversen Schow, a former pupil of Heyne in Göttingen.[70] The first scholar who tried to read the original text of a palimpsest was Jean Boivin, assistant librarian in the Royal Library of Paris, who in 1692 discovered a Greek transcription of sacred scripture dating to the fifth century.

The use of chemical reagents to help analyze numerous Latin palimpsests grew throughout the first half of the nineteenth century. Angelo Mai (1782–1854) was a pioneer of such new methods. He even used gall from gallnuts, formations that occur on oak trees, to help clarify letters hidden beneath the surface of a palimpsest.[71] Working from 1810 to 1819 in the Biblioteca Ambrosiana in Milan and from 1820 in the Vatican library—libraries that by then already housed the greater part of the palimpsests of Bobbio—Mai discovered the epistles of Fronto, the opening sections of Cicero's *De re publica*, and the remains of the *Vidularia* of Plautus, among other texts. Mai's finding of Cicero earned him considerable fame, as witnessed by Leopardi's poem *Ad Angelo Mai* (1820). In 1838 Pope Gregory XVI appointed Mai to the College of Cardinals.[72]

Another remarkable find, that of the palimpsest containing the *Institutiones* of the second-century AD jurist Gaius, occurred at the same time that Mai was making his discoveries. In this case, the Dane Barthold Georg Niebuhr (1776–1831), while traveling to Rome as a Prussian ambassador, stopped in Verona in 1816 and detected the palimpsest in the Biblioteca Capitolare.[73] In the years that followed, Niebuhr worked closely with Mai in the study of palimpsests in the Vatican library and later identified fragments of the fifth-century poet and rhetorician Merobaudes in a palimpsest of St. Gall.[74]

HISTORIOGRAPHY

The literary historiography of the sixteenth and seventeenth centuries grew in the wake of humanistic thought. The reconstructions advanced by literary historiographers came not from a single period but developed over time, from ancient authors all the way up to the age of the rediscovery of historiography. Over the centuries the discipline of historiography grew and changed. Some approaches were but piecemeal, as is evident in Pietro Crinito's *De Poetis Latinis Libri Quinque* (1505), not unlike Voss's contribution on Greek and Latin historians.

In his publication of *Historica Critica Latinae Linguae* (1716), Johann Georg Walch studies Latin authors from various periods, distinguishing between prose writers and poets. As regards the periodization of literary history, humanistic historiography already tended to stop considering literary sources beyond the imperial period. For example, we saw that Valla placed a terminus at a rather early date, as early as Quintilian. Guarino, however, in his epistle to Leonello d'Este (1449) had readdressed and embraced the Isidorian periodization, which regarded the imperial literature as the beginning of the downward spiral in the corruption of the Latin language. Interpreting this metaphorically, Isidorus called the archaic phase of Latin literature its "childhood," the second (classical) period its "adulthood," and its final period, "elderly."[75]

Another metaphor, involving "organic" forms, enjoyed good fortune in the Early Modern period. Other designations were based on metals, a motif dating as far back as the Archaic Greek poet Hesiod. The distinction between the "Golden" age, covering late Republican and Augustan literature, and the "Silver" age for the next stage is frequently found in books written as recently as a few decades ago. Among the most elaborate uses of this schema is that of the German philologist Caspar Schoppe (Scioppius, 1576-1649), who posits six different periods, characterized (always assuming a progressive decay) first by "gold," then "silver," followed by "bronze," "iron," "wood," and finally "terracotta."[76]

Schematizations of this kind were not incompatible with the continual model of literary history. In his *Poetics* (1561), Julius Caesar Scaliger proposed that the dawn of a *nova pueritia* (new childhood) followed the atrophy of classical literature, and was represented by Petrarch and by Renaissance humanism in general. The antiquarian emphasis characteristic of this era led some to focus all their attention on ancient authors alone. Johann Albert Fabricius (1668-1736), with the *Bibliotheca Latina* (1697) and then the *Bibliotheca Graeca* (1705-1728), proposed for ancient literature temporal boundaries ranging from the early imperial to the Antonine to the early Christian periods.

In Italy a continual model was then adopted by Girolamo Tiraboschi, a Jesuit and the successor of Muratori, who was then serving as director of the Biblioteca Estense in Modena. In his monumental *History of Italian Literature* (1772-1781), he devotes the first sections to classical Latin literature. Yet the Italian writer Ugo Foscolo criticized the work for its lack of discrimination, for he saw in it only an "archive of materials, histories, documents, and arguments reasoned and ordered to serve the literary history."[77] What Foscolo lamented as absent in the work of Tiraboschi was, essentially, the notion of literary history that in those same years was being developed in Germany by Wolf.

THE "SCIENCE" OF ANTIQUITY

In the modern period, classical culture no longer enjoyed the pervasiveness and sway that it had been endowed with by the humanists. Indeed, by the end of the eighteenth century, it had become a subject area, well reflected in the German term *Altertumswissenschaft* ("science of antiquity"), first used by Friedrich August Wolf. In doing so he created a bright-line rule for classical studies. As humanism passed from Italy to France, and later to the Netherlands and England, what we now call classics developed into a robust area of study. As a discipline it found its home in Germany, which now assumed a leading role.

Friedrich August Wolf (1759-1824) began his academic career in 1777 as a student in Göttingen, where he incurred the displeasure of a professor, Christian Gottlob Heyne. Heyne disdained Wolf's notion that an Ur-text (that is, a much older archetype) of Homer might be discerned. Nevertheless, by 1787 Wolf had become professor of philology and pedagogy at the University of Halle, a training center for teachers and officials of the Prussian state. In 1795 he published his famous *Prolegomena ad Homerum*, in which he attempted to show the growth of the Homeric tradition from the genesis of the rhapsodic poems to the Alexandrian exegesis. To do this, he availed himself of the scholia of the Codex Marcianus Graecus 454 (now in the British Museum; also known as *Venetus A*), small portions of which had been published a few years earlier. The novelty of Wolf's method, compared to the more traditional approaches of previous studies, is most apparent in Wolf's emphasis on the historical context, which he considered essential for a correct reading of the text itself.[78]

In the years of the Napoleonic Wars, Wolf wrote his foundational work, dedicated to Goethe, entitled "Darstellung Altertumswissenschaft nach der

Begriff, Umfang, Zweck und Wert" ("Presentation of the Science of Antiquity according to Its Concept, Scope, Purpose, and Value"). He published this work in the first issue of the journal, the *Museum der Alterthums-Wissenschaft* (1807), of which he became the director along with Philip Buttmann. The program outlined in his "Darstellung" was realized in concrete form when, in 1810, the University of Berlin was founded. Designed by Wilhelm von Humboldt, according to the wishes of the king of Prussia, the new university brought Wolf to its faculty, as he was by then among the leading educators of his time. The so-called science of antiquity that he had established became one of the university's two key pillars, alongside Hegelian philosophy.[79]

An important development in German education was destined to mark the next stage in the growth of European universities. This advance can be seen in the separation of university training, such as that sponsored by the Prussian state, from the kind of education that the Lutheran Church had sponsored. Until now, humanism had never demanded a classical culture dissociated from Christian ideals. On the contrary, from Petrarch to Erasmus the goal had been a synthesis of the spiritual and intellectual. Yet, as Rudolf Pfeiffer observes, Wolf was "a good classicist and had no interest in the Greek Bible; instead, he inclined toward Plato."[80] Such a turning toward a secularized classicism coincided with a further change in the humanistic tradition as it moved toward a fresh interest in Greek antiquity as opposed to Roman. Greek antiquity was now seen as the matrix of classical culture, alien not only to Christian tradition but to Jewish as well. The Wolfian model marked the decline of the formative project that Erasmus had distinguished in the trilingual college of Leuven, where the three key languages of European culture, Latin, Greek, and Hebrew, were taught with equal dignity.

The concept of literature was profoundly influenced by emerging national identities. This was exemplified in Germany by Johann Gottlieb Fichte's *Discourses to the German Nation* (1807–1808). In the cultural context of Romanticism, literature came to be understood as an expression of a people and of a national identity. Seizing upon this hypothesis, Wolf programmatically tied the *opera* of Latin authors to the Roman historical and cultural context. The task that he envisioned for the history of Latin literature, therefore, was the tracing out of the Roman *Volksgeist*, or "spirit of the people."[81]

To achieve this goal, Wolf distinguished between what he dubbed "external history" (*äussere Geschichte*)—the collection of information about Latin authors and their work—from "internal history" (*innere Geschichte*)—the elements that constitute Roman literature and its development, especially as concerns the history of the language and cultural institutions. The external history coincided largely with traditional historiography, while the internal history was something new, requiring that the historian bring to light deeply spiritual literature and the people whose expression it is. In academic lessons published posthumously, Wolf expresses this view clearly in his claim that to know antiquity one must transfer oneself to the days in which its people were thriving. One would have an inaccurate vantage point if one only judged from the things in the present that did not exist. One of the consequences of this assumption is that history should examine "not only the writers whose works are preserved intact, but also the most excellent, whose works have been lost."[82]

Wolf's idea marked a clear shift from the tradition of literary historiography. His departure from his predecessors is highlighted by the fact that he wrote in German instead of the traditional scholarly language, Latin. Already in 1787 in Halle, Wolf had written *Geschichte der römischen Literatur* (*History of Roman Literature*) in German. Though he returned to Latin in his *Prolegomena* (1795), he wrote again in German in 1807 in his "Darstellung." It is significant that already in his *Geschichte* he would use the name "Roman" in place of "Latin," putting aside definitively the continuity between ancient and modern that the previous historiography had inherited from humanism, which had anchored "Latin" literature to the nation that engendered it, that is, Rome. Further evidence of Wolf's establishment of a new direction can be seen in his discrete periodization of ancient literature, which had previously depended on a judgment about decadence caused by the barbarian invasions. In his new vision, Wolf's demarcation coincides with the history of Rome. He sets as a *terminus ante quem* for Roman literature the natural division of the fall of the Roman Empire (AD 476).

The formal distinction between "external" and "internal" history was later diluted or dropped, but the overall structure imposed on the history of Latin literature remains to this day. Literary development was now closely connected to the history of Rome. Individual authors would continue to be considered within the ambit of global history, although examined with varying interpretative methodologies. The periodization that Wolf introduced remained in effect, essentially breaking the connection with the production of neo-Latin that humanism had considered fundamental.

The historical approach outlined by Wolf in *Prolegomena ad Homerum* implied close cooperation between the different disciplines outlined in previous centuries. Wolf lists fourteen of them in the appendix to his "Darstellung," which form a sort of encyclopedic compendium for the scholar of the ancient world: language, grammar, interpretation and philological criticism, metrical studies, geography, history, chronology, mythology, literary history, art history, rhetoric, archeology, numismatics, and epigraphy. Indeed, only two disciplines are missing from Wolf's list, paleography and papyrology, which would be added to the curriculum of ancient studies in the course of the nineteenth century. With these additions, the Wolfian schematization prevailed and greatly influenced the organization of the discipline of classical studies in European universities, in which the designation "Altertumswissenschaft" enjoyed considerable fortune.

In the new configuration of these ancient scientific studies, one particular area of discrete studies emerged, not unrelated to other disciplinary areas. Wolf seemed to be aware of the fact that contemporary science was growing separate from ancient culture and therefore was coming to be considered on its own. Thus Wolf wrote, "The dominant scientific studies of today, removed far from the influence of Greece and Rome, could not await further discoveries coming from that quarter."[83] Wolf's assertion implicitly acknowledged the disagreements between those who placed a premium on ancient thought and those who embraced modern ideas. That debate had in fact taken place more than a century earlier, singling out only a few exceptions, such as Protestant theology and jurisprudence, which had maintained a direct relationship with the ancient texts. Wolf warns against abandoning the study of ancient culture: "When one fails to find anything more to learn from the ancients, he then too often forgets that one can learn many things from them."[84]

This statement accepts the premise that classical culture can have a formative role for education, separate from whatever any individual discipline might encompass, and it thus harks back to the function of the humanities advanced by the Renaissance. Thus Wolf propounds further, "Languages, the first artistic creations of the human spirit, involve the full range of general ideas and forms of our thought that have been acquired and perfected in the cultural evolution of the nations."[85]

Such classical training does not necessarily coincide with specialized pursuit of the scientific studies of antiquity. Study of specific sciences remains reserved for those who choose to specialize. Nevertheless, classical training requires knowledge of the two classical languages as well as Greco-Roman culture and literature. In the case of Latin, that knowledge also included the capacity to compose in the language. Although Latin composition was a traditional practice, Wolf envisioned it acting no longer as a form of communication, as this role had been taken over by vernacular languages.

Rather, Wolf saw in Latin composition its own "intrinsic value for the practice of developing facility with the language" and thus the classical scholar of his day "expresses in modern languages more sensibly a knowledge of antiquity better than someone trying to make a trickle of words and Latin phrases."[86] It is a paradox of sorts that he is best remembered for his Latin work, *Prolegomena ad Homerum*, when in fact he mostly used German for his essays, as he believed that each young scholar then could habituate himself "also to the improvement of language of his own country."[87] Wolf closes his "Darstellung" with a quote from Jean Paul, the writer and educator of German Romanticism: "Humanity today would sink into an inscrutable abyss if the youth should not take a whirl on the carousel of life through the quiet shrine of the great ancient times and of its men."[88]

THE RISE OF PHILHELLENISM

The notion of philhellenism or what is sometimes called neohellenism was propelled to the forefront of the German academy by Johann Joachim Winckelmann (1717–1768), a historian of classical art whose research on antiquity had a major impact on the culture of the eighteenth century. He derived his idealization of aesthetics from his study of Greek statuary, piquing the interest of authors such as Lessing and Goethe.[89] At the same time, Romantic appreciation of the Homeric poems waxed greater. The oral expression of those poems, viewed by many as faithful to the Greek spirit, was compared to several sagas and medieval cultural traditions that were rediscovered in the early nineteenth century, a topic that would be reprised just over a hundred years later by Milman Parry and Albert Bates Lord.

In the nineteenth century, the positive reevaluation of Greek culture was essentially a reversal of the perspective of the humanists, who had preferred Rome's dominant position in their appraisal of ancient culture; they had regarded the recovery of Greek classics merely as a bonus. Wolf viewed the Greeks essentially as the first people "in whom the impulse to perfect itself in a multiplicity of ways sprang from innermost needs, spirit, and feeling, and where the passionate inclination quickly to switch from one object to another begat a well ordered circle of arts and of knowledge, raising humankind's existence to selfless commitment based on its nobler strengths."[90] Wolf, on the other hand, did not see the Romans as a people of original talent, but rather believed that they imitated the Greeks so magnificently that they enriched what they received with practical doctrine. Thus Wolf espouses that "their direct impact upon legal theory was almost equal to that of the Greeks in the field of philosophy."[91]

Wolf's view represented a telling contrast. His revision of Greece as antiquity's more valuable contributor would remain dominant in European culture and German philology throughout the nineteenth century and beyond. One of the consequences of this idealization of the Greeks was the diminished value placed on other ancient cultures. Egyptians, Hebrews, and Persians were viewed only as undeveloped forms of civilization, without access "to a higher, truly spiritual culture" that belonged only to the Greeks and Romans. The peoples of the East, by contrast, were not viewed as going beyond the basic spontaneous spiritual formation, "especially the Jews as can be seen in their sacred texts; also, among peoples of the east as a whole, there is a character that differs starkly from the Greeks."[92]

With this statement, Wolf offers a prelude to the anti-Semitism and racism that spread to Europe in the nineteenth and twentieth centuries, as is reflected in the sympathy he shows for a viewpoint of exclusivity—indeed a marginalization of the "other"—resulting in a palpable elitism. "Allow me," he writes, "in the spirit of the ancients, who looked upon the Barbarians, from top to bottom, as a less noble branch of humanity, to restrict the name 'antiquity,' in its exceptional significance, to two peoples of refined spiritual culture, scholarship, and art."[93]

This "scientific" viewpoint, which really is an opinion rather than a point of science, was viewed by Wolf as essentially historical in nature. Such a notion dovetails with an idea that he puts forth in particular in his "Darstellung." There he distinguishes between a documentary philology, based on manuscripts, and a philology "superior in form, which one might call rather 'divination,'" that "clarifies, through internal proofs, something for which there are not any obvious testimonies, testimonies that sometimes may not even exist." He continues, "History has to bear the blame of being in fact narrative, not mathematics."[94] This "scientific" perspective allows, of course, a good deal of latitude for the insertion of opinions about the supremacy or inferiority of ancient cultures based on preconceived prejudices.

Wolf's historic orientation was reprised by his student, August Boeckh (1785-1867), who systematized a framework for "Altertumswissenschaft"[95] in his *Enzyklopädie und Methodenlehre der philologischen Wissenschaften* (*Encyclopedia and Methodology of Philological Science*), a work that demonstrates by its very title a connection in spirit with the Hegelian encyclopedia. The influence of Hegelian historicism is very strong in Boeckh's work.[96] This is perhaps not surprising, because Boeckh had been a colleague of Hegel at the University of Berlin.

In the 1820s Boeckh found himself at the center of a lively controversy with Johann Gottfried Hermann (1772-1844). Hermann, whose *Weltanschauung* was essentially Kantian, had studied under Friedrich Wolfgang Reitz (1733-1790), a student of metrics, an editor of Aristotle, and an admirer of Richard Bentley's approach to textual philology. As an heir to this tradition, Hermann favored education in ancient languages as the foundation of philological and ancient studies in general. In fact he was antipathetic to any hint of historicism or Romanticism. His opposition emerged in particular in a clash with the Greek scholar Karl Otfried Müller (1797-1840), who was interested in ancient Greek culture and religion as a whole, not merely on the basis of linguistic study.[97]

The controversy born of the tension between the philological and the cultural considerations of antiquity began, in a sense, with these two scholars. In any case, it was certainly enhanced by their personal idiosyncrasies. The representatives of each approach learned to coexist. In many ways, they complemented each other, as they both were, in different ways, deeply involved in the project of *Altertumswissenschaft*. In the nineteenth century the linguistic approach, by then dubbed "philological," became dominant, owing in part to important developments in textual philology, exemplified in the methodology of a scholar named Karl Lachmann.[98]

LACHMANN'S "METHOD"

One might properly place quotes around the entire phrase "Lachmann's Method," for it is a historiographic myth: the principles that are ascribed to Lachmann's name had already made their appearance in large part before the work of this one individual.[99] Nevertheless, their attribution to him, stemming from his edition of Lucretius published in 1880, is traditional and therefore significant.[100]

Regardless of the actual role or investigative merits of Lachmann, the "method" attributed to him led to a turning point in the history of philology. This approach affected not only classical texts but also novels, legal codes, and anything involving handwritten tradition. The popularity of Lachmann's method stems from a system of "scientific" documentation prevalent during the nineteenth and twentieth centuries and related to positivism, a product of the Enlightenment that asserted that all truth is scientifically discernible.[101] This style of nineteenth-century philology follows the path forged by Hermann in that it was more concerned with the regularity of phenomena and form than with historical reconstruction.

The success of Lachmann's method was also enhanced by related factors, chief among which was the increased accessibility of codices, starting during the French Revolution. After centuries of dispersion in private collections, these codices were now often assembled in large public libraries. In addition, the consolidation of paleographic methods led to more accurate dating of manuscripts. Further, the development and employment of new photographic techniques allowed philologists to collate manuscripts without having to travel throughout Europe.

Karl Lachmann (1793–1881) studied in Leipzig and Göttingen. Beginning in 1818 he taught in Königsberg and, from 1828, in Berlin. In addition to his work on classical authors, he engaged in philological studies of German texts and of the New Testament.[102] Among the numerous editions published by Lachmann, his edition of Lucretius' *De rerum natura* (1880), commonly referred to as the first critical edition of that author, demonstrated the method of new philology.

The programmatic idea that Lachmann pursued is equivalent to a reconstruction of the mechanics of the text through a methodical examination of manuscripts. We find it already formulated in his preface to the New Testament published in 1842, where he writes almost capriciously, "We can and must establish the text without reference to its interpretation."[103] Yet it was impossible for scholars who followed in his wake to practice such a "pure" method without any interpretation; thus it is not without reason that Timpanaro once called out Lachmann's boastful positivistic stance as being quite different from Lachmann's own practice of the method.[104]

The claim, despite its emptiness, does point out the cultural exigency to which the method responds. There was a shift from an empirical practice of textual criticism, whereby an *emendatio* was derived apart from the *textus receptus*, to a procedure strongly influenced by the inventiveness and *ingenium* of the philologist. By employing a methodical and scientific procedure, the philologist could perform his work based on clear rules and transparent procedures.

Questioning the elevated status of the *textus receptus* was, in fact, nothing new. Bentley had already criticized the practice of overvaluing manuscript readings, as had Wolf after him. Even the principal innovation of the Lachmannian method, the *recensio*, had precedent. That central feature, elucidated below, in short derives from the examination of the internal tradition of handwritten texts, elimination of insignificant witnesses, and the reconstruction of the relationship between those witnesses, with a view to determining which manuscripts are of greater and which of lesser merit. The principle of *eliminatio codicum descriptorum* was anticipated as early as Politian.[105]

Other elements of the Lachmannian method date to the first half of the nineteenth century: these include close scrutiny of the codices to establish correspondence between them, prefiguring the *stemma codicum*. This procedure was already proposed in 1831 by Carl Zumpt for the tradition of Ciceronian texts and was perfected in the case of the Verrine orations later by a student of Wolf, Friedrich W. Ritschl, *Doktorvater* of, among others, Friedrich Nietzsche and Basil Gildersleeve, who twenty years before the century's end would found the *American Journal of Philology*.[106]

Another important contribution was made by Thomas Magister, whose primary focus of research was establishing the texts of Dionysius of Halicarnassus and Plautus. Yet such philological analysis was still in its infancy. Even the *stemma* of the tradition of the text of Lucretius was not traced by Lachmann but by Jacob Bernays, another of Ritschl's students.[107] Finally, the concept of the archetype was developed by Nicolai Madvig, who in 1439 had published an edition of Cicero's *De finibus*. Nevertheless, because these disparate practices came together in Lachmann's edition of the *De rerum natura*, his name rather than that of Bernays, Ritschl, or Madvig is associated with the method of establishing a text that involved the key editorial concepts of archetype, *stemma*, and *recensio*.

The last of these concepts, *recensio*, involves recognizing the witnesses of a manuscript text and defining their mutual relationships. The latter is not a problem, of course, if there is only one witness: in this case, the editor will pass directly to the practice of *examinatio*. But for the majority of classical texts there are multiple witnesses and in some cases more than one hundred such manuscripts. Once the witnesses have been compiled, the first procedure of *recensio* is the *eliminatio codicum descriptorum*, wherein a copied manuscript (*codex descriptus*)—specifically, a copy of a codex already in hand—is eliminated. If, for example, two codices, manuscripts A and B, contain the same mistakes but manuscript B also contains errors not present in A, B can be assumed to be a copy of A. When making this copy of manuscript A, the copyist of manuscript B inherited the first text's errors while also introducing his own mistakes into the later text. Codices thus transmitted have no independent value for the reconstruction of the text, as they do not carry information beyond that provided by the *antigraph*, and therefore should not be taken into account in the next phase of *recensio*.[108]

The second stage of *recensio* consists in defining the relationships between the codices that remain. These are now grouped into families based on the mistakes they have in common. These clusters are not based upon every error in a text but only the most significant ones, known as "Leitfehler," which when translated from German means a "guiding error." Such mistakes include those that cannot be emended by conjecture by the next copyist and therefore allow the philologist to posit relationships of dependency between various witnesses. If one codex includes a certain copyist error, it is inevitable that later codices in the same family would have the same error. This applies not only to errors of transcription, but also to *lacunae*, or "gaps," in the text. Thus, if a line of text is missing in one manuscript because the copyist failed to include it, all codices that depend on that manuscript will also omit that line.

The German scholar Paul Maas distinguished between "separative" and "subjunctive" mistakes, that is, those subordinate within the tradition. The first type, called "Trennfehler," establishes the independence of two witnesses: if B does not have an error found in A, then B cannot depend on A. The second, "Bindefehler," establishes mutual dependence of at least two witnesses upon one common source: if A and B both have the same error, but neither is the original *apograph*, then, assuming the mistake was not made independently, they both must depend on another manuscript (C), which must previously have introduced the shared error. A reading in the text in question, as an heir of C's bloodline, is thus "subjoined" to C and cognate with the manuscript that shares its error.

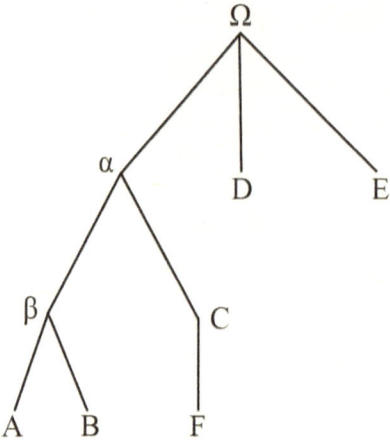

FIGURE 5.1. *Sample* stemma 1.

The set of relationships between codices is represented by what is known as the *stemma codicum*. Thus, the establishment of the descent of manuscripts is known as the "stemmatic method."[109] In the so-called Lachmannian method, the *stemma*'s apex is the archetype, the ancestor from which the tradition derives.[110] From the archetype lines are drawn representing branches that bear individual known codices or manuscript families. In the case of such families, the founder of a single-family branch is called a "subarchetype." The lines from that source correspond to the relationships of dependence within the family. In the case of a lost codex, the manuscripts that branch from the source within a family act as witnesses to the family and its source.

The current practice is to represent the source manuscript with Latin capital letters, such as A, B, or C, sometimes choosing the initial of the name of the current codex, such as V for Vaticanus or M for Mediceus. The subarchetypes are normally marked with Greek letters in lowercase, such as α, β, or γ, while the archetype is designated with the capital omega (Ω). A single family can comprise subfamilies; for example, if three codices, A, B, and C, share a number of significant errors, but two of them, say A and B, have certain errors in common not shared by C, they would represent a subfamily. In that case, the subfamily of A and B presupposes a common subarchetype, say β, which descends, as the C codex does, from the subarchetype α, the source of the entire family.

We offer an example of a *stemma* in figure 5.1. Manuscripts A, B, C, and F, which are in the same family because of significant errors they share, depend on the subarchetype α. The errors shared by manuscripts A and B descend from the subarchetype β. Manuscripts D and E present errors that are separate from A,

B, and C, and therefore depend on an archetype independent of α. Manuscript F is *descriptus* insofar as it depends on codex C — that is, aside from its own errors, it has the same mistakes as C.

The archetype is the reconstructed and thus virtual specimen from which an entire manuscript family arises. The term *archetypus* is used by ancient authors to mean the original copy or authorial autograph. The paleographer, however, refers to the archetype as a less ideal form, that is, the place where manuscript traditions could theoretically coincide with the original, were the codices or branches of the tradition to depend directly on the original. Such a circumstance, however, which can occur in the case of medieval or Renaissance texts, is unlikely for classical texts, as the documentation that survives was itself assembled many centuries after the creation of the original document. The randomness of available testimony can result in a reconstruction of the archetype that may not be entirely accurate.[111] Indeed, identifying a new witness often necessitates changes to the configuration of the archetype. It is not always the case that the reconstruction is inaccurate. There are instances in which the archetype, reconstructed on the basis of tradition, accurately matches a manuscript that existed at some point. This can be confirmed, for example, when a codex descends directly from a manuscript that had been rediscovered and copied, but was lost afterward.

The entire process that permits the construction of the *stemma* assumes that the transmission of the text occurs strictly in a vertical direction, that is, each scribe used a single specimen and inherited its errors. But it is not uncommon for such a vertical transmission also to have had a horizontal element (the distinction between vertical and horizontal transmission is suggested by the configuration of the *stemma*, in which branches of the tradition descend from above instead of below). Horizontal transmission occurs when the copyist uses not only the *antigraph* but also another witness, one that usually belongs to a branch of the tradition from higher up in the *stemma*.

Vertical transmission generally characterizes texts that had less than optimal fortune during the Middle Ages. It was unlikely for a scribe to have access to more than one copy of a work that was not very widespread. For certain works of heightened popularity, however, it was entirely possible that there would be multiple copies. In such cases, the copyist confronted with passages or words too difficult to read might resort to one or more specimens, interpolating as often as necessary. The resulting text, in which readings from two or more *antigraphs* might turn up, would not have a clear lineage. This phenomenon is called *contaminatio*.[112]

Maas was well aware of the consequences that *contaminatio* entailed as far as the stemmatic method was concerned. Beginning in his second edition of

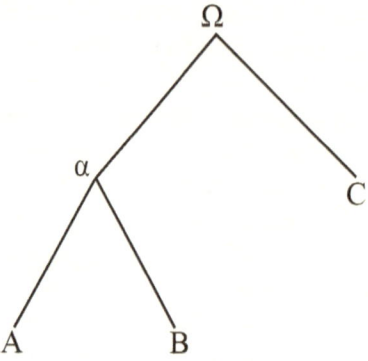

FIGURE 5.2. *Sample* stemma 2.

Textkritik, one finds at the conclusion of his discussion a statement characterized by an inauspicious, even ponderous tone: "Gegen die Kontamination ist kein Kraut gewachsen" ("no remedy for *contaminatio* has been discovered").[113]

For the philologist, the step after *recensio* is *examinatio*. This process provides the basis for careful examination of the document. The first step is to identify errors or variations present in individual manuscripts, but not in the subarchetype on which they depend.[114] These errors or changes are attributable to the copyist of the codex, and they are not taken into account for the formation of the text. This process of eliminating exceptional readings is called *eliminatio lectionum singularium*.

Next, the philologist must attempt to determine the reading of the archetype. In the case of the *stemma* traced above, the reconstruction of the archetype is not difficult, as it comprises only three branches (A and B, C, and D), each of which presents a number of errors. The reading of the archetype is the one on which the two branches converge, while the third is characterized by error. One speaks, in such cases, about a reconstruction of the mechanics of the text. In many cases, however, the stemma is divided into two branches, as in figure 5.2. A *stemma* of this type does not allow a linear reconstruction to spring from an individual archetype. For cases in which the codices provide two different readings, or variants, the editor must decide whether the correct reading is that of branch A and B or that of C.

There are two main criteria for choosing between variants of a text. The first criterion, the *usus scribendi*, assumes that the language conventions and author style will remain consistent throughout the text. Where errors run counter to these, the editor may hypothesize corrections that match what we would expect.[115] The second criterion, the *lectio difficilior*, looks at what types of mistakes could produce the error. With both of these criteria in mind, when there are mistakes, the editor should tend to assume that the most difficult reading is correct, as this is most conducive to copyist error.

The reconstruction of the text does not preclude the reconstruction of the archetype. Although the archetype is equivalent to the original, it may be corrupted. In these cases, the editor has two options. The first is simply to report the corruption, preserving the tradition in the text but accompanying it with a sign, the so-called *crux desperationis* (†).[116] This solution is generally adopted only when the text of the manuscripts presents grammatical or syntactical irregularities so extreme as not to appear solvable with a convincing conjecture. The second solution is to amend the text by replacing the corrupt portion with a conjecture. This solution obviously incorporates the philological practice of *emendatio ope ingenii*, an exercise common well before Lachmann. The delicacy of this operation is underscored by its definition of *divinatio*, which combines the art of emendation with the ancient custom of predicting the future.

The criteria that the editor must take into account in this difficult task are in part those already followed for the choice of variants, that is, those followed in the *usus scribendi*. Based on this notion, the editor's conjecture must be consistent with the language and style of the author. Further, the adoption of any *lectio difficilior* must take into account the mistakes that the scribe could have more easily committed. The editor must also consider the kind of errors present throughout the text, the kind of writing used, as each script has its own errors that easily occur, and, finally, the copyists who were transcribing the text even before the archetype, whose mistakes could have affected the tradition of the text in question.

The transmission of the text may involve not only errors in transcribing single words, but even larger textual omissions and additions. The reconstruction of the manuscript tradition can remedy such phenomena, if they occurred only in some branches of the tradition. If the gap, however, is already present in the archetype, this reconstruction will not help correct the error. Instead, the editor can sometimes compensate by integrating appropriate and logical conjectures. In such a case, the portion of text that fills the gap is presented between angled brackets: <...>.

More often, especially if the *lacuna* is sizable, the editor cannot help but offer a signal as to the difficulty in the text, by placing one or more asterisks in this place to indicate that there is a textual problem of some kind: ***. In the case of interpolations—that is, portions of text not belonging to the original work but introduced at a later stage of the tradition—it is customary to retain the reading of the text in question rather than to expunge it, but in square brackets: [...]. The result of the processes described is the critical edition, an edition of the text in Greek or Latin accompanied by an *apparatus criticus* (critical apparatus), generally printed in smaller type at the bottom of the page. The apparatus informs the reader of the variants provided by the codices for each portion of the text.

For poetic texts, the postponement of the *apparatus criticus* to the bottom of the text is dependent on the number of verses. For prose texts, this is determined by line numbers. The apparatus will contain, for each word or phrase, the various alternate readings that derive from the manuscript tradition. That tradition is referred to via the abbreviations or initials of its individual codices or manuscripts, generally the same as those used in the *stemma*, which is itself either portrayed graphically (as above) or discussed in detail in the critical edition's preface. In that *praefatio*, itself sometimes written in Latin, the editor gives an account of the manuscript tradition, and, indirectly, of the results of *recensio* and criteria in *examinatio*.[117] Both the text and the apparatus are preceded by a list of the codices alongside their related acronyms (as indicated in the *conspectus siglorum*, the list of available manuscripts with their standard abbreviations) and, when possible, by the *stemma*.

An apparatus may be positive or negative.[118] For a positive apparatus, all the variants present in the manuscripts are listed, including those adopted in the text. For a negative apparatus, only the variants discarded are included. The reader will deduce by exclusion the witnesses of the variant adopted in text. When a conjecture is adopted in the text, the apparatus will give notice of the reading (or readings) of the manuscripts and the name of the scholar who proposed the conjecture. The apparatus may provide other indications for the individual codices, such as the presence of *lacunae*, glosses, or corrections.[119]

6 / TOOLS FOR THE MODERN SCHOLAR

DRAMATIC PERFORMANCES IN ANCIENT ATHENS WERE COMpetitive, with prizes awarded by the city to the victors. Because these competitions were an expenditure for the city, records of the victorious plays, poets, and actors for each festival were kept in the city's archives. These records attained a new meaning ca. 335 BC, when Aristotle compiled his *Didaskaliai*, a chronologically organized compilation of victorious poets and their plays, a list similar in content to the financial records of the archons, but with a much different purpose — preserving literary history.[1] It is impossible to know what precisely Aristotle envisioned for his work, but whether foreseen or not, it became an important tool for Hellenistic scholars interpreting and preparing fresh editions of Athenian drama. Unfortunately only fragments of Aristotle's groundbreaking work are extant.

Scholarship has always benefited from such tools as the *Didaskaliai*. What to most was mere record keeping became for Alexandrian scholars an indispensable means of accurately contextualizing their discussions of Athenian drama. Over time numerous other such aids to classical scholarship were assembled, without which many fruitful strands of enquiry would have been impossible. The number and availability of research tools for classics has increased dramatically in recent years thanks in large part to the convenience of electronic databases and the easy access to such data that the Internet provides.

Compilations of documents that once required many large volumes and supplements have become immensely more efficient through the transition to an electronic format, and the openness of the Internet provides individual scholars the opportunity to make their own data available to the scholarly community. Unquestionably, electronic tools have indelibly altered methods of modern research. In what follows we will consider some of the more prominent tools for scholarly inquiry in the field of classics, including a brief background. Some of the most important trace their origins back centuries before the advent of the computer. As we will see, however, the path to the current set of electronic research tools has not always been clear, and digital resources bring with them their own set of challenges and controversies.

For millennia, scholars have relied on archives like Aristotle's compilation of dramatic victories, but the history of modern research and the tools that it encompasses dates most immediately to the nineteenth century and leads up to the current age of digital resources. The speed and accessibility of online databases has been revolutionary. In 1993 one eminent scholar put the matter vividly in his review of Jon Solomon's *Accessing Antiquity: The Computerization of Classical Studies*: "The powers of memory are vastly over-rated. It is now eleven years since I bought a computer, nine years since TLG announced its CD, four years since I found the Internet, and I simply cannot remember how we ever lived without these tools. What dark hovels did we dwell in? What vast caverns of ignorance did we patrol in search of elusive scraps of information that now leap to our call and dance in serried ranks before us?"[2] The tools providing these "serried ranks" of data are in fact the unimaginably powerful successors of methods and collections honed in the modern era and ultimately derived from a long tradition of classical scholarship going back to Aristotle.

Before embarking upon our survey of scholarly resources, we should note that while classics and its related disciplines have made great use of tools both in print and online to grant all scholars access to the literary and material culture of antiquity, the methods of presenting that information are often particularized in each subfield and not necessarily intuitive for a newcomer. This chapter aims in some ways to fill that gap by providing an introduction to and historical context for some valuable scholarly tools.

We begin with the important disclaimer that our work necessarily stands not only *on* the shoulders of giants, but among them as well. In this vein, we should mention here the highly practical and useful guide by David Schaps for those beginning to undertake research in classics and its related fields. His *Handbook for Classical Research*[3] considers, among other things, such topics as how to construct a bibliography, read an archaeological report, and make sense of a published inscription. Schaps's clear and concise approach provides an excellent point of departure for those setting out to do research in just about any subfield pertaining to classical antiquity.

Perhaps the nineteenth century's most significant contribution is the *Corpus Inscriptionum Latinarum* (*CIL*). As noted in chapter 2, interest in recording Latin inscriptions dates as far back as the Carolingian period. The *Codex Einsidlensis* n. 326 is a compilation of Greek and Latin inscriptions organized geographically and topographically, as they were encountered by a traveler to Rome. Though the first systematic collection of inscriptions can be attributed to Poggio Bracciolini in 1429, the most significant advance in epigraphic studies belongs to Cyriacus of Ancona. In the first half of the fifteenth century, Cyriacus traveled through Italy, Greece, and the Levant and in addition to recording inscriptions, also made observations about their monuments that are now standard practice for epigraphers.[4]

Work on epigraphy continued throughout the sixteenth century. At the turn of the next century, Jan Gruter, with the help of Joseph Justus Scaliger, produced a comprehensive collection containing twelve thousand inscriptions (1602). This work was carefully indexed and became the standard source for inscriptions until the formation of Mommsen's monumental *CIL*.[5] Theodor Mommsen (1817–1903) determined that the venerable work of Gruter and Scaliger was no longer adequate to meet the needs of scholars.[6] He presented a plan for *CIL* to the Royal Prussian Academy of Sciences, but his project was initially not well received; August Boeckh in particular, whose incomplete *Corpus Inscriptionum Graecarum* provided a model for Mommsen, was hostile to Mommsen's plans for *CIL*.[7] Nevertheless, Mommsen set to work on *CIL* in 1853 and continued in this endeavor until his death.[8]

Mommsen's great codification of Latin inscriptions is itself a monument of scholarship that persists as the standard collection for Roman epigraphy. *CIL* currently contains about 180,000 inscriptions and is still a work in progress that continues to be updated by scholars. The seventeen volumes of *CIL* (with work on an eighteenth underway) are, generally speaking, arranged geographically, but there are some exceptions. The first volume (now in its second edition, cited as *CIL* I²) contains inscriptions from the Roman Republic to the death of Julius Caesar. The geographic arrangement of the collection begins with *CIL* II, which encompasses inscriptions from Spain (specifically the Roman provinces of Hispania Citerior, Baetica, and Lusitania), while *CIL* III contains inscriptions from the eastern Mediterranean and provinces along the Danube.[9] The fourth volume lists wall inscriptions from Pompeii, Herculaneum, and Stabiae, with a supplement that includes wax tablets and vases.[10] *CIL* V gathers inscriptions from Cisalpine Gaul (northern Italy),[11] while *CIL* VI is devoted to inscriptions from Rome, whose total of 54,000 represents the largest of the *CIL* volumes.[12]

In each volume, every inscription is preceded by a heading known as a *lemma*, which is a unique number used to identify the inscription. Following the *lemma* comes information about the context of the inscription, which can include details about the size or condition of the stone and a statement of where the stone was found, its "provenance." The unique number attached to inscriptions in the *lemma* provides easy identification of each entry; it is standard to refer to inscriptions from *CIL* according to their volume and *CIL* number (e.g., *CIL* IX 1514). Because *CIL* is comprehensive, it includes many inscriptions that have been published elsewhere, and it is therefore not uncommon for a single inscription to be cited in multiple collections. One might consider, for example, the *Senatus consultum de Bacchanalibus*, a senatorial decree dating from 186 BC that restricts the practice of the cult of Bacchus. This inscription is number 581 in *CIL* I², but also appears as inscription 18 in another important collection known as *Inscriptiones Latinae Selectae* (*ILS*). One is likely to find this inscription referred to in scholarship, then, as *CIL* I² 581 = *ILS* 18.

The simplest way to locate inscriptions is by *CIL* number. If the *CIL* number is unknown or a topical search of *CIL* is required, the indices must be consulted. Each volume of *CIL* has its own set of indices, which allow for searches by name, location, grammatical features, and other topics. To date, the ability to consult all of *CIL*'s thousands of inscriptions requires access to hard copies of its many volumes and fascicles. *CIL* has a useful webpage with a wealth of information on the history of the collection and a searchable database that provides information on provenance, bibliography, photos, and cross-references with other collections,[13] but transcriptions of the inscriptions are not yet presented there.[14]

Although the *CIL* is the most extensive collection of Latin inscriptions, the number and size of its volumes make it inaccessible outside of a well-equipped research library. Though efforts are underway to make the *CIL* available online, the constraints of so large a collection were recognized immediately, and solved in part through another collection, the above-mentioned *Inscriptiones Latinae Selectae*, the first volume of which was published by Hermann Dessau in 1892. The *ILS* is a collection of Latin inscriptions in five volumes, and thus forms a far smaller and more manageable dataset than *CIL*. The *ILS* was designed to be a corpus of inscriptions for the benefit of scholars who could not get access to *CIL*—Dessau observes that few scholars have whole volumes of *CIL* at home[15]—and those who are new to the study of inscriptions. Dessau also sought to make epigraphy accessible to nonexperts, an intention reflected in the collection's size and the criteria for the inscriptions it includes; accordingly, he selected "those inscriptions whose contents he deemed most worth knowing,"[16] assembling inscriptions representative of numerous different types.[17]

The organization of Greek inscriptions, however, is sadly not as neat as those in Latin. The largest collection of Greek inscriptions is the *IG* (*Inscriptiones Graecae*). This collection grew out of August Boeckh's *Corpus Inscriptionum Graecarum* (*CIG*), which sought to gather all known Greek inscriptions, but could not keep pace with new findings and fresh discussions of known epigraphic texts. By the time the indices were published in 1877 (the first part of the collection appeared in 1825), the collection was already outdated and a new project, the *IG*, had been undertaken to replace it.[18] The *CIG*'s inauspicious beginning shaped its successor in some important and unfortunate ways, especially since the *CIL* provides a model for what it might have been. The *IG* is geographically oriented, like the *CIL*, but the region it encompasses is limited to Europe. Volumes planned for Greek inscriptions outside mainland Greece were never realized, since a great portion of their contents had already been taken up in other collections (such as the *Tituli Asiae Minoris* or *Inscriptiones Creticae*).[19] Greek inscriptions are thus scattered over a number of collections, making both the inscriptions and the collections more difficult to use.

In spite of the *IG*'s obvious limitations, new editions of some volumes have been produced. The first and second volumes of the *IG* received second editions (frequently referred to as the *editio altera*), and have replaced *IG* I–III.[20] A few other volumes (IV and IX) have received second editions as well, and the challenge of keeping the *IG* up-to-date continues.[21]

The Packard Humanities Institute (PHI) provides an online, searchable database of the *IG*'s inscriptions; the website is kept current, and even includes fascicles of the third edition, published in 2012.[22] Though the PHI search engine is extremely useful, it does not replace the hard copies of the *IG*, whose most recent editions include photographs of the inscriptions. The advantage to the PHI version, however, is that the commentary is in English as opposed to Latin, as it is in the print edition.

In addition to the collections of inscriptions noted above, other important tools include *L'Année épigraphique* (*AE*), an annual publication of newly discovered inscriptions that provides revisions of older texts based on new scholarship for texts up through the seventh century AD. *AE* has undergone numerous changes since its inception in 1888 as a supplement to *CIL* under the lengthy title *Revue des publications épigraphique relatives à l'Antiquité romaine*.[23] It has grown in scope and usefulness, as it now includes Greek inscriptions, but only those that pertain to Rome. This publication features transcriptions of new inscriptions, citations for scholarship that treats particular inscriptions, and summaries of the works it cites. Its numerous indices and geographic organization make it easily accessible.[24]

Another important periodical essential for finding the most recent epigraphic research for Greek inscriptions is the *Supplementum Epigraphicum Graecum* (*SEG*). The *SEG* publishes new inscriptions and corrects older inscriptions and also summarizes articles (and even books, in its *varia* section) that substantially address Greek inscriptions, with the goal of "mak[ing] available all texts published in learned periodicals and books."[25] The *SEG* is geographically organized and includes indices in each of its annual editions, and thus resembles *AE* in its form and uses. The *SEG* is published by Brill, and is available online through a subscription service. The online version includes all volumes of the *SEG* and so is an extremely convenient way to search the entirety of *SEG* for information on an inscription or region.

Epigraphers since the nineteenth century have made herculean efforts to centralize both Greek and Latin inscriptions in large collections. Though *CIL* is the most successful outcome of these efforts, the *IG*, too, is indispensable. Although the majority of ancient Greek and Latin inscriptions are therefore easy to locate, there nevertheless exist many smaller collections of value to the epigrapher, especially those that gather inscriptions outside the chronological bounds of *CIL* and *IG*. These collections can most easily be found using the *Guide de l'épigraphiste*, now in its fourth edition. The *Guide* is organized according to criteria used for collections; there are sections organized by language and chronology (Greek inscriptions up to 1453 and Latin inscriptions up to the Merovingian era, for example) and sections for collections organized by type (religious texts, *instrumenta domestica*, etc.). As collections do for individual inscriptions, the *Guide* provides a *lemma* for each collection with a description of its contents, and indicates when older collections have been superseded by new ones. The *Guide*'s bibliography includes monographs that treat inscriptions as well. Unlike the other resources discussed, the *Guide* is not limited to ancient texts and is therefore useful for locating collections of Christian and other late inscriptions not found in the large collections of *CIL* and *IG*.

Epigraphic collections are not limited to updates of older projects. New, web-based initiatives to generate scholarly databases of original collections have begun to appear in recent years. As interest in complete online collections of ancient material was being articulated in the 1990s, so too was the insistence that the encoded data be transferable across platforms. Tom Elliot responded to this need with the creation of EpiDoc, which is, in essence, a set of guidelines for encoding scholarly material in a broadly accessible digital format (XML).[26] Originally aimed at epigraphic material, EpiDoc standards have been expanded to include online papyrological material as well. EpiDoc has benefited from the support of the numerous scholars who anticipated the need for standards for encoding ancient material.[27] The foresighted creation of EpiDoc has surely augmented the usefulness of online databases, which, while not replacing the print copies of collections, have made scholarly fields, especially epigraphy and papyrology, more accessible to students and scholars alike.

For epigraphical resources that employ the EpiDoc format, one excellent example is the U.S. Epigraphy Project,[28] which was begun in 1995, and is currently housed at Brown University under the direction of John Bodel. The U.S. Epigraphy Project collects data on inscriptions held by institutions in the United States. The database currently contains just over 2,400 Latin and Greek inscriptions, many with images. In addition to the searchable database, the site contains a wealth of helpful links to online epigraphic resources.

Navigating the thousands of extant inscriptions is, perhaps, initially daunting, but ultimately quite manageable, thanks to the diligent and ongoing work of cataloguing and digitizing inscriptions old and new. The situation for papyri, however, is unfortunately a bit more complicated. No systematic, unified collection of papyri exists on the scale of *CIL*. For a discussion of available papyrological resources it may be best to begin with the two major sources of extant papyrus rolls and fragments—Herculaneum and Egypt—which have provided the bulk of our extant papyri. The Herculaneum papyri come from a single villa and represent evidence for a personal library well stocked with works by Philodemus, while the papyri from Egypt are not so narrow in their genre or geography. The latter collection paints a vivid picture of the society of the ancient world, and Roman Egypt in particular.[29]

One might recall from chapter 1 that the Herculaneum papyri, carbonized (but not burned) in the eruption of Vesuvius in AD 79, were first discovered between 1752 and 1754.[30] In the early part of the twentieth century various attempts were undertaken to unravel and read these rolls, and the result too often was merely damage to the papyri. Antonio Piaggio, a friar who invented a machine to open the carbonized rolls, also catalogued those that he believed could be read,[31] as most of the papyri did not seem to him legible. Recent advances in technology, however, have increased the legibility of even some of the most damaged rolls. In particular the use of multispectral imaging is noteworthy for allowing scholars to see otherwise invisible text.[32] The result of these technological advances has been a marked increase in the rate of publication of Herculaneum papyri in recent decades.

The dissemination of the content of these papyri requires a different approach from inscriptions or even other fragments of papyrus. Unlike inscriptions, which tend to be relatively brief, the papyrus rolls can be quite extensive, requiring entire monographs for the publication of fragments from a single roll. Also rather unique is that the rolls are all of the same provenance and type (that is, they are all literary papyri from Herculaneum), unlike the Egyptian papyri, which are spread across different sites and of varying types. Thus, geography and genre are moot points in organizing the papyri, which are instead grouped according to author and work. The most prominent author found in the papyri is Philodemus, and many of the papyri that contain his works have been published in the series titled "La Scuola di Epicuro," under the direction of M. Gigante. Currently the series has produced seventeen volumes and five supplements.

"La Scuola di Epicuro" is, however, but one source for the texts from Herculaneum. Several other authors are represented in the papyri, even if only in fragmentary form. Many of the papyri from these disparate authors have been published in the journal *Cronache Ercolanesi*, the bulletin of the Centro Internazionale per lo Studio dei Papiri Ercolanesi.[33] Others are spread across various journals and monographs. Images of the papyri can be difficult to locate.

Although they can typically be found as part of the publication of the papyri, to date they have not been made available online, but efforts are underway to fill this gap.

The website administrated by the Friends of Herculaneum Society provides an extremely helpful bibliography grouped primarily by author.[34] The individual Herculaneum papyri, however, are classified by the abbreviation *P.Herc.*, followed by a number unique to each individual fragment. These have not been systematically published, although they tend to appear in print, and the pace of publication has required frequent updates.[35]

Unlike the papyri from Herculaneum, Egyptian papyri, while in need of conservation, fortunately were well preserved by the dry climate of northern Africa, and so do not require the same kind of technological intervention necessary for reading carbonized papyri. Rather they are normally found in a state of preservation adequate for immediate transcription and interpretation. As mentioned in chapter 1, the two most prominent figures for the early publication of Egyptian papyri are Bernard Grenfell and A. S. Hunt, who discovered a trove of papyri in the rubbish heaps of Oxyrhynchus. Grenfell and Hunt's first expedition to Oxyrhynchus in 1897 was financed by the Egypt Exploration Fund. The expedition yielded a trove of some 2,300 papyri.[36]

At the end of their excavation season, Grenfell and Hunt took the papyri they had discovered back to Oxford and soon produced the first volume of *Oxyrhynchus Papyri (P.Oxy.)*, which was published as early as 1898. It contained 158 papyri chosen for their legibility and because they were deemed generally representative of the variety of material preserved in the papyri.[37] *Oxyrhynchus Papyri* was published by the Graeco-Roman branch of the Egypt Exploration Society, which had been founded a mere six years earlier. The fund that this society administers continues to allow for the production of volumes in the series, but the society is now known simply as the Egypt Exploration Society. Volumes of *Oxyrhynchus Papyri* are currently published as part of its Graeco-Roman Memoirs series.

Each volume of *Oxyrhynchus Papyri* contains a transcription, commentary, and, where appropriate, translation of the included papyri; a full set of indices can be found at the end of each. In contrast to Grenfell and Hunt's original publications of *Oxyrhynchus Papyri*, recent installations tend to be thematically focused. The latest volume, for example, is a collection of medical texts from Oxyrhynchus.[38] The most convenient way to locate relevant papyri in the Oxyrhynchus collection is to use the database of Oxyrhynchus Online,[39] which is searchable for all eighty volumes by author, title, genre, date, *P.Oxy.* volume, and publication number. Though the database is quite helpful and equipped with useful metadata (such as date, provenance, genre, etc.), its chief function lies in directing users to the relevant volumes of *P.Oxy.*

It has been convenient to single out for discussion the papyri from Herculaneum and Oxyrhynchus both because they are famous collections and because their excavation histories have influenced their publication. Important as these findspots are for papyrus, however, they are hardly the sum of papyrology, as papyri continue to turn up in various places in the Mediterranean basin and beyond, and in private collections as well. As the field of papyrology continues to flourish, it also faces challenges, one of which is simply attending to the tremendous amount of unpublished papyri in collections throughout the world. By one estimate, there are currently between one million and one and a half million unpublished papyri in known collections.[40] Although there is not a pressing need for all of them to be published, as many are but fragmentary and offer only slender bits of information, nevertheless the mere number is daunting.[41]

Efforts to increase accessibility to the texts of extant papyri through the use of online platforms have met with considerable success. Numerous databases gathering information about and images of papyri have been established, such as the Advanced Papyrological Information System (APIS), begun in 1995, the Duke Data Bank of Documentary Papyri (DDbDP), founded as early as 1982, and Trismegistos, a metadata catalogue established in 2005.[42] Each of these databases, and others as well, sought to provide increased access to papyri within a particular scope.

In 2006 and 2007, a plan was proposed that would unite the resources of various papyrological databases into a single website.[43] Over the next five years, the database Papyri.info was constructed. The new website was structured around the DDbDP and required converting the coding to EpiDoc. The information provided by the DDbDP is augmented by the data collected by APIS, Trismegistos, HGV (Heidelberger Gesamtverzeichnis der Griechischen Papyrusurkunden Ägyptens), BP (Bibliographie Papyrologique), and APD (the Arabic Papyrology Database). For any given papyrus the information available from each of the participating databases is displayed on a single electronic page. Thus, while there is sometimes overlap in the metadata provided, the contribution of each data pool is discernible in a single place. Especially noteworthy are the transcriptions provided by the DDbDP,[44] where some papyri are accompanied by translations.[45]

In addition to the cataloguing of published papyri, online tools are also being utilized to speed work on unpublished papyri. One such effort, the Ancient Lives Project,[46] under the direction of Dirk Obbink, is a collaborative website that relies upon crowdsourcing to measure and transcribe unpublished fragments of Oxyrhynchus papyri. The data on transcriptions and measurements gathered through the website are analyzed by professional papyrologists before being published in the *Oxyrhynchus Papyri* volumes of Graeco-Roman Memoirs. The process of preparing papyri for publication remains time-consuming, but resources such as the Ancient Lives Project represent great strides both in making more papyri available to the public and in familiarizing a greater number of scholars with this traditionally specialized discipline.

Inscriptions and papyri are not the only ancient texts gathered in collections. Countless fragments of authors whose works are otherwise lost are preserved as quotations and paraphrases in literature and scholia. These brief citations have been assembled into a number of collections based on genre. Like epigraphical compilations, the number of those that gather literary fragments is extensive, and only a few will be noted here. An early and highly significant compilation of ancient texts is August Nauck's *Tragicorum Graecorum Fragmenta* (*TrGF*), first published in 1856, and followed by a second edition in 1889. *TrGF* collects fragments and testimonia for Greek tragedy, including minor tragedians and anonymous fragments (*adespota*) as well as the major tragedians (Aeschylus, Sophocles, and Euripides). Volume 1 notably includes *didaskaliai* and other sources. Nauck's five-volume collection has undergone substantial revision in recent years, beginning with a new edition of volume 1 edited by Bruno Snell and Richard Kannicht in 1971, with another edition following in 1986.[47]

The volumes of *TrGF* are arranged by ancient author.[48] Each volume's fragments (F) are arranged by title of the tragedy and are preceded by testimonia (T). No commentary is provided, but each volume contains a full *apparatus criticus*, which often provides references to both ancient sources and modern scholarly discussions. Currently *TrGF* is not available online.

In addition to fragments of Greek tragedy, those of Greek comedy have also received extensive attention. As early as 1826, August Meineke envisioned a complete collection of comic fragments, but the enormity of the task led him to focus solely on fragments of Attic comedy. The grand vision of a complete collection that included Doric fragments was renewed by Georg Kaibel, who in 1899 produced a single volume of *Comicorum Graecorum Fragmenta* before his death.[49] Kaibel's text gathered the remnants of Doric comedy and the mimes of Phylaces, and became the basis for Rudolf Kassel and Colin Austin's *Poetae Comici Graeci* (*PCG*), a compendium that fulfills Meineke's original vision for a complete collection.

All of *PCG* is contained in eight volumes, which are organized alphabetically by ancient author[50] and subdivided by comedy. The entry for each author is headed by testimonia, and fragments are provided with an *apparatus criticus*. A commentary follows each fragment, though, as with *TrGF*, it is highly abbreviated. *PCG* is not currently available online.

Another important collection is Hermann Diels's *Die Fragmente der Vorsokratiker*, published in three volumes in 1903. Diels sought to make his compendium of pre-Socratic texts comprehensive, and consequently took an inclusive approach to the subjects encompassed by ancient philosophy. In particular, Diels was careful to include texts related to mathematics and medicine.[51] Diels continued work on his collection, publishing three subsequent editions. A fifth and sixth edition, edited by Walther Kranz, kept Diels's great collection up to date. The sixth edition, published in 1951-1952, was the last to incorporate significant enhancements.

Diels's *Fragmente*, typically referred to by the editors' names, Diels-Kranz (D-K), is arranged by author, each of which is numbered. The section devoted to particular authors are further partitioned into A (testimonia) and B (fragments) sections. Each entry in the A and B sections are numbered as well. Thus a reference to an entry in D-K would provide the author number, lettered section, and entry number (e.g., 28 B2 D-K for Polycleitus fragment 2). Care must be taken when using D-K citations since the numbering for pre-Socratic authors was altered between the fourth and fifth editions. In addition to the Greek testimonia and fragments, D-K provides a German translation of the fragments (not the testimonia). Anglophones may wish to consult Kathleen Freeman's *Ancilla to the Pre-Socratic Philosophers*, an English translation of the fragments based on the fifth edition of D-K. Unfortunately, D-K is not currently available online.

Felix Jacoby's *Die Fragmente der griechischen Historiker* (*FGrHist*) began as Jacoby's dissertation in 1900 under Hermann Diels and from there became his life's work. The first volume of *FGrHist* was published in 1923, and Jacoby continued his work on the project until his death in 1959.[52] *FGrHist* eschews strict chronological or alphabetical ordering in favor of organization by genre. Jacoby divided the collection as a whole into three parts: (1) genealogy and mythography; (2) histories covering a span of time (*Zeitgeschichte*); and (3) history of cities and cultures. At Jacoby's death, the three parts were contained in fifteen volumes, although part 3 was still incomplete. A three-volume set of indices was published in 1999 by Pierre Bonnechère.[53] Work on *FGrHist* is ongoing thanks to an international group of scholars. The scope of the collection has expanded to include a fourth and a fifth part — (4) biography and antiquarian literature; (5) historical geography — with fascicles of part 4 already beginning to appear.[54]

The organization of *FGrHist* places text and commentary in separate volumes. Volumes that contain text are organized by genre with a number assigned to each author. As in Diels's collection, Jacoby divides each author's section into testimonia (T) and fragments (F), each of which is also numbered. Thus, according to Jacoby's system, a particular fragment would be identified by reference to author and fragment numbers (e.g., Pherecydes of Athens 3 F 23). A complete electronic version of *FGrHist* is available from Brill.[55]

Compendia of literary fragments and inscriptions are currently still primarily accessible through volumes in print, but that is not the case for a number of scholarly tools. The most recent developments in standard philological resources, which now are in some cases over a century old, are directly tied to robust online platforms. The impulse to convert the works of antiquity into digital form was felt as early as the 1960s, and much work to bring classics into the digital age was carried out in that decade.[56] The first great success in creating a significant digital resource for classicists, however, began in the early 1970s. Far prior to the computer age, there had been a desire to make searching the corpus of Greek authors easier. After several failed attempts to systematize Greek literature, the advent of digitization provided the impetus to try again. Following a few years of planning, the work to digitize and catalogue every extant word of the Greek language began in 1973. Theodore Brunner was selected to direct the project, a position he held for twenty-five years. The resulting database was named *Thesaurus Linguae Graecae* (*TLG*), an homage to Estienne's sixteenth-century work. The *TLG* has since evolved with new technological developments and remains indispensable for anyone who works with Greek texts. The task of compiling the *TLG* was enormous, but the intensive labor of coding all of Greek literature was not the only obstacle. In the 1970s no computer software had yet been designed that could read Greek characters, a problem tackled by David Packard, both a classicist and part of the family that founded the computer manufacturer Hewlett-Packard.

In 1981 Packard, a pioneer in many respects in applying computers to the needs of research in classics, modified the operating system of a Hewlett-Packard computer to enable the processing and display of Greek text. Named "Ibycus," Packard's system and the computer on which it ran debuted in 1985. It used CD-ROM technology, very advanced at the time, for searching the assembled database of the *TLG*.[57] The Ibycus computer soon became obsolete as Unix-based software able to produce Greek characters became more widespread.[58] By the 1990s *TLG*'s technology was outdated, and the decision was made to move the database to the Internet. By 2001 the *TLG* was available online, where it continues to receive significant updates.

In its current state, the *TLG* provides limited but valuable open-source access to its corpus of texts. For full access to the texts of the *TLG*, a subscription is required. The most recent iteration of the *TLG* gathers a significant amount of data, and only some of its features are mentioned here. Each word in the texts of its extensive corpus is linked to a morphological analysis that parses the word and further links to entries from a variety of lexica, including the LSJ, the standard Greek–English lexicon, as well as specialized lexica (such as Powell's *Lexicon to Herodotus* and Cunliffe's *Lexicon of the Homeric Dialect*). Statistical analyses can display the frequency of a word over different time periods and identify the authors who most frequently use it. One of the *TLG*'s most interesting features is its ability to search by "N-grams."⁵⁹ This is essentially a method of performing textual searches that identify phrases with significant overlap. The software accounts for variations in orthography, and is a useful, but not perfect, tool for identifying parallel passages among authors. Finally, the *TLG* enables parallel browsing of two authors at once; that is, the site is designed, at the user's choice, to display two authors' texts simultaneously.

The *TLG* has undergone many changes, from magnetic tapes to CD-ROMs to the Internet. Another important, early web-based platform for classical studies is the venerable (and still valuable) Perseus Project.⁶⁰ Begun in the late 1980s, the Perseus Project, under the direction of Greg Crane, long an important voice in the challenge of bringing classics into the digital world, marked a departure from previous efforts at digitizing classical resources. Crane himself recounts his experience when, as a graduate student at Harvard, he advanced the idea of a Unix-based approach to classical resources at the 1983 meeting of the American Philological Association. Crane's approach was at odds with that of David Packard, who was championing his Ibycus system.⁶¹ Crane's support for Unix, which would be readable by different machines, ultimately prevailed, for as advanced as Packard's Ibycus system was at the time, it nonetheless was tailored to specialized equipment, the Ibycus scholarly computer, dedicated to searching classical texts and thus having otherwise limited application. As Crane notes, however, the Ibycus system laid the groundwork that would be vital for the future of online classics databases.⁶²

The Perseus Project contains an impressive database of classical literature. The original Latin or Greek text is accompanied by a translation (the *TLG* directs to some of these), and each word of the original language is linked to a page that contains a set of useful data, including a basic definition, all possible grammatical interpretations of verb forms, and a statistical analysis showing the likelihood of each grammatical form fitting the context of the sentence.⁶³ The *TLG* and the Perseus Project are notable as digital resources designed as unique projects.

Many other projects entailed the digitization of collections that had appeared in print form prior to the advent of the personal computer. It is to a summary of these venerable resources that we now turn. Perhaps the first resource encountered by the student of antiquity is the lexicon. For Greek, the standard lexicon is that produced by Henry George Liddell and Robert Scott, first published in 1843. Liddell and Scott's work was indebted to earlier lexica compiled by Franz Passow and Carl Benedict Hase. Hase's lexicon drew upon Estienne's *Thesaurus Linguae Graecae*, improving upon the sixteenth-century work not only with additions and corrections, but also by converting the etymological arrangement of entries to an alphabetical format. The lexicon was revised eight times before the death of Liddell in 1898. In 1911 Henry Stuart

Jones undertook the revision of the work, which is now in its ninth edition.[64] The most recent version of Liddell and Scott's lexicon, which includes a supplement from 1996, is now known as the "LSJ," and has been available in digital format since the 1990s, on the Perseus Project and, most recently, the *TLG* websites. For the newcomer, LSJ can be intimidating. The work was intended to meet the needs of scholars and so contains far more detail than can be offered in a smaller dictionary.

For example, an average entry for a verb in the LSJ includes a set of the verb's conjugated forms, each with a reference to where the form can be found in literature. This (often lengthy) set of citations and verbal forms precedes the definition of the word, which might seem odd to one who considers such a translation the primary purpose of the lexicon. The possible meanings of the word follow these initial citations, but even this can cause confusion. For words that cover a large semantic range, broad meanings are subdivided into more specific ones, each with citations from literary or documentary sources. The grammatical forms and textual citations are described by abbreviations, which are grouped into lists and explained in the front matter of the lexicon.

Unlike other reference works that may be too rare or expensive to be found even in institutional collections, the LSJ is still widely used in its print form. The version found on the *TLG* website debuted in 2011. The Perseus Project version is also quite usable but notably lacks the list of *lemmata* that allow one to take in multiple entries at once and make searches easier for those uncertain of a word's orthography (often due to a morphological difference between how a word appears in the text versus the *lemma*). Both the Perseus Project and the *TLG* closely replicate the format of the LSJ's text (including the LSJ page number for each entry) and, what is especially useful, link each citation to its full entry in the *TLG* (for those with a subscription, at least).

Though early efforts to create a print version of the *TLG* were stymied by the morphological and dialectical problems of cataloguing the Greek language, similar efforts for the Latin language were more successful. The *Thesaurus Linguae Latinae* (*TLL*) is, in essence, an exhaustive dictionary of the Latin language that seeks to account for all nuances in meaning of every Latin word that appears in literature up to AD 200, with additional selections up to AD 600. With the support of academic institutions in Berlin, Göttingen, Leipzig, Munich, and Vienna, work on the project began in 1894. Numerous scholars would compile "slips," which were handwritten entries for all extant Latin words.[65] Early on in the project the work was proceeding well (the first fascicle was published in 1900), and the *TLL* was scheduled to be completed by 1915, but that date had to be pushed back to 1930 after a series of delays.[66] The *TLL* remains incomplete, although work has continued over the century or more since it began. Perhaps less than a third of the *TLL* remains to be compiled, and fascicles continue to appear in print.[67]

Naturally, the overall structure of the *TLL* is alphabetical, with each volume collecting entries for words from one or two letters of the alphabet; some of these volumes are subdivided into more manageable "parts." Publication of the *TLL* now occurs in the form of fascicles that appear about once each year, but the fascicles do not appear in strict alphabetical order because the rate of completion of each is uneven. Currently, the *TLL* is complete up through the letter "O," except for the letter "N" (part 1 of volume 9), of which only two fascicles have appeared, the most recent in 2014. Even though the letter "N" is incomplete, the first fascicle of volume 11, published in 2012, has appeared and covers the beginning of the letter "R." The task of compiling the *TLL* is divided among a team of Latinists who are at work on different letters, which explains why the most recent fascicles are staggered across the two letters. Furthermore, work on the letter "Q" has been postponed due to the particular difficulty of words that begin with that letter.[68] The process is logical, but potentially confusing for the student approaching the *TLL* for the first time.

Although it is incomplete, the *TLL* is a highly valuable resource for Latinists. It is different in concept from its Greek counterpart, for the *TLG* was designed to take advantage of computer search capabilities. The *TLL*, by contrast, was designed to be, and still is, a lexicon, albeit one containing far more information that we might typically expect. A subscription version of the *TLL* is available online; it essentially reproduces the same data found in the print editions, and thus provides numerous references to ancient authors but not the full text of those authors. Each entry is divided into a heading followed by definitions. Depending on the word, the heading may contain such information as modern etymological analysis, ancient etymological accounts, variations in spelling, epigraphic abbreviations, brief chronologies of the word in prose and poetry, and issues of textual reliability.[69] As mentioned above, the print version of the *TLL*, begun in the nineteenth century, was composed in Latin and the electronic version exactly reproduces the print editions, which continue to be published in Latin.

Not all resources for classics and its related fields are centered on texts. Material culture, including art from pottery, walls, coins, and other artifacts, provide valuable evidence for understanding ancient culture and literature. One of the most prominent aspects of both visual and literary art is the incorporation of figures from myth, and so through the initiative of Lilly Kahil, an archeologist, efforts to compile a record of all known representations of mythological figures in the visual arts began. The new collection was to be called the *Lexicon Iconographicum Mythologiae Classicae* (*LIMC*), and a foundation to oversee its creation was formed in 1973.[70] It was recognized from the beginning that the enormity of the task would require an international effort, a fact that is reflected by the multilingual nature of the volumes, in which one will find articles in German, French, English, and Italian. The *LIMC* consists of eight volumes (seven regular volumes and a supplement). The first volume was published in 1981, with the final volume appearing in 1997 (the indices followed in 1999, at which point the project was considered complete, but another supplement would appear in 2009). Each volume is divided into two parts, the first of which contains the articles for each entry, while the second contains plates with the images referenced in that volume's articles.

The *LIMC* is an excellent resource for those concerned with how figures from literary myths are depicted in the plastic arts and contains much more than a catalogue of images. The collection is arranged alphabetically; each article devoted to a mythological character consists of an introduction, a bibliography, a catalogue of images, and a systematic commentary on each individual image therein.[71]

No online source replicates the full set of information found in the print version of *LIMC*, but data from the *LIMC* are available electronically in various forms. IconicLIMC[72] contains a repository of *LIMC*'s images but is not an intuitive site to navigate. Far superior is the database organized by the French *LIMC* team.[73] Although *LIMC* France provides a limited number of images, it hosts a wealth of metadata derived from *LIMC*'s catalogue of images, as well as updated bibliography.

As work on the *LIMC* was coming to a close, many involved with the project desired to continue work on the extensive archive they had built for the *LIMC*. Even as work on *LIMC* continued, meetings were being held in preparation for the next project, which ultimately became known as the *Thesaurus Cultus et Rituum Antiquorum* (*ThesCRA*).[74] As the name suggests, *ThesCRA* is a resource devoted to compiling articles on ancient Greek and Roman cult practices. Specifically, *ThesCRA* is intended to present a "comprehensive account of all substantial aspects of Greek, Roman, and Etruscan religion, apart from any assessment of the purely spiritual or philosophical, and only incidentally with the historical."[75] Its subject matter differs from that of the *LIMC*, but the organization of the articles and overall format of the collection owe much to its predecessor. Thus, articles are still composed by an international team of scholars and are arranged in a format similar to *LIMC*, with an initial discussion of the subject of the *lemma*, followed by bibliography and testimonia, which contain references to inscriptions, artifacts, and other ancient sources relating to the topic under discussion. The testimonia are formatted in much the same way as the catalogue section of *LIMC* articles. Each volume contains a set of plates showing images of many of the artifacts listed in the testimonia. Once again, an asterisk indicates an image in the plates, and a bold dot marks an illustration set alongside the text. Work on the *ThesCRA* yielded its first volume by 2005, and the collection was completed with an index in 2014.[76]

In addition to resources on ancient texts and material culture, important bibliographical tools are also available to scholars of classics. The most significant such tool is *L'Année Philologique* (*APh*), founded in 1927 by Jules Marouzeau with the assistance of Juliette Ernst.[77] Prior to Marouzeau's establishment of *APh*, bibliographic notices for classics were compiled in *Bibliotheca Scriptorum Classicorum*, edited by Wilhelm Engelmann. As a result of the First World War, however, bibliographic compendia were either discontinued or published in diminished form.[78] Nevertheless, Marouzeau was instrumental in organizing the bibliography for classical scholarship in the wake of the war both through his efforts with *APh* and with a separate, two-volume work titled *Dix années de bibliographie classique*, a compendium of bibliography for the years 1914–1924. First published in 1928 for the years 1924–1926, *APh* continued from where *Dix années* left off and has since continued to produce an annual volume meant to compile an index of scholarship from the previous year. Owing to the wide dissemination of learning, volumes now often also encompass bibliography from two years earlier.

APh is an essential bibliographical tool for research in classics and its related fields. The first part of each volume is arranged alphabetically according to ancient author or as categories of texts (such as *carmina epigraphica*). The breadth of research the volume reports on is perhaps more evident in its second half, which is arranged by subject. The list of subject headings is robust; these include ancient literature, linguistics, grammar, paleography, papyrology, history, and numismatics. Some of these topics, such as history, are further subdivided. Searches are further supported by multiple indices, including an index of rubrics (subject headings) in each volume, an index of ancient names (*index nominum antiquorum*), a geographic index, an index of scholars (*index nominum recentiorum*, for scholars from the medieval era to the present who are mentioned in published works), and an index of modern authors.

Entries in *APh* follow a regular format. The author and title of the work begin the entry, followed by the relevant bibliographic information. For monographs this includes the place of publication, press, publication date, and number of pages; for journals one finds the abbreviation for the journal's title (listed in the front matter of each volume, these abbreviations are standard for all bibliographic references in scholarship), the publication date, volume number, sometimes followed by a fascicle number in parentheses, and finally the pagination. Other indications—such as whether the work is a supplement or part of a collection—are to be found when appropriate. Many monograph entries are followed by references to scholarly reviews of the book. The main entry is set off from the book reviews by double vertical bars (| |), and references to individual reviews are set apart by a single bar (|). Some entries receive a summary or description of contents; these are set off by a diamond (♦). Each entry can also be identified by an individual record number, which consists of the volume number followed by a unique identifier (e.g., 41-04225). *APh* does not duplicate entries in its volumes, preferring instead to refer readers to entries by *APh* number.

In its print form, *APh* has long been a standard research tool for fields in any way concerned with ancient Greece and Rome. Despite the valuable information it collects, it does have some shortcomings. The most significant impediment to using *APh* is the fact that entries for specific topics may be distributed across multiple (potentially many) rubrics.[79] In addition, each volume accounts for a single year of scholarship, and setting up a multi-rubric search across multiple volumes to account even just for recent scholarship can quickly become time-consuming. Fortunately, technological advances have provided a convenient solution; as was the case with other established collections, however, the process was gradual.

APh is produced from a number of offices (currently there are six, in the United States, Italy, Germany, Spain, France, and Switzerland) that work together to assemble each volume.[80] The transition toward electronic data entry began in the 1970s at the American office of *APh*.[81] At about the same time, Dee Clayman began creating the Database of Classical Bibliography (DCB) from previous volumes of *APh*. The first volume of the DCB came out on CD-ROM in 1997 and included data from volumes 47–58 of *APh*, as well as its own search software (something lacking on the original *TLG* CD-ROM).[82] The online version of *APh* replicates the data, largely including the formatting, of entries in the print version, but with vastly superior means of searching and sorting information. Volumes of *APh* continue to be put out in print, but the website, which benefits from the same data, makes the print version outdated.[83]

Less complete, but nonetheless useful and more up-to-date is the database search engine TOCS-IN. Begun in 1992, TOCS-IN[84] gathers bibliographic details from nearly two hundred journals in classics and related fields into a searchable database. The contents of available journals date most reliably to 1992; some content before 1992 is included, but has not been systematically compiled. The site relies entirely on volunteers, who are asked to submit the table of contents for assigned journals as coded files. The details available from TOCS-IN are thus limited to citations of articles, and gaps are likely to be found due to shortfalls in volunteers. Nevertheless, TOCS-IN is a useful tool. Because the works it covers and the details it provides are more modest in scope than *APh*, TOCS-IN's database for the most recent journals is likely to be updated more quickly than *APh*, which is currently running a year behind.

An important research tool broadly applicable to all fields concerned with classical antiquity is an encyclopedia of Greek and Roman culture and its reception known as *The New Pauly (NP)*. As the name implies, *The New Pauly* (in German *Der neue Pauly*) is based on an older collection titled the *Realencyclopädie der classischen Altertumswissenschaft* (referred to as either the *RE* or Pauly-Wissowa). The *RE* was begun by August Friedrich Pauly, who published the first volume in 1839 but died before the project could be completed. The task of compiling the *RE* was then taken over by Georg Wissowa in 1890 and, finally, by Konrat Ziegler.[85] As with many such projects, the final product became much larger and more time-consuming than originally planned. The *RE* did not attain its finished state until 1980, when it reached sixty-eight volumes, fifteen supplements, and indices. The completion of the encyclopedia took well over a century, during which time some of the entries became outdated. The enormity of the collection made it expensive and unwieldy.

To make the *RE* more accessible, a smaller version was produced even before the completion of the original *RE*. Dubbed *Der kleine Pauly*, this new version was published from 1964 until 1975 and represents a selection of the much larger *RE*, narrowed down to five volumes. The *Kleine Pauly* was not produced as a mere copy of certain *RE* entries; instead, it presented updated material and new entries that, in some instances, made it superior to the larger *RE*.[86] Of course, at a mere five volumes, it lacks the breadth of the *RE*.

Today, the preferred version of Pauly's great work is *Der neue Pauly*, or, for Anglophones, the English translation known as *The New Pauly*. The English version of the *NP* saw its first volume published in 1996. The project produced twenty volumes with two volumes of indices, the last of which was published in 2011. The volumes of the *The New Pauly* are divided into two sections: the first group contains alphabetically arranged articles dedicated to all aspects of Greco-Roman culture and civilization; the second contains articles on the classical tradition. These are also alphabetically arranged but are not integrated with articles on culture and civilization. *The New Pauly* is also available online from Brill in both its English and German versions, but requires a subscription.

Another encyclopedia of classical studies undertaken in the latter half of the nineteenth century is Iwan von Müller's *Handbuch der klassischen Altertumswissenschaft (HdA)*. Eschewing the alphabetized entries of the *RE*, Müller's *Handbuch* offers extensive analyses of all aspects of Greco-Roman society (and beyond) organized according to topic.[87] Work on the *HdA* has continued unabated since Müller's death, first under Walter Otto and Hermann Bengtson, and most recently under the editorship of Hans-Joachim Gehrke and Bernhard Zimmermann. New and revised volumes continue to be produced—the most recent in 2014—as the scale of the project has continued to expand. Volumes of the *HdA*, which cover subjects as diverse as Greek and Latin grammar and

Celtic history, are akin to stand-alone volumes in a scholarly series; one will not find the *HdA* gathered on a single shelf in the library. The volumes can therefore be difficult to locate if one does not know the title or editor of a particular text. Due to the stunning scope of the project and the many volumes it encompasses, some volumes are unavoidably out-of-date. Most, however, have been updated and *HdA* remains a vital resource for classical scholars. Particularly noteworthy are the volumes on Greek and Latin grammar, *Griechische Grammatik* (2.1) and *Lateinische Grammatik* (2.2).

Finally, for research in medieval and Renaissance manuscripts, the database known as "Digital Scriptorium" (DS) is a vital recent contribution.[88] As with papyri, information and images associated with specific manuscripts are often provided online by the institutions that own the collection. DS aggregates data and images from numerous libraries and collections, and so, while far from exhaustive, it is an excellent resource for manuscript studies.

IN CLOSING, IT SHOULD BE NOTED THAT THE RESOURCES discussed in this chapter, whether the print collections of the nineteenth century or the new online databases, are painstaking endeavors. Many a scholar has forsaken a more glorious career—perhaps becoming a top name in Pindaric studies—in order to provide a resource for others. In any case, they have all, in various ways, had a significant role in advancing research in classical antiquity and medieval studies. The research tools assembled here were chosen for being the most useful to the broadest possible group of students and scholars; many discipline-specific collections have necessarily been omitted. Of course, as online platforms are utilized for research at a seemingly exponential rate, the items on a list of this sort will only grow. Indeed, it may be that the next challenge to face scholars is not finding information, but synthesizing the hordes of data now available at our fingertips.

NOTES

FOREWORD

1. The classic article on this subject is Anthony Grafton, "On the Scholarship of Politian and Its Context," *Journal of the Warburg and Courtauld Institutes* 40 (1977): 150–188.
2. Sebastiano Timpanaro, *The Genesis of Lachmann's Method*, ed. and trans. Glenn W. Most (Chicago: University of Chicago Press, 2005).
3. L. D. Reynolds and N. G. Wilson, *Scribes and Scholars: A Guide to the Transmission of Greek and Latin Literature*, 4th ed. (Oxford: Clarendon, 2014).
4. From a letter to his friend Francesco Vettori, dated December 10, 1513, in J. B. Atkinson and D. Sices, trans., *Machiavelli and His Friends: Their Personal Correspondence* (De Kalb: Northern Illinois University Press, 1996), 262–265.
5. Gilbert Highet, *The Classical Tradition: Greek and Roman Influences on Western Literature* (repr. of 1949 ed.; Oxford: Oxford University Press, 2015).
6. Charles Martindale, *Redeeming the Text: Latin Poetry and the Hermeneutics of Reception*, Roman Literature and Its Contexts (Cambridge: Cambridge University Press, 1993), 7.
7. Jerome McGann, *A Critique of Modern Textual Criticism* (Chicago: University of Chicago Press, 1983).
8. For an introduction to the issues here, see Lorna Hardwick and Christopher Stray, eds., *A Companion to Classical Receptions* (Oxford: Blackwell, 2008).
9. This point is well understood by David Scott Wilson-Okamura (*Virgil in the Renaissance* [Cambridge: Cambridge University Press, 2010], 20–27, 252–281), but his efforts to produce reliable figures were hampered by the limited data that were available at the time he was working. I am preparing a survey of the early printed commentaries on Virgil for the Catalogus Translationum et Commentariorum, but right now the figures in this paragraph are preliminary.
10. Ibid., 4.
11. This process is explained in detail in my "Virgil in the Renaissance Classroom: From Toscanella's *Osservationi . . . sopra l'opere di Virgilio* to the *Exercitationes rhetori-*

cae," in *The Classics in the Medieval and Renaissance Classroom: The Role of Ancient Texts in the Arts Curriculum as Revealed by Surviving Manuscripts and Early Printed Books*, ed. J. F. Ruys, J. Ward, and M. Heyworth, Disputatio 20 (Turnhout: Brepols, 2013), 309–328.

12. Craig Kallendorf, "Commentaries, Censorship, and Printed Books: Neo-Latin in a Transnational World," in *Acta Conventus Neo-Latini Vindobonensis*, Proceedings of the Sixteenth International Congress of Neo-Latin Studies, ed. Astrid Steiner-Weber and Franz Römer (Leiden: Brill), in preparation.

CHAPTER 1

1. William V. Harris, *Ancient Literacy* (Cambridge, MA: Harvard University Press, 1989), vii.

2. L. H. Jeffery, "Greek Alphabetic Writing," in *The Cambridge Ancient History*, vol. 3, part 3, ed. John Boardman, I. E. S. Edwards, N. G. L. Hammond, and E. Sollberger (Cambridge: Cambridge University Press, 1982), 831.

3. Jeffery (1982) 819; B. S. J. Isserlin, "The Earliest Alphabetic Writing," in *The Cambridge Ancient History*, vol. 3, part 3, ed. John Boardman, I. E. S. Edwards, N. G. L. Hammond, and E. Sollberger (Cambridge: Cambridge University Press), 816.

4. The term *boustrophedon*, which refers to the turning (*strophe*) of an ox (*bous*) as it draws the plow across a field, indicates a method of writing in which the text alternates between retrograde (right to left) and orthograde (left to right) writing styles (the text could begin with either). Further, see Barry Powell, *Homer and the Origin of the Greek Alphabet* (Cambridge: Cambridge University Press, 1991), 11.

5. Isserlin (1982) 816; John Healey, *The Early Alphabet* (Berkeley: University of California Press, 1990), 37–38.

6. Herodotus 5.58.

7. Powell (1991) 13; Jeffery (1982) 827–828.

8. Jeffery (1982) 822; cf. Powell (1991) 15.

9. Powell (1991) 13–14.

10. Powell (1991) 10–11; Jeffery (1982) 822.

11. A. Kirchhoff, *Studien zur Geschichte des griechischen Alphabets* (Berlin: Ferd, 1863), 168.

12. Calvert Watkins, "Greece in Italy outside Rome," *Harvard Studies in Classical Philology* 97 (1995): 37.

13. Harris (1989) 46.

14. Merle K. Langdon, "A New Greek Abecedarium," *Kadmos* 44 (2005): 179.

15. Homer, *Iliad* 6.152–170.

16. πέμπε δέ μιν Λυκίην δέ, πόρεν δ' ὅ γε σήματα λυγρά / γράψας ἐν πίνακι πτυκτῷ θυμοφθόρα πολλά ("He sent him to Lycia, and gave him woeful signs having engraved many life-destroying marks on a folded tablet").

17. Cf. Rufus Bellamy, "Bellerophon's Tablet," *Classical Journal* 84 (1989): 293: "writing tablets were not only known, but well known in eighth-century Greek courts." Bellamy also provides a good summary of scholarly views on the passage (290n3).

18. Barry Powell, "Homer and Writing," in *A New Companion to Homer*, ed. Ian Morris and Barry Powell (Leiden: Brill, 1997), 27.

19. B. L. Ullman, "The Etruscan Origins of the Roman Alphabet," *Classical Philology* 22 (1927): 372.

20. The *praenomen* was a Roman's first name. Romans also had a family name (*nomen*) and a surname (*cognomen*), often derived from a personal trait.

21. Plutarch, *Quaestiones Romanae* 54.

22. John Edwin Sandys, *Latin Epigraphy*, 2nd ed. (Chicago: Ares, 1927), 35–36.

23. Suetonius, *Vita divi Claudi* 41.3. The Romans used *V* and *I* as both vowels and semi-consonants; *U* and *J* would not be introduced until the sixteenth century. Cf. Leila Avrin, *Scribes, Script, and Books: The Book Arts from Antiquity to the Renaissance* (Chicago: American Library Association, 1991), 60–61.

24. Lawrence Keppie, *Understanding Roman Inscriptions* (Baltimore: Johns Hopkins University Press, 1991), 29. For the appearance of these letters in inscriptions, see *ILS* 210.

25. Pliny the Elder, *Naturalis historia* 13.69. See Edward Maunde Thompson, *An Introduction to Greek and Latin Paleography* (Oxford: Clarendon, 1912), 8.

26. Philip Mayerson, "The Role of Flax in Roman and Fatimid Egypt," *Journal of Near Eastern Studies* 56 (1997): 202.

27. For the magistrate lists and the temple of Juno Moneta, see Livy, *Ab urbe condita* 4.7, 13, 20, 23; for the Samnite ritual, see 10.38.

28. L. B. van der Meer, *Liber Linteus Zagrabiensis: The Linen Book of Zagreb. A Comment on the Longest Etruscan Text*, Monographs on Antiquity 4 (Leuven: Peeters, 2007), 4.

29. Pliny the Elder, *Naturalis historia* 13.69: *postea publica monumenta plumbeis voluminibus, mox et privata linteis confici coepta aut ceris*.

30. Giulia Piccaluga, "La specificità dei libri lintei romani," *Scrittura e Civiltà* 18 (1994): 7–8.

31. Fronto, *Ad M. Caesarem et invicem* 4.4: *Nullus angulus fuit, ubi delubrum aut fanum aut templum non sit; praeterea multi libri lintei, quod ad sacra adtinet*.

32. Piccaluga (1994) 10.

33. A. K. Bowman and J. D. Thomas, *Vindolanda: The Latin Writing Tablets*, Britannia Monograph Series (London: Society for the Promotion of Roman Studies, 1983), 34–36.

34. These tablets were initially diptychs until a senatorial decree in AD 62 required the use of triptychs. Cf. Gregory Rowe, "The Roman State: Laws, Lawmaking, and Legal Documents," in *The Oxford Handbook of Roman Epigraphy*, ed. Christer Bruun and Jonathan Edmondson (Oxford: Oxford University Press, 2014), 310.

35. Cf. Herodas, *Mime* 3.14–16.

36. Bernhard Bischoff, *Latin Palaeography: Antiquity and the Middle Ages*, trans. Dáibhí Ó Cróinín and David Ganz (Cambridge: Cambridge University Press, 1990), 14.

37. Bowman and Thomas (1983) 468.

38. Rowe (2014) 310.

39. Pliny the Younger, *Letters* 1.6.

40. Pliny the Younger, *Letters* 4.16.

41. Edward Capps Jr., "The Style of Consular Diptychs," *Art Bulletin* 10 (1927): 61.

42. O. M. Dalton, *Byzantine Art and Archaeology* (New York: Dover, 1961), 202.

43. Vindolanda (modern Chesterholm).

44. Bowman and Thomas (1983) 36–37.

45. Harris (1989) 46.

46. Rosalind Thomas, "Writing, Reading, Public and Private 'Literacies': Functional Literacy and Democratic Literacy in Greece," in *Ancient Literacies: The Culture of Reading in Greece and Rome*, ed. William Johnson and Holt Parker (Oxford: Oxford University Press, 2009), 18.

47. While women did not enjoy the same level of literacy as men, the fact that some women could read and write is evidenced not only by the tone and quality of Sappho's work but also simply by the very existence of her oeuvre. Her influence on Greek and Roman literature, particularly polymetric and elegiac poetry, was substantial.

48. Harris (1989) 62–64.

49. Harris (1989) 49. Harris's work on ancient literacy has been highly influential, but its claims for a low level of ancient literacy have been frequently challenged. For example, cf. Leonard Curchin, "Literacy in the Roman Provinces: Qualitative and Quantitative Data from Central Spain," *American Journal of Philology* 116 (1995): 461; J. H. Humphrey, ed., *Literacy in the Ancient World* (Ann Arbor: University of Michigan Press, 1991); and John Bodel, "Epigraphy and the Ancient Historian," in *Epigraphic Evidence: Ancient History from Inscriptions*, ed. John Bodel (New York: Routledge, 2001), 15. Though Harris's percentages remain a subject of debate (and may well underestimate the prevalence of literacy), his sensitivity to differing levels of reading proficiency provides an important complication to the simplistic notion of ancient literacy. Cf. Thomas (2009) 14.

50. As demonstrated by the existence of ceramic nonsense inscriptions, which give the impression of sophistication without communicating anything. Cf. Henry Immerwahr, "Nonsense Inscriptions and Literacy," *Kadmos* 45 (2006): 136–172.

51. Harris (1989) 49.

52. Aristophanes, *Clouds* 18–24; Xenophon, *Oeconomicus* 9.10.

53. Isaeus 6.7; 9.2; and 11.8 evidences written testaments. Documents recording sacral manumissions attest the increased role of writing. From Delphi alone, twelve hundred such inscriptions, dating from the second century BC to the first century AD, have been found. Cf. Harris (1989) 70–71.

54. Harris (1989) 78–77.

55. That writing was done by the court is suggested by Aristophanes, *Clouds* 769–772, where Strepsiades proposes to escape a lawsuit by using a piece of glass to focus sunlight and melt the letters as the court clerk inscribes them on a wax tablet. For the submission of written complaints by the litigants in legal cases, see Aeschines, *In Ctesiphontem 217 and 219*, who uses the expression τὴν γραφὴν ταύτην ἀπενεγκεῖν to mean "bring an indictment." Many more examples of writing in fifth- and fourth-century Athenian courts are provided by George Miller Calhoun, "Oral and Written Pleading in Athenian Courts," *Transactions and Proceedings of the American Philological Associa-*

tion 50 (1919): 180–188. Further, F. D. Harvey, "Literacy in the Athenian Democracy," *Revue des Études Grecques* 79 (1966): 593.

56. Written record keeping seems to require an explanation: θαυμάσῃ δὲ μηδεὶς ὑμῶν εἰ ἀκριβῶς ἴσμεν· οἱ γὰρ τραπεζῖται εἰώθασιν ὑπομνήματα γράφεσθαι ὧν τε διδόασιν χρημάτων ("None of you should be amazed if I know this accurately, for bankers are accustomed to writing notes concerning monies they lend" [Demosthenes 49.5]). Cf. Thomas (2009) 17.

57. Harris (1989) 77.

58. Aristophanes playfully but tellingly evokes a suspicion and disapproval of reading, which he associated with novel figures like Euripides. Cf. *Frogs* 52–54, 1113; T. J. Morgan, "Literate Education in Classical Athens," *Classical Quarterly* 49 (1999): 48.

59. Plato, *Republic* 376e.

60. Morgan (1999) 49–50.

61. Harris (1989) 121.

62. Harris (1989) 144–146.

63. Watkins (1995) 45.

64. Graeme Barker and Tom Rassmussen, *The Etruscans* (Oxford: Blackwell, 2000), 97. Cf. M. Bonghi Jovino, *Gli etruschi di Tarquinia* (Modena: Panini, 1986), 172.

65. Of the many extant fragments of the Twelve Tables, most come from assorted works of Cicero and from the Roman jurist known only as Gaius (fl. AD 160). Cf. P. R. Coleman-Norten, "Cicero's Contribution to the Text of the Twelve Tables," *Classical Journal* 46 (1950): 51–52, esp. his n. 4; and Harris (1989) 152.

66. Greg Woolf, "Literacy or Literacies in Rome?" in *Ancient Literacies: The Culture of Reading in Greece and Rome*, ed. William Johnson and Holt Parker (Oxford: Oxford University Press, 2009), 51–52.

67. Cato, *De agricultura* 2.5–6.

68. Whether literacy was *typical* of freedmen is another, more difficult matter. Cf. Woolf (2009) 52. Cf. Harris (1989) 7–8, for different types of literacy.

69. Cf. Edward E. Best Jr., "Attitudes toward Literacy Reflected in Petronius," *Classical Journal* 61 (1965): 72–76. There could also be economic advantages to educating slaves. Cf. Joseph Vogt, "Alphabet für Freie und Sklaven: Zum sozialen Aspekt des antiken Elementarunterrichts," *Rheinisches Museum* 116 (1973): 137.

70. John A. North, "The Books of the *Pontifices*," in *La mémoire perdue: Recherches sur l'administration romaine*, ed. C. Moatti (Rome: Collection de l'École francaise de Rome, 1998), 54.

71. Michael Affleck, "Priests, Patrons, and Playwrights: Libraries in Rome before 168 BC," in *Ancient Libraries*, ed. Jason König, Katerina Oikonomopoulou, and Greg Woolf (Cambridge: Cambridge University Press, 2013), 130.

72. North (1998) 46–50.

73. Affleck (2013) 127. Livy (*Ab urbe condita* 1.20.5–6) mentions written instructions for ritual acts existing as far back as the regal period.

74. Harris (1989) 152, 206; Woolf (2009) 48–49.

75. Bodel (2001) 16–18.

76. Alison Cooley, "From Document to Monument: Inscribing Official Documents

in the Greek East," in *Epigraphy and the Historical Sciences*, ed. John Davies and John Wilkes (published to British Academy Scholarship Online: January 2012), 159.

77. Harris (1989) 206.
78. Harris (1989) 217.
79. Cooley (2012) 159–162.
80. For an introduction to Greek and Roman numismatics, see William E. Metcalf, *The Oxford Handbook of Greek and Roman Coinage* (Oxford: Oxford University Press, 2012).
81. Cf. Bodel (2001) 2–3.
82. Bodel (2001) 30.
83. A. G. Woodhead, *The Study of Greek Inscriptions* (Norman: University of Oklahoma Press, 1992), 44. An example is *IG* 7.581. Epitaphs in verse are collected in Werner Peek, *Griechische Vers-Inscriften* (Berlin: Akademie, 1955).
84. Woodhead (1992) 43.
85. Sandys (1927) 94.
86. Sandys (1927) 118.
87. Keppie (1991) 54.
88. *M(arcus) Agrippa L(uci) f(ilius) co(n)s(ul) tertium fecit.* Cf. *ILS* 129. Ironically, this inscription was actually posted by Hadrian when he refashioned the temple after a fire and was meant to honor Agrippa, the temple's original builder, as an important figure in the Augustan principate and Roman history generally. Further, Mary T. Boatwright, "Hadrian and the Agrippa Inscription of the Pantheon," in *Hadrian: Art, Politics, and Economy*, ed. T. Opper, British Museum Research Publications 175 (London: Cambridge University Press, 2013), 19–30.
89. Sandys (1927) 57.
90. Cooley (2012) 159–163.
91. Woodhead (1992) 95.
92. Woodhead (1992) 97.
93. Bodel (2001) 166.
94. Further on the *CIL* can be found in chapter 6 below.
95. Christer Bruun and Jonathan Edmondson, eds., "The Epigrapher at Work," in *The Oxford Handbook of Roman Epigraphy* (Oxford: Oxford University Press, 2014), 7.
96. In the first half of the twentieth century a rubber substance was used less successfully.
97. Bruun and Edmondson (2014) 8.
98. Woodhead (1992) 6; cf. Bruun and Edmondson (2014) 12.
99. Keppie (1991) 107.
100. Woodhead (1992) 58.
101. Woodhead (1992) 58.
102. Rudolf Pfeiffer, *History of Classical Scholarship: From the Beginnings to the End of the Hellenistic Age* (Oxford: Oxford University Press, 1968), 51.
103. Olli Salomies, "Names and Identities: Onomastics and Prosopography," in *Epigraphic Evidence: Ancient History from Inscriptions*, ed. John Bodel (New York: Routledge, 2001), 74–76.

104. Cf. Salomies (2001) 77.

105. Bodel (2001) 30-31.

106. ἔδοξεν τῆι βουλῆι καὶ τῶι δήμωι / Θεμιστοκλῆς, 2-3. The text can be found in Michael H. Jameson, "A Decree of Themistokles from Troizen," *Hesperia* 29 (1960): 199-200.

107. Jameson (1960) 206.

108. Herodotus 8.40. Cf. Charles W. Fornara, "The Value of the Themistocles Decree," *American Historical Review* 73 (1967): 425-433.

109. Jameson (1960) 203. In lines 40-44, the decree orders a hundred ships to meet the enemy at Artemisium. Cf. Mikael Johansson, "The Inscription from Troizen: A Decree of Themistocles?," *Zeitschrift für Papyrologie und Epigraphik* 137 (2001): 71-72.

110. Bodel (2001) 48.

111. Jameson (1960) 204.

112. Fornara (1967) 426-427.

113. Pliny the Elder, *Naturalis historia* 13.68: *Chartae usu maxime humanitas vitae constet, certe memoria.*

114. Pliny the Elder, *Naturalis historia* 13.69.

115. Orsolina Montevecchi, *La Papirologia* (Milan: Vita e Pensiero, 2008), 9-10. Cf. also Paul Schubert, "Editing a Papyrus," in *The Oxford Handbook of Papyrology*, ed. Roger Bagnall (Oxford: Oxford University Press, 2009), 197-215.

116. Cf. Montevecchi (2008) 7, and Roger S. Bagnall, *Reading Papyri, Writing Ancient History* (New York: Routledge, 1995), 14-15, for approximations of numbers of papyri.

117. The term "literary" in this context is convenient but imprecise. One might further distinguish certain genres, such as school texts or magical papyri, as "subliterary."

118. Bagnall (1995) 9. Cf. Montevecchi (2008) 7.

119. This conventional wisdom has not gone unchallenged. Cf. Todd M. Hickey, "Writing Histories from the Papyri," in *The Oxford Handbook of Papyrology*, ed. Roger Bagnall (Oxford: Oxford University Press, 2009), 500.

120. Bagnall (1995) vii-viii, 19-21.

121. Naphtali Lewis, *Papyrus in Classical Antiquity* (Oxford: Clarendon, 1974), 84. The last extant work written on papyrus dates from the twelfth century. Cf. Lewis (1974) 94.

122. George Houston, *Inside Roman Libraries* (Chapel Hill: University of North Carolina Press, 2014), 98.

123. Bagnall (1995) 10-11.

124. Montevecchi (2008) 32.

125. E. G. Turner, "Oxyrhynchus and Its Papyri," *Greece and Rome* 63 (1952): 128.

126. See Bagnall (1995) 9-13.

127. Bagnall (1995) 14.

128. Lewis (1974) 87-88.

129. Aristophanes, *Frogs* 52-54, and Plato, *Apology* 26d-26e. Cf. Felix Reichmann, "The Book Trade at the Time of the Roman Empire," *Library Quarterly* 8 (1938): 40-41.

130. Lewis (1974) 87-88.

131. W. Muss-Arnolt, "On Semitic Words in Greek and Latin," *Transactions of the*

American Philological Association 23 (1892): 125. A blank roll, however, was known as a *chartes*, whence the Latin *charta*. See Skeat (1995) 76–77.

132. 458 Skutsch.

133. 798 Krenkel.

134. Cassius Hemina, fr. 37 Peter.

135. Livy, *Ab urbe condita* 10.31.10.

136. Bagnall (1995) 14.

137. Bagnall (1995) 14. Cf. Harris (1989) 144–145.

138. Pliny the Elder, *Naturalis historia* 13.89.

139. On *ostraka*, see Lewis (1974) 129–134; Skeat (1995) 88.

140. Pliny the Elder, *Naturalis historia* 13.74–76.

141. Cf. Lewis (1974) 42–57.

142. Johnson (1993) 47–49.

143. Pliny the Elder, *Naturalis historia* 13.74–76.

144. William Johnson, "The Posidippus Papyrus: Bookroll and Reader," in *The New Posidippus*, ed. Kathryn Gutzwiller (Oxford: Oxford University Press, 2008), 72.

145. Johnson (2008) 72–73.

146. John Van Sickle, "The Book-Roll and Some Conventions of the Poetic Book," *Arethusa* 13 (1980): 7. Cf. Frederic G. Kenyon, *Books and Readers in Ancient Greece and Rome* (Oxford: Clarendon, 1932), 61–62. William A. Johnson, *Bookrolls and Scribes in Oxyrhynchus* (Toronto: University of Toronto Press, 2004), 149, notes that roll length is likely to have been highly variable.

147. Johnson (2004) 264. See also Richard Janko, "The Herculaneum Library: Some Recent Developments," *Estudios Clásicos* 121 (2002): 27–28.

148. T. C. Skeat, "The Length of the Standard Papyrus Roll and the Cost-Advantage of the Codex," *Zeitschrift für Papyrologie und Epigraphik* 45 (1982): 172. Cf. Turner (2015) 4.

149. Galen, *On Grief*, 28–29.

150. Nicholls (2010) 379.

151. Theodor Birt, *Das Antike Buchwesen* (Berlin: Wilhelm Herz, 1882), 132.

152. Van Sickle (1980) 8–9.

153. Rex Winsbury, *The Roman Book* (London: Duckworth, 2009), 46–47.

154. Kenyon (1932) 58.

155. Johnson (2008) 74.

156. Kenyon (1932) 59–60.

157. Houston (2014) 9–10.

158. Kenyon (1932) 59.

159. Mario Capasso, "*Omphalos/Umbilicus*: Dalla Grecia a Roma," in *Volumen: Aspetti della tipologia del rotolo librario antico*, Cultura 3 (Napoli: Procaccini, 1995), 75.

160. Montevecchi (2008) 21.

161. Papyri of reduced quality were not necessarily of value even for informal uses. For example, *ostraka* served for mundane, ephemeral documents, e.g., receipts or election ballots. T. C. Skeat, "Was Papyrus Regarded as 'Cheap' or 'Expensive' in the Ancient World?" *Aegyptus* 75 (1995): 75–93.

162. Lewis (1974) 46. Cf. Skeat (1995) 87–90.

163. Skeat (1995) 82–83; cf. Kenyon (1932) 60.

164. E. G. Turner, "Writing Material for Businessmen," *Bulletin of the American Society of Papyrologists* 15 (1978): 164.

165. Skeat (1995) 82–83.

166. Martial, *Epigrams* 8.62.

167. Martial, *Epigrams* 4.86.7–11.

168. Kenyon (1932) 61.

169. Pliny the Elder, *Naturalis historia* 13.79.

170. Examples of literary works written on the *verso* are extant; perhaps the most famous of these is the London Papyrus, which contains a significant fragment of Aristotle's *Constitution of the Athenians* on its *verso*. The *recto*, written about AD 79, bears farm accounts. Like most other literary examples of opisthographic writing, the *recto* of the London Papyrus contains a documentary text that was repurposed to hold a literary text. Further, Kenyon (1932) 61.

171. T. C. Skeat, "The Use of Dictation in Ancient Book-Production," *Proceedings of the British Academy* (1956): 179–208. Line-by-line coping of exemplars did occur on occasion; cf. Johnson (2004) 49.

172. Johnson (2004) 57.

173. On book rolls, see Johnson (2004) 91–93.

174. David Diringer, *The Book before Printing: Ancient, Medieval, and Oriental* (New York: Dover, 1982), 172.

175. C. H. Roberts and T. C. Skeat, *The Birth of the Codex* (Oxford: Oxford University Press, 1983), 6–8.

176. E. G. Turner, *Greek Papyri: An Introduction* (Princeton: Princeton University Press, 1968), 9.

177. Jesse Meyer, "Parchment Production: A Brief Account," in *Scraped, Stroked, and Bound*, ed. Jonathan Wilcox (Turnhout: Brepols, 2013), 93.

178. R. Reed, *Ancient Skins, Parchments, and Leathers* (New York: Seminar, 1972), 53.

179. Meyer (2013) 94.

180. Roberts and Skeat (1983) 8–10. Cf. Lewis (1974) 129–134, and Skeat (1995) for discussions of the cost of writing materials.

181. Pliny the Elder, *Naturalis historia* 13.70.

182. Richard Johnson, "Ancient and Medieval Accounts of the 'Invention' of Parchment," *California Studies in Classical Antiquity* 3 (1970): 116–118.

183. Pliny the Elder, *Naturalis historia* 13.70: "afterwards the use of parchment, through which the immortality of men persists, spread extensively" (*postea promiscue patuit usus rei qua constat immortalitas hominum*). Cf. Kenyon (1932) 90.

184. Johnson (1970) 119.

185. The English "parchment" is also derived from the word, which descends from the Latin via the French "parchemin." Diringer (1982) 170.

186. Roberts and Skeat (1983) 5.

187. Horace, *Ars poetica* 386–390.

188. One may recall Pliny the Younger, who speaks of composing on wax tablets while hunting (*Epistulae* 1.6).

189. Quintilian, *Institutio oratoria* 10.3.31.

190. 2 Timothy 4.13. Cf. Roberts and Skeat (1983) 22.

191. Martial, *Epigrams* 1.2; 14.184, 186, 188, 190, and 192. Cf. Roberts and Skeat (1983) 24–28; Kenyon (1932) 89–91.

192. Galen, *Peri alypias* 33; Nicholls (2010) 381.

193. Nicholls (2010) 380–383.

194. Roberts and Skeat (1983) 35.

195. Roberts and Skeat (1983) 37.

196. E. G. Turner, *The Typology of the Early Codex* (Eugene, OR: Wipf and Stock, 1977), 37.

197. Turner (1977) 37–40.

198. Turner (1977) 40. Though Roberts and Skeat (1983) 49–51 dispute these advantages.

199. Skeat (1982) 172–175. Cf. T. C. Skeat, "The Origin of the Christian Codex," *Zeitschrift für Papyrologie und Epigraphik* 102 (1994): 265.

200. Turner (1977) 73. The potential for confusion in the ordering of loose pages led to the development of pagination. Cf. Roberts and Skeat (1983) 51–52.

201. Turner (1977) 44–46.

202. Roberts and Skeat (1983) 38–39.

203. Skeat (1994) 263–264. Cf. Kenyon (1932) 99–100.

204. Ulpian, *Ad Sabinum* 24; cf. *Digesta Justiniani* 32.52pr.

205. Roberts and Skeat (1983) 32.

206. For a discussion of how the conception of the *bibliotheke* progresses, see Wolfam Hoepfner, "Zu griechischen Bibliotheken und Bücherschränken," *Archäologischer Anzieger* 1 (1996): 25–27.

207. Strabo, *Geography* 13.1.54.

208. Christian Jacob, "Fragments of a History of Ancient Libraries," in *Ancient Libraries*, ed. Jason König, Katerina Oikonomopoulou, and Greg Woolf (Cambridge: Cambridge University Press, 2013), 69.

209. Thomas Hendrickson, "The Invention of the Greek Library," *Transactions of the American Philological Association* 144 (2014): 379–380; Roger Bagnall, "Alexandria: Library of Dreams," *Proceedings of the American Philosophical Society* 146 (2002): 349.

210. Tzetzes, *Prolegomena de comoediae Aristophanis* 2, places the book count at 490,000; Gellius (7.17.3) claims the number is 700,000. Further, Bagnall (2002) 349.

211. *Letter of Aristeas* 9–11; Bagnall (2002) 349–354.

212. P. M. Fraser, *Ptolemaic Alexandria* (Oxford: Oxford University Press, 1985), 1:324.

213. Jacob (2013) 64–65.

214. Judith S. McKenzie, Sheila Gibson, and A. T. Reyes, "Reconstructing the Serapeum in Alexandria from the Archaeological Evidence," *Journal of Roman Studies* 94 (2004): 99–100; Epiphanius, *On Weights and Measures* 11.

215. In Turkey, the remains of a structure have been identified by Conze as those of the Pergamene library, which itself was adjoined to another structure despite the space supposedly dedicated to the study of texts, though some have questioned this identification. Cf. Gälle Coqueugniot, "Where Was the Royal Library of Pergamum? An Insti-

tution Found and Lost Again," in *Ancient Libraries*, ed. Jason König, Katerina Oikonomopoulou, and Greg Woolf (Cambridge: Cambridge University Press, 2013), 122. Cf. Hendrickson (2014) 380-385.

216. Known from a decree of 117/6 BC. Cf. Hendrickson (2014) 397-398.

217. Morgan (1999) 51-52.

218. Cf. Annette Harder, "From Text to Text: The Impact of the Alexandrian Library on the Work of Hellenistic Poets," in *Ancient Libraries*, ed. Jason König, Katerina Oikonomopoulou, and Greg Woolf (Cambridge: Cambridge University Press, 2013), 96-108.

219. For the problems, see Bagnall (2002) 350-351.

220. Tzetzes, *Prolegomena in Aristophanis comoediae* 2. Cf. Rudolf Blum, *Kallimachos: The Alexandrian Library and the Origins of Bibliography*, trans. Hans Wellisch (Madison: University of Wisconsin Press, 1991), 126.

221. Affleck (2013) 126.

222. Strabo, *Geography* 13.1.54.

223. Suetonius, *Augustus* 29; Cassius Dio 53.1.3.

224. Temples in Rome from an early period were repositories of documents accessible to the public; such collections may have influenced early public library development. Cf. Ewen Bowie, "Libraries for the Caesars," in *Ancient Libraries*, ed. Jason König, Katerina Oikonomopoulou, and Greg Woolf (Cambridge: Cambridge University Press, 2013), 239-242.

225. Pliny the Elder, *Naturalis historia* 7.115, 35.10; Suetonius, *Julius Caesar* 44.

226. Suetonius, *Augustus* 29, notes that the Temple of Apollo held senate meetings. Tacitus, *Annales* 2.37, further describes Marcus Hortalus gazing upon an image of his famous ancestor during one of these meetings. Ovid, *Tristia* 3.1.60-68, describes the structure's elaborate exterior. Cf. Matthew Nicholls, "Roman Libraries as Public Buildings in the Cities of the Empire," in *Ancient Libraries*, ed. Jason König, Katerina Oikonomopoulou, and Greg Woolf (Cambridge: Cambridge University Press, 2013), 262-263.

227. Pliny the Elder, *Naturalis historia* 35.10.

228. Suetonius, *Tiberius* 74; for Verres' failure to remove the statue, see Cicero, *In Verrem* 2.4.119.

229. Bowie (2013) 241-242; Nicholls (2013) 266.

230. Galen, *Peri alypias* 2-5; Matthew Nicholls, "Galen and Libraries in the *Peri Alupias*," *Journal of Roman Studies* 101 (2011): 129.

231. Fr. 327 in R. Kassel and C. Austin, *Poeti Comici Graeci* (Berlin: de Gruyter, 2001).

232. Aristophanes, *Frogs* 52-54.

233. See the scholium to Aristophanes, *Frogs* 53.

234. Richard Moorton, "Euripides' *Andromeda* in Aristophanes' *Frogs*," *American Journal of Philology* 108 (1987): 434-436.

235. Aristophanes, *Frogs* 1114.

236. L. E. Rossi, "I generi letterari e le loro leggi scritte e non scritte nelle letterature classiche," *Bulletin of the Institute of Classical Studies of the University of London* 18 (1971): 78.

237. E. G. Turner, *Athenian Books in the Fifth and Fourth Centuries B.C.* (London: H. K. Lewis, 1952), 20-21.

238. Plato, *Apology* 26d–26e.

239. Strabo, *Geography* 13.1.52.

240. Cf. Raymond J. Starr, "The Circulation of Literary Texts in the Roman World," *Classical Quarterly* 37 (1987): 222; and Reichmann (1938) 42.

241. Sander Goldberg, *Constructing Literature in the Roman Republic: Poetry and Its Reception* (Cambridge: Cambridge University Press, 2005), 49–50.

242. Starr (1987) 213–215, notes that, once finished, the author would make more copies for distribution to friends, but only to friends; the dedicatee would be the first recipient, followed by others in the author's closed circle.

243. Cicero, *Ad Atticum* 4.10.1.

244. Cicero, *Ad Atticum* 16.11.1.

245. Starr (1987) 214.

246. Propertius, *Elegies* 2.34b.41–42.

247. Larry Richardson Jr., *A New Topographical Dictionary of Ancient Rome* (Baltimore: Johns Hopkins University Press, 1992), 44–45, s.v. "Auditorium Maecenatis." S. B. Platner and T. Ashby, *A Topographical Dictionary of Ancient Rome* (London: Oxford University Press, 1929), 61, s.v. "Auditorium Maecenatis." Recently, Alexander G. Thein, "Auditorium Maecenatis," in *Digital Augustan Rome* (http://digitalaugustanrome.org, Philadelphia, 2011), no. 313, suggests that "the location confirms the long-established identification with Maecenas," and notes that a graffito citing the poet Callimachus may suggest that the auditorium provided "a roofed theater for literary performances," although other functions are also possible. He concludes, "it was probably a multifunctional space." Further, see M. de Vos, s.v. "Horti Maecenatis. 'Auditorium,'" *Lexicon Topographicum Urbis Romae* 3 (1983): 74–75; Silviana Rizzo, "L'Auditorium di Mecenate," in *Archeologia in Roma capitale, 1870–1911* (Venice: Marsilio, 1983), 228. For a detailed discussion of the excavation that also considers early photographs of the auditorium and the surrounding landscape, see Ruth Christine Häuber, "Zur Topographie der Horti Maecenatis und der Horti Lamiani auf dem Esquilin in Rom," *Kölner Jahrbuch* 23 (1990): 15, 29 fig. 13, 62, and map 3. Cf. also John Bodel, *Graveyards and Groves: A Study of the Lex Lucerina* (Cambridge, MA: Harvard University Press, 1994), 13–18.

248. L. Richardson Jr., *Pompeii: An Architectural History* (Baltimore: Johns Hopkins University Press, 1988), 131–134; Paul Zanker, *Pompeii: Public and Private Life*, trans. D. L. Schneider (Cambridge, MA: Harvard University Press, 1998), 65–68.

249. B. A. van Groningen, *"Ekdosis,"* *Mnemosyne* 16 (1963): 5.

250. Richard Sommer, "T. Pomponius Atticus und die Verbreitung von Ciceros Werken," *Hermes* 61 (1926): 392. Cf. Starr (1987) 215.

251. E.g., Catullus, *Carmina* 1.1; see Starr (1987) 214.

252. Van Groningen (1963) 3–4.

253. Harry Y. Gamble, *Books and Readers in the Early Church: A History of Early Christian Texts* (New Haven: Yale University Press, 1995).

254. Cicero, *Ad Quintum fratrem* 3.4.5; see Sommer (1926) 392–393.

255. Catullus, *Carmina* 14.17.

256. Gamble (1995) 87.

257. Martial, *Epigrams* 2.1.

258. Houston (2014) 118.

259. An example is Ovid's citation at *Metamorphoses* 10.475 of *Aeneid* 10.475; see R. Alden Smith, *Virgil* (Chichester: Wiley-Blackwell, 2011), 9.

260. M. B. Parkes, *Their Hands before Our Eyes: A Closer Look at Scribes* (Burlington, VT: Ashgate, 2008), 4.

261. Johnson (2004) 158-159. See also Gamble (1995) 87.

262. Peter White, "Positions for Poets in Early Imperial Rome," in *Literary and Artistic Patronage in Ancient Rome*, ed. Barbara Gold (Austin: University of Texas Press, 1982), 50. See also Gamble (1995) 83.

263. Martial, *Epigrams* 11.3.

264. White (1982) 51-54.

265. The best discussion of this phenomenon is that of White (1982) 54-55. Cf. Martial, *Epigrams* 3.38, for the penury of those who practice law or write poetry for profit.

266. Pliny the Younger, *Letters* 1.2.6.

267. The introduction to Quintilian's *Institutiones oratoriae* contains a letter to his publisher, Trypho, who has encouraged Quintilian to send him his books: *Efflagitasti cotidiano convicio ut libros quos ad Marcellum meum de institutione oratoria scripseram iam emittere inciperem* ("Every day you've urged me eagerly to begin sending to you immediately the books I wrote to Marcellus on the instruction of oratory"). Cf. Reichmann (1938) 44-45.

268. Starr (1987) 223.

269. Pliny the Younger, *Letters* 9.11.

270. Pliny the Younger, *Letters* 1.2.

271. James Zetzel, *Latin Textual Criticism in Antiquity* (New York: Arno, 1981), 235. Cf. Fronto, *Ad M. Caesarem et Invicem* 1.7.4, and the subscription in the H and P manuscripts of Solinus.

272. S. West, "Chalcenteric Negligence," *Classical Quarterly* 20 (1970): 290.

CHAPTER 2

1. Cf. Gugliemo Cavallo and Herwig Maehler, eds., *Hellenistic Bookhands* (Berlin: de Gruyter, 2008), 19.

2. M. B. Parkes, *Pause and Effect: Punctuation in the West* (Berkeley: University of California Press, 1993), 10.

3. Seneca, *Epistles* 40.11.

4. William Johnson, "The Ancient Book," in *The Oxford Handbook of Papyrology*, ed. Roger Bagnall (Oxford: Oxford University Press, 2009), 262.

5. Parkes (1993) 23.

6. Cf. Otha Wingo, *Latin Punctuation in the Classical Age* (The Hague: Mouton, 1972), 20-22.

7. Calvert Watkins, "Observations on the 'Nestor's Cup' Inscription," *Harvard Studies in Classical Philology* 80 (1976): 34-35.

8. Kevin Robb, *Literacy and Paideia in Ancient Greece* (Oxford: Oxford University Press, 1994), 68 n. 1.

9. A. G. Woodhead, *The Study of Greek Inscriptions* (Norman: University of Oklahoma Press, 1992), 30.

10. Cavallo and Maehler (2008) 5.

11. John Bodel, "Epigraphy and the Ancient Historian," in *Epigraphic Evidence: Ancient History from Inscriptions*, ed. John Bodel (New York: Routledge, 2001), 27.

12. Cavallo and Maehler (2008) 4-5.

13. L. H. Jeffery, *The Local Scripts of Archaic Greece* (Oxford: Clarendon, 1990), 67, 90, 96, 116, 133.

14. Jeffery (1990) 67.

15. Woodhead (1992) 28.

16. Alison Cooley, *The Cambridge Manual of Latin Epigraphy* (Cambridge: Cambridge University Press, 2012), 309; Bodel (2001) 16.

17. Bodel (2001) 24.

18. Cavallo and Maehler (2008) 18-19.

19. Johnson (2009) 262-263.

20. Johnson (2009) 263.

21. Raffaella Cribiore, *Gymnastics of the Mind: Greek Education in Hellenistic and Roman Egypt* (Princeton: Princeton University Press, 2001), 134.

22. Parkes (1993) 10-11.

23. The absence of punctuation placed a good deal of responsibility upon the reader for interpretation. Cf. Parkes (1993) 11.

24. Johnson (2009) 263.

25. Josef Balogh, *"Voces paginarum,"* *Philologus* 82 (1927): 220, asserts that the ancients read silently "as a rule." Against this, see Bernard Knox, "Silent Reading in Antiquity," *Greek, Roman, and Byzantine Studies* 9 (1968): 421-435.

26. Augustine, *Confessions* 6.3; Plutarch, *Brutus* 5.

27. Horace, *Satires* 1.6.122-123; Suetonius, *Augustus* 39; see Knox (1968) 423-424, 428.

28. Cf. Alberto Manguel, *A History of Reading* (New York: Viking, 1996).

29. Parkes (1993) 9-10.

30. Johnson (2009) 261; William Johnson, "The Function of the Paragraphus in Greek Literary Prose Texts," *Zeitschrift für Papyrologie und Epigraphik* 100 (1994): 65.

31. Johnson (1994) 65-67.

32. Isocrates, *Antidosis* 59. Cf. E. G. Turner, *Athenian Books in the Fifth and Fourth Centuries B.C.* (London: H. K. Lewis, 1952), 7.

33. Cavallo and Maehler (2008) 19-20.

34. Gwendolen Stephen, "The Coronis," *Scriptorium* 13 (1959): 5.

35. Richard Janko, "The Derveni Papyrus: An Interim Text," *Zeitschrift für Papyrologie und Epigraphik* 141 (2002): 1; Cavallo and Maehler (2008) 7.

36. Stephen (1959) 4; Ulrich von Wilamowitz-Moellendorff, *Timotheos: Die Perser* (Leipzig: J. C. Hinrichs'sche Buchhandlung, 1903), 8.

37. *Palatine Anthology* 12.257. Cf. Stephen (1959) 4.

38. Stephen (1959) 5.

39. Stephen (1959) 9-10.

40. Cavallo and Maehler (2008) 22.

41. Rudolf Pfeiffer, *History of Classical Scholarship, 1300–1850* (Oxford: Clarendon, 1976), 79.
42. Pfeiffer (1976) 93–94.
43. Pfeiffer (1976) 90–91.
44. Cf. P. M. Fraser, *Ptolemaic Alexandria* (Oxford: Oxford University Press, 1985), 1.330–331.
45. Fausto Montana, "Hellenistic Scholarship," in *Brill's Companion to Ancient Scholarship*, ed. Franco Montanari, Stephanos Matthaios, and Antonios Rengakos (Leiden: Brill, 2015), 90–91.
46. Franco Montanari, "Correcting a Copy, Editing a Text: Alexandrian *Ekdosis* and Papyri," in *From Scholars to Scholia: Chapters in the History of Ancient Greek Scholarship*, ed. Franco Montanari and Lara Pagani (Berlin: de Gruyter, 2011), 1–2.
47. Montanari (2011) 2–3.
48. Pfeiffer (1976) 115 n. 4.
49. Pfeiffer (1976) 178.
50. Cavallo and Maehler (2008) 22.
51. Pfeiffer (1976) 178; Monica Berti and Virgilio Costa, *La bibliotheca di Alessandria: Storia di un paradiso perduto* (Rome: Tored, 2010), 148.
52. Philomen Probert, *Ancient Greek Accentuation: Synchronic Patterns, Frequency Effects, and Prehistory* (Oxford: Oxford University Press, 2006), 19–21.
53. Probert (2006) 48–50.
54. Tzetzes, *Prolegomena in Aristophanis comoediae* 1; Cf. Pfeiffer (1968) 105–107.
55. Montana (2015) 112; on the complicated meanings of *kritikos*, *grammatikos*, and *philologos*, see Pfeiffer (1976) 158–159.
56. The librarians, in order, were Zenodotus, Apollonius Rhodius, Eratosthenes of Cyrene, Aristophanes of Byzantium, Apollonius Eidographus (the classifier), and Aristarchus of Samothrace.
57. Pfeiffer (1976) 218; Berti and Costa (2010) 153.
58. Rudolf Blum, *Kallimachos: The Alexandrian Library and the Origins of Bibliography*, trans. Hans Wellisch (Madison: University of Wisconsin Press, 1991), 150–153.
59. Pfeiffer (1976) 184–185; Berti and Costa (2010) 148.
60. Pfeiffer (1976) 185. On *cola*, see Paul Maas, *Greek Metre* (Oxford: Clarendon, 1962), 38, and M. L. West, *Greek Metre* (Oxford: Clarendon, 1982), 5–6.
61. Pfeiffer (1976) 217.
62. Berti and Costa (2010) 154.
63. Eleanor Dickey, *Ancient Greek Scholarship* (Oxford: Oxford University Press, 2007), 5–6.
64. Pfeiffer (1976) 235.
65. Montana (2015) 149.
66. Esther Hansen, *The Attalids of Pergamon* (Ithaca, NY: Cornell University Press, 1971), 410.
67. Montana (2015) 151.
68. Hansen (1971) 413–414.
69. Dickey (2007) 78.

70. Pfeiffer (1976) 266.

71. Dionysius Thrax, *Techne grammatike* 1.

72. Dionysius Thrax, *Techne grammatike* 2.

73. Possibly an interpolation. Cf. David Blank, "Remarks on Nicanor, the Stoics, and the Ancient Theory of Punctuation," *Glotta* 61 (1983): 51–52.

74. Cf. Blank (1983) 48–51.

75. Dickey (2007) 11 n. 25.

76. N. G. Wilson, "A Chapter in the History of Scholia," *Classical Quarterly* 17 (1967): 244–256.

77. Dickey (2007) 18–19.

78. Stephanos Matthaios, "Philology and Grammar in the Imperial Era and Late Antiquity: An Historical and Systematic Outline," in *Brill's Companion to Ancient Scholarship*, ed. Franco Montanari, Stephanos Matthaios, and Antonios Rengakos (Leiden: Brill, 2015), 258.

79. Matthaios (2015) 261; Dickey (2007) 75.

80. Dickey (2007) 75.

81. Dickey (2007) 94.

82. Eleanor Dickey, "The Sources of Our Knowledge of Ancient Scholarship," in *From Scholars to Scholia: Chapters in the History of Ancient Greek Scholarship*, ed. Franco Montanari and Lara Pagani (Berlin: de Gruyter, 2011), 471.

83. N. G. Wilson, *Scholars of Byzantium* (Baltimore: Johns Hopkins University Press, 1983), 145–147; Dickey (2007) 90.

84. Suetonius, *On Grammarians and Rhetoricians* 2.1–2.

85. Robert Kaster, ed. and trans., *De grammaticis et rhetoribus* (Oxford: Oxford University Press, 1995), 62; Erich Gruen, *The Hellenistic World and the Coming of Rome* (Berkeley: University of California Press, 1984), 1:255.

86. Kaster (1995) 61–67.

87. Cf. Kaster (1995) 66 for the interpretive problem with the passage.

88. Gabriele Costa, *Sulla preistoria della tradizione italica* (Florence: Olschki, 2000), 32–33.

89. Costa (2000) 43.

90. Cicero, *Brutus* 75. Cf. Sander Goldberg, *Constructing Literature in the Roman Republic: Poetry and Its Reception* (Cambridge: Cambridge University Press, 2005), 2–3.

91. Jay Fisher, *The Annals of Quintus Ennius and the Italic Tradition* (Baltimore: Johns Hopkins University Press, 2014), 1. Jackie Elliott, *Ennius and the Architecture of the Annales* (Cambridge: Cambridge University Press, 2013).

92. Burkhart Cardauns, *Marcus Terentius Varro: Einführung in sein Werk* (Heidelberg: Winter, 2001), 9–11.

93. Suetonius, *Julius Caesar* 44.2.

94. Cardauns (2001) 10.

95. Cicero, *Academica* 9. Varro's works include the *De antiquitate litterarum* (On the Antiquity of Literature), the *De origine linguae latinae* (On the Origin of the Latin Language), *De similitudine verborum* (On the Similarity of Words), *De utilitate sermonis* (On the Utility of Speech), and the *Quaestiones Plautinae* (Plautine Questions). In addition to

his interest in etymology and literature, Varro was an antiquarian and philosopher. He wrote a *De philosophia*, while his antiquarian works include the *Antiquitates divinae et humanae*, a book on Roman life that adduced evidence from its language, literature, custom, and other sources. Cf. Arnaldo Momigliano, "Ancient History and the Antiquarian," *Journal of the Warburg and Courtauld Institutes* 13 (1950): 288.

96. Goldberg (2005) 67–68.

97. It was written in twenty-five books, of which only books 5–10 survive, in a tradition dating back to an eleventh-century manuscript from Monte Cassino. The first book of the *De lingua Latina* contained a dedication to Cicero and a general introduction. Books 2–7 were broadly about etymology and covered both theory and practical applications. Books 8–13 treated the inflection and morphological changes of Latin words, again both in theory and with attention to practical application. The remaining books addressed issues of syntax. Cf. Roland Grubb Kent, "On the Text of Varro, *de Lingua Latina*," *Transactions and Proceedings of the American Philological Association* 67 (1936): 64; Cardauns (2001) 30–31.

98. Aulus Gellius, *Noctes Atticae* 10.11.1. Other works of Nigidius Figulus are eclectic, including one work on the gods, another on the wind, and another still on dreams.

99. Lucan, *De bello civili* 1.639–645; Cassius Dio 45.1.3.

100. Much of this discussion relies on E. G. Turner, *The Typology of the Early Codex* (Eugene, OR: Wipf and Stock, 1977), 55–71.

101. Turner (1977) 55.

102. Turner (1977) 56.

103. Turner (1977) 57–58; J. A. Szirmai, *The Archaeology of Medieval Bookbinding* (Brookfield, VT: Ashgate, 1999), 12.

104. Ibid.

105. Turner (1977) 58.

106. Szirmai (1999) 13.

107. Szirmai (1999) 15.

108. Turner (1977) 62.

109. Michael Gullick, "How Fast Did Scribes Write? Evidence from Romanesque Manuscripts," in *Making the Medieval Book: Techniques of Production*, ed. Linda Brownrigg (Los Altos Hills, CA: Anderson-Lovelace, 1992), 40–41. Instructions for treating parchment are recorded in Conrad of Mure's thirteenth-century work *De naturis animalium* 3.1.4.15.1 (*item de pella, qualiter de ea fiat carta*). Cf. R. Reed, *Ancient Skins, Parchments, and Leathers* (New York: Seminar, 1972), 134.

110. Leslie Jones, "Pricking Manuscripts: The Instruments and Their Significance," *Speculum* 21 (1946): 389.

111. Leslie Jones, "Ancient Prickings in Eighth-Century Manuscripts," *Scriptorium* 15 (1961): 14.

112. Jones (1946) 389.

113. Jones (1961) 14.

114. Cf. Jones (1946) 395 for skepticism about the use of wheels.

115. Jones (1961) 389–395.

116. Gullick (1992) 41.

117. Eltjo Buringh, *Medieval Manuscript Production in the Latin West: Exploration with a Global Database* (Leiden: Brill, 2011), 16.

118. "Diplomatics" derives from the Latin *diploma*, which is a document drawn up by a magistrate. Henry Van Hoesen, *Roman Cursive Writing* (Princeton: Princeton University Press, 1915), 5. See further on Jean Mabillon in chapter 5.

119. Van Hoesen (1915) 5–8. For more on Mabillon and a description of majuscule and minuscule, see below.

120. Albert Gruijs, "Codicology or the Archaeology of the Book? A False Dilemma," *Quaerendo* 2 (1972): 94–95.

121. François Masai, "Paléographie et codicologie," *Scriptorium* 4 (1950): 292; Gruijs (1972) 96–97.

122. Albert Derolez, "Codicologie ou archéologie du livre?," *Scriptorium* 27 (1973): 49. Karl Löffler, *Einführung in die Handschriftenkunde* (Leipzig: K. W. Hiersemann, 1929) and Joachim Kirchner, *Germanistische Handschriftenpraxis: Ein Lehrbuch für die Studierenden der deutschen Philologie* (Munich: C. H. Beck, 1950).

123. Willy Clarysse, "Egyptian Scribes Writing Greek," *Chronique d'Égypte* 68 (1993): 188–189; Adam Bülow-Jacobson, "Writing Materials in the Ancient World," in *The Oxford Handbook of Papyrology*, ed. Roger Bagnall (Oxford: Oxford University Press, 2009), 18.

124. B. A. Van Groningen, *Short Manual of Greek Palaeography* (Leiden: A. W. Sijthoff, 1940), 21. See also Raymond Clemens and Timothy Graham, *Introduction to Manuscript Studies* (Ithaca, NY: Cornell University Press, 2007), 18.

125. Clarysse (1993) 189.

126. Bülow-Jacobson (2009) 18.

127. The English term "cursive" derives from the Latin *currere*, meaning "to run."

128. Cavallo and Maehler (2008) 4–6.

129. Cavallo and Maehler (2008) 5–9.

130. Cavallo and Maehler (2008) 11–17.

131. Jonathan Edmondson, "Inscribing Roman Texts: *Officinae*, Layout, and Carving Techniques," in *The Oxford Handbook of Roman Epigraphy*, ed. Christer Bruun and Jonathan Edmondson (Oxford: Oxford University Press, 2014), 122.

132. Edmondson (2014) 123.

133. Edmondson (2014) 122–125.

134. Benet Salway, "Late Antiquity," in *The Oxford Handbook of Roman Epigraphy*, ed. Christer Bruun and Jonathan Edmondson (Oxford: Oxford University Press, 2014), 370–371.

135. Bernhard Bischoff, *Latin Palaeography: Antiquity and the Middle Ages*, trans. Dáibhí Ó Cróinín and David Ganz (Cambridge: Cambridge University Press, 1990), 65–66.

136. Bischoff (1990) 61–62; Bowman and Thomas (1983) 53.

137. The connection, if any, between ORC and NRC is a matter of debate; the current direction of studies of Roman cursive owes much to the twentieth-century paleographer Jean Mallon, who argued, contrary to then-unanimous scholarly opinion, that NRC did not develop from ORC, but rather is related to a change in the manner of writ-

ing book-hand script. Cf. Jean Mallon, *Paléographie romaine* (Madrid: Consejo Superior de Investigaciones Científicas, Instituto Antonio de Nebrija de Filología, 1952), 106. Bowman and Thomas (1983) 53–56.

138. Augustine, *Confessions* 6.10.16.

139. As evidenced by a *colophon* composed by Statilius Maximus.

140. Parkes (2008) 5.

141. Lynn Thorndike, "Copyists' Final Jingles in Mediaeval Manuscripts," *Speculum* 12 (1937): 268.

142. Lynn Thorndike, "More Copyists' Final Jingles," *Speculum* 31 (1956): 323. Morris Bishop, *The Middle Ages* (Boston: American Heritage, 1968), 258.

143. Thorndike (1956) 325–326.

144. Martin Steinmann, "Lesen und Schreiben in den Klöstern des frühen Mittelalters," in *Teaching Writing, Learning to Write. Proceedings of the XVIth Colloquium of the Comité International de Paléographie Latine*, ed. Pamela Robinson (London: King's College London, 2010), 26.

145. Steinmann (2010) 26.

146. Parkes (2008) 6.

147. Steinmann (2010) 28–29. A similar notion can be found in the thirteenth-century Dominican order. See Cynthia Cyrus, *The Scribes for Women's Convents in Late Medieval Germany* (Toronto: University of Toronto Press, 2009), 74.

148. Cyrus (2009) 72.

149. Walter Horn and Ernest Born, "The Medieval Monastery as a Setting for the Production of Manuscripts," *Journal of the Walters Art Gallery* 44 (1986): 34.

150. Gullick (1992) 48.

151. Horn and Born (1986) 16.

152. Janos Szirmai, "Carolingian Bindings in the Abbey Library of St. Gall," in *Making the Medieval Book: Techniques of Production*, ed. Linda Brownrigg (Los Altos Hills, CA: Anderson-Lovelace, 1992), 157.

153. Horn and Born (1986) 22.

154. Horn and Born (1986) 22.

155. This figure includes scribes of both books and documents. Cf. Horn and Born (1986) 33.

156. On the difficulty of the scribe's labor, see Gullick (1992) 43; Horn and Born (1986) 33–34.

157. Gullick (1992) 39–40.

158. Parkes (2008) 6.

159. For the example of Melania the Younger, see Parkes (2008) 6. Cf. Cyrus (2009) 18–28.

160. C. P. Hammond, "A Product of a Fifth-Century Scriptorium Preserving Conventions Used by Rufinus of Aquileia," *Journal of Theological Studies* 29 (1978): 367.

161. James Thompson, *The Medieval Library* (New York: Hafner, 1957), 38–39; cf. James O'Donnell, *Cassiodorus* (Berkeley: University of California Press, 1979), 177–222.

162. C. P. Hammond, "Products of Fifth-Century Scriptoria Preserving Conventions Used by Rufinus of Aquileia: Script," *Journal of Theological Studies* 35 (1984): 347.

163. Parkes (2008) 6. Cf. *Regula magistri* cap. 85.

164. H. I. Marrou, "La technique de l'édition a l'époque patristique," *Vigiliae Christianae* 3 (1949): 215.

165. Gullick (1992) 40.

166. Cyrus (2009) 32–33.

167. Gullick (1992) 40.

168. Gullick (1992) 52.

169. Gullick (1992) 47.

170. Graham Pollard, "The *Pecia* System in the Medieval Universities," in *Medieval Scribes, Manuscripts, and Libraries. Essays Presented to N. R. Ker*, ed. M. B. Parkes and Andrew G. Watson (London: Oxford University Press, 1978), 156.

171. Cyrus (2009) 32–33.

172. Gullick (1992) 41.

173. Aliza Cohen-Mushlin, "A School for Scribes," in *Teaching Writing, Learning to Write. Proceedings of the XVIth Colloquium of the Comité International de Paléographie Latine*, ed. Pamela Robinson (London: King's College London, 2010), 61.

174. Cohen-Mushlin (2010) 63–68.

175. Gullick (1992) 39.

176. Cohen-Mushlin (2010) 64.

177. Cyrus (2009) 82.

178. Florence Edler de Roover, "The Scriptorium," in *The Medieval Library*, ed. James Westfall Thompson (New York: Hafner, 1957), 598.

179. Cyrus (2009) 83.

180. Sylvia Huot, "'Ci parle l'aucteur': The Rubrication of Voice and Authorship in *Roman de la Rose* Manuscripts," *Substance* 17 (1988): 42.

181. Cohen-Mushlin (2010) 66.

182. De Roover (1957) 599.

183. De Roover (1957) 604.

184. Cyrus (2009) 85.

185. Szirmai (1999) 103.

186. Gary Frost, "Material Quality of Medieval Bookbindings," in *Scraped, Stroked, and Bound: Materially Engaged Readings of Medieval Manuscripts*, ed. Jonathan Wilcox (Turnhout: Brepols, 2013), 129–130.

187. Szirmai (1999) 23–25.

188. Frost (2013) 130.

189. Szirmai (1999) 103–119.

190. Frost (2013) 130.

191. Jean Destrez, *La pecia dans les manuscrits universitaires du XIIIe et XIVe siècles* (Paris: Vautrain, 1935), 6.

192. T. C. Skeat, "The Use of Dictation in Ancient Book-Production," *Proceedings of the British Academy* 42 (1956): 200–205.

193. William A. Johnson, *Bookrolls and Scribes in Oxyrhynchus* (Toronto: University of Toronto Press, 2004), 39–40.

194. James Willis, *Latin Textual Criticism* (Urbana: University of Illinois Press, 1972), 47.

195. Manuscripts were written in vernacular languages as well as Latin, and it may be that, when possible, scribes were assigned texts that corresponded to their fluencies. Cf. Cyrus (2009) 85.

196. Willis (1972) 53-54, 57.

197. Willis (1972) 57.

198. Other frequent errors are caused by the interchange of *l* and *i* (e.g., *iubet/lubet*) and *d/cl* (*clipeo/dibeo*; *Clemens/demens*). For extensive examples of confusion between Greek letters, see Douglas Young, "Some Types of Scribal Error in Manuscripts of Pindar," *Greek, Roman, and Byzantine Studies* 6 (1965): 247-252.

199. M. L. West, *Textual Criticism and Editorial Technique Applicable to Greek and Latin Texts* (Stuttgart: Teubner, 1973), 27; Willis (1972) 68.

200. Willis (1972) 68.

201. West (1973) 27.

202. *Isthmean* 1.47: Rome, Biblioteca Apostolica Vaticana, ms. Graecus 1312. See Young (1965) 254.

203. *Pythian* 8.65: Göttingen, ms. Philologus 29. See Young (1965) 250.

204. Roger Pack, "Scribal Errors in an Autograph Manuscript," *American Journal of Philology* 101 (1980): 459-460.

205. An example can be seen in a manuscript of Aulus Gellius (Rome, Biblioteca Apostolica Vaticana, ms. Reginensis Latinus 1646) reading *quibus frons . . . ut mons pons frons* instead of the correct *mons, pons, fons*. Cf. P. McGushin, review of P. K. Marshall, *A. Gellii, Noctes Atticae, Mnemonsyne* 25 (1972): 96.

206. Ward Briggs Jr., "Housman and Polar Errors," *American Journal of Philology* 104 (1983): 269.

207. Willis (1972) 98.

208. Young (1965) 267.

209. Willis (1972) 99-100.

210. Codex Andethannensis (Paris, Bibliothèque Nationale, ms. Latinus 10314) and Montpellier, Bibliothèque de la Faculté de Médecine, ms. H 113. See Willis (1972) 99.

211. The line is from Manilius' *Astronomica* in Codex Gemblacensis (Brussels, ms. 10012) and Leipzig, Universitätsbibliothek, ms. 1465. Cf. E. Courtney, "Housman's Manilius," in *A. E. Housman: Classical Scholar*, ed. D. J. Butterfield and Christopher Stray (London: Duckworth, 2009), 38.

212. West (1973) 16.

213. Madrid, El Escorial, ms. Q.1.1 reads *eumque*, while Florence, Biblioteca Medicea-Laurenziana, ms. Plutei 90, sup. 25, reads *huncque*. See Willis (1972) 122.

214. Martial archetypum codicum *HTR*. See West (1973) 18.

215. Milan, Biblioteca Ambrosiana, ms. S.P. 10/27.

216. Cf. M. D. Reeve, "The Tradition of *Consolatio ad Liviam*," *Revue d'Histoire des Textes* 6 (1978): 79-98.

217. Galen, *In Hippocratis librum De natura hominis commentarius*, I 127.

218. Heinz-Günther Nesselrath, "Language and (in-)Authenticity: The Case of the (Ps.-)Lucianic *Onos*," in *Fakes and Forgers of Classical Literature*, ed. Javier Martinez (Leiden: Brill, 2014), 195.

219. H. Rushton Fairclough, "The Poems of the Appendix Vergiliana," *Transactions and Proceedings of the American Philological Association* 53 (1922): 5–6.

220. Niklas Holzberg, "Impersonating Young Virgil: The Author of the *Catalepton* and His *Libellus*," *Materiali e Discussioni per l'Analilsi dei Testi Classici* 52 (2004): 30. Cf. Irene Peirano, *Rhetoric of the Roman Fake: Latin Pseudoepigraphia in Context* (Cambridge: Cambridge University Press, 2012), 74–115.

221. Pliny the Younger, *Epistulae* 2.10.1.

222. Suetonius, *De grammaticis et rhetoribus* 3.3.

223. Suetonius, *De grammaticis et rhetoribus* 5.1.

224. Pliny the Elder, *Naturalis historia* 21.

225. Walter Lapini, "Philological Observations and Approaches to Language in the Philosophical Context," in *Brill's Companion to Ancient Scholarship*, ed. Franco Montanari, Stephanos Matthaios, and Antonios Rengakos (Leiden: Brill, 2015), 1038.

226. Diogenes Laertius 8.84–85.

227. Hermann Diels, *Die Fragmente der Vorsokratiker* (Berlin: Weidmann, 1903), Protagoras B 5.

228. Cicero, *Ad Atticum* 13.21a.1–2.

229. Quintilian, *Institutio oratoria* intro. 7.

230. Quintilian, *Institutio oratoria* 7.2.24.

231. Rex Winsbury, *The Roman Book: Books, Publishing, and Performance in Classical Rome* (London: Duckworth, 2009), 132.

232. Galen, *De libris propriis* 19 (8 Kuhn).

233. Matthew Nicholls, "Galen and Libraries in the *Peri alupias*," *Journal of Roman Studies* 101 (2011): 129.

234. Cicero, *Ad Atticum* 12.6a.1.

235. Rome, Biblioteca Apostolica Vaticana, ms. Latinus 5757.

236. Aristophanes, *Clouds* 518–526.

237. Zachary Biles, *Aristophanes and the Poetics of Competition* (Cambridge: Cambridge University Press, 2011), 169.

238. Another work to appear in multiple editions is Callimachus' *Aetia*. In the case of the *Aetia*, however, the second edition was an expansion of the original work, not a replacement for it. The *Aetia* is a collection of etiological stories that was first published in Callimachus' youth as a two-book work in which the poet imagines himself in discussion with the Muses. Years later, books 3 and 4 were added, along with a new introduction to the collection as a whole. Further, Annette Harder, *Callimachus: Aetia*, vol. 1 (Oxford: Oxford University Press, 2012), 2–3.

239. Jerome, *Commentary on Matthew* prologue 77. See Hammond (1984) 391.

240. Hammond (1984) 391.

241. Marrou (1949) 211.

242. Marrou (1949) 218–219.

243. The Latin text is in C. Lambot, "Lettre inédite de S. Augustin relative au *De civitate Dei*," *Revue Bénédictine* 51 (1939): 112–113.

CHAPTER 3

1. Hans Robert Jauss, *Toward an Aesthetic of Reception* (Minneapolis: University of Minnesota Press, 1982). For a concise and focused application of Jaussian theory to Latin texts, see Lowell Edmunds, *From a Sabine Jar: Reading Horace Odes 1.9* (Chapel Hill: University of North Carolina Press, 1992).
2. T. S. Eliot, "Tradition and the Individual Talent," in *The Sacred Wood: Essays on Poetry and Criticism* (New York: Knopf, 1921), 45.
3. Charles Martindale, "Introduction: Thinking through Reception," in *Classics and the Uses of Reception*, ed. Charles Martindale and Richard F. Thomas (Chichester: Wiley-Blackwell, 2006), 2.
4. Suetonius, *De grammaticis et rhetoribus* 2.2.
5. Aulus Gellius, *Noctes Atticae* 3.3.4.
6. This is not the case for Fabius Pictor, whose work in Greek was well known even in imperial times.
7. Cicero, *Brutus* 65.
8. On the crisis in theatrical production, see Mario Citroni, "Poetry in Augustan Rome," in *A Companion to Ovid*, ed. Peter E. Knox (Chichester: Wiley-Blackwell, 2009), 20. One example of such a canon is that of Volcatius Sedigitus, whose list included ten comic authors, giving first place to Caecilius Statius, second to Plautus, only sixth to Terence, and to Ennius a mere tenth place.
9. Horace, *Epistles* 2.1.157–163.
10. Our translation of Quintilian, *Institutio oratoria* 10.88: "*Ennium sicut sacros uetustate lucos adoremus.*"
11. Plutarch, *Theseus* 20.2.
12. Plutarch, *Numa* 22. Other sources include Pliny the Elder, *Naturalis historia* 13.84–87; Livy 40.29; Valerius Maximus 1.1.12. Further see the outdated, but nonetheless valuable, article of Clarence A. Forbes, "Books for the Burning," *Transactions and Proceedings of the American Philological Association* 67 (1936): 118.
13. Livy 39.6.8.
14. Suetonius, *Augustus* 31.1.
15. This work (ca. 43 BC), enjoyed considerable success and likely had a remarkable distribution in light of a papyrus fragment found in 1978 in Egypt, which comes from a copy of Gallus' work and may even date to the period in which Gallus was serving as governor of Egypt, at the opening of the 20s BC. Cf. R. D. Anderson, P. J. Parsons, and R. G. M. Nisbet, "Elegiacs by Gallus from Qaṣr Ibrîm," *Journal of Roman Studies* 69 (1979): 125–127.
16. Ovid, *Tristia* 3.1.71–72.
17. Suetonius, *Vita divi Iulii* 56.7: "All of these books Augustus forbade to be pub-

lished in a short and straightforward letter sent to Pompeius Macrus, whom he had tasked with putting the libraries in order" (our translation).

18. Tacitus, *Annales* 4.34–35.

19. Cassius Dio, *Historiae Romanae* 57.24.4: "They were later republished [ἐξεδόθη τε αὖθις], since others, particularly his daughter Marcia, had hidden copies of them" (our translation).

20. Suetonius, *Vita Caligulae* 34.2.

21. Tacitus, *Agricola* 2.1. Cf. Charles W. Hedrick Jr., *History and Silence: Purge and Rehabilitation of Memory in Late Antiquity* (Austin: University of Texas Press, 2000), 163–164.

22. Aulus Gellius, *Noctes Atticae* 5.4.1–2.

23. Aulus Gellius, *Noctes Atticae* 2.14.1.

24. Aulus Gellius, *Noctes Atticae* 9.14.3: "I also recall having found in the Tibur library both *facies* and *facii* written in the same book of Claudius."

25. Aulus Gellius, *Noctes Atticae* 2.3.5.

26. Aulus Gellius, *Noctes Atticae* 18.5.11; cf. Suetonius, *De grammaticis et rhetoribus* 2.

27. Aulus Gellius, *Noctes Atticae* 1.7.1.

28. The adjective "Tironian" could possibly also merely indicate "careful." Further, J. E. G. Zetzel, "*Emendavi ad Tironem*: Some Notes on Scholarship in the Second Century A.D.," *Harvard Studies in Classical Philology* 77 (1973): 225–243. More recently, Alan Cameron, *The Last Pagans of Rome* (Oxford: Oxford University Press, 2011), 450; and Hedrick (2000) 188.

29. Wytse Hette Keulen, *Gellius the Satirist: Roman Cultural Authority in* Attic Nights (Leiden: Brill, 2009), 313–315.

30. Peter Brown, *The World of Late Antiquity: From Marcus Aurelius to Muhammad* (London: Thames and Hudson, 1974).

31. This event, recorded in the *Acta martyrum scillitanorum*, is one of many stories of martyrdom found in these texts that represent the oldest Christian Latin texts. They both record key events of that period and document the rise of Christianity.

32. See Joseph Farrell, *Latin Language and Latin Culture: From Ancient to Modern Times* (Cambridge: Cambridge University Press, 2001), 10.

33. This upheaval is exhibited in the East with Zenobia's conquests and in the West with the various disturbances caused by Marcus Cassius Latinas Postumius. Cf. Javier Teixidor, *The Pagan God: Popular Religion in the Greco-Roman Near East* (Princeton: Princeton University Press, 1977), 109.

34. Oronzo Pecere, "La tradizione dei testi latini tra IV e V secolo attraverso i libri sottoscritti," in *Società romana e impero tardoantico*, vol. 4: *Tradizione dei classici, trasformazioni della cultura*, ed. Andrea Giardina (Bari: Laterza, 1986), 75.

35. See *Papyrus Herculaneum* 817.

36. Codex Bembinus: Rome, Biblioteca Apostolica Vaticana, ms. Latinus 3226.

37. Naples, Biblioteca Nazionale, ms. Latinus 2.

38. Rome, Biblioteca Apostolica Vaticana, ms. Latinus 3256.

39. St. Gall, Stiftsbibliothek, ms. 1394.

40. The appellation of "uncial" for this new font was based on a passage of Jerome and introduced by Mabillon in the seventeenth century.

41. "A cursive majuscule script called uncials (meaning 'inch-high') developed in the 3rd century (probably in Africa, and possibly under the influence of patristic authors writing in Greek), in which the pen would not have to be lifted as often" (David C. Greetham, *Textual Scholarship: An Introduction*, Garland Reference Library of the Humanities 1417 [London: Routledge, 2013], 117); "The M, regarded as characteristic of uncial, apparently began originally with a straight shaft to which were added two equal arches one after the other; its later customary twin-arched structure, however, appears already in third-century African inscriptions and in many fourth-century manuscripts" (Bernhard Bischoff, *Latin Palaeography: Antiquity and the Middle Ages*, trans. Dáibhí Ó Cróinín and David Ganz [Cambridge: Cambridge University Press, 1990], 68–69). See E. A. Lowe, *Codices Latini Antiquiores* 12 (supplement) (Oxford: Clarendon, 1971), plate 7; Stanley Morison, *Politics and Script: Aspects of Authority and Freedom in the Development of Graeco-Latin Script from Sixth Century B.C. to Twentieth Century A.D.*, Lyell Lectures (Oxford: Oxford University Press, 1972), reprods. 46–48.

42. Moreover, there are several important codices containing Christian texts, such as St. Petersburg ms. Q.v.i.3 of the fifth century, an important witness of the works of Saint Augustine.

43. Bernard P. Grenfell and Arthur S. Hunt, *The Oxyrhynchus Papyri*, part 4 (London, 1898), 90–116; see "Epitome of Livy XXXVII–XL and XLVIII–LV," plate VI (col. VIII).

44. St. Gall, Stiftsbibliothek, ms. 1395: *Veterum Fragmentorum Manuscriptis Codicibus Dectractorum Collection Tom. II*, 25.

45. Cf. Jan Assmann, *The Price of Monotheism* (Stanford: Stanford University Press, 2009).

46. On the importance of this conversion as a pivotal historical event, see Paul Veyne, *When Our World Became Christian (312–394)* (Cambridge: Polity, 2010).

47. Cf. 1 Thessalonians 4 and 1 Peter 4.

48. E.g., 1 Corinthians 5, 10, and 12.

49. Pliny, *Epistulae* 10.46.

50. James MacGregor, *The Apology of the Christian Religion: Historically Regarded with Reference to Supernatural Revelation and Redemption* (Edinburgh: T. and T. Clark, 1891), 188.

51. Justin Martyr, *First Apology, Second Apology, Dialogue with Trypho*.

52. Tatian, *Address to the Greeks* 19; see also Eusebius, *Historia Ecclesiastica* 4.16.7–8.

53. Further, see Candida R. Moss, *Ancient Christian Martyrdom: Diverse Practices, Theologies, and Traditions* (New Haven: Yale University Press, 2012), 95.

54. David N. Greenwood, "The *Alethes logos* of Celsus and the Historicity of Christ," *Anglican Theological Review* 96 (2014): 705–713.

55. Tertullian, *De praescriptione haereticorum* 7.

56. Tertullian, *Ad martyras*, part 4.

57. Augustine, *De christiana doctrina* 2.43.

58. Augustine, *De civitate Dei* 8.11.

59. Augustine, *De christiana doctrina* 2.60 (our translation).

60. Joseph Tixeront, *A Handbook of Patrology* (St. Louis: Herder, 1920), 73.

61. Craig Kallendorf, *The Protean Virgil* (Oxford: Oxford University Press, 2015), 49–58.

62. Jerome, *Epistulae* 22.30: "*Ciceronianus es, non Christianus.*" Such appreciation of Cicero did not end with Jerome, of course. Cf. Valery Rees, "Ciceronian Echoes in Marcilio Ficino," in *Cicero Refused to Die: Ciceronian Influence through the Centuries*, ed. Nancy van Deusen (Leiden: Brill, 2013), 146.

63. As Augusto Fraschetti ("Trent'anni dopo 'Il conflitto fra paganesimo e cristianesimo nel secolo IV,'" in *Pagani e cristiani da Giuliano l'Apostata al sacco di Roma. Atti del convegno internazionale di studi: Rende, 12–13 novembre 1993*, ed. F. E. Consolino [Soveria Mannelli: Rubbettino] 5–14) points out, the title of the volume edited by Momigliano shows there is no monolithic scholarly consensus on this topic. See Arnaldo Momigliano, *The Conflict between Paganism and Christianity in the Fourth Century* (Oxford: Clarendon, 1963).

64. Jerome, *Apologia adversus libros Rufini* 2.11.

65. W. R. Inge, "The Permanent Influence of Neoplatonism upon Christianity," *American Journal of Theology* 4 (1900): 328–344.

66. Cf. Ausonius, *Epistulae* 21.62–72. Ausonius' concern is veiled in mythological references, but, in light of his response, Paulinus seems to have taken his master's meaning (*Carmina* 10.189-195). Cf. Dennis E. Trout, *Paulinus of Nola* (Berkeley: University of California Press, 1999), 69–71.

67. By the fourth century, the altar of Victory was an item of considerable antiquity, as it had remained in the Roman Senate house from 29 BC until Constantine removed it 356, only to be restored by the Emperor Julian.

68. Symmachus, *Relationes* 3.10 (our translation).

69. Ambrose, *Epistulae* 18.

70. Ambrose, *Epistulae* 17.

71. *CIL* 6.1710; cf. Hedrick (2000) 233.

72. Augustine, *De civitate Dei* 5.26.

73. Orosius, *Historiae adversus paganos* 7.35.

74. Harris (1989) 206 and Woolf (2009) 48–49.

75. A witness of this text remains (Paris, Bibliothèque Nationale, ms. Latinus 8084, written in rustic capitals, the script used for pagan authors).

76. R. Alden Smith, *Virgil* (Chichester: Wiley-Blackwell, 2011), 2–5.

77. Paulinus of Nola, *Epistulae* 28.6.

78. Sidonius Apollinaris, *Epistulae* 2.9.4.

79. Franco Cardini, *Cristiani perseguitati e persecutori* (Rome: Salerno, 2011); Giovanni Filoramo, *La croce e il potere: I cristiani da martiri a persecutori* (Bari: Laterza, 2011).

80. An island near the coast of Provence, between Cannes and Antibes.

81. Cyprian of Toulon, *Vita Caesarii* 1.9.

82. Robert A. Kaster, *Guardians of Language: The Grammarian and Society in Late Antiquity*, Transformation of the Classical Heritage 11 (Berkeley: University of Cali-

fornia Press, 1988), 93. See also Danuta Shanzer, "Latin Literature, Christianity, and Obscenity in the Later Roman West," in *Medieval Obscenities*, ed. Nicola McDonald (Woodbridge: Boydell and Brewer, 2006), 185.

83. An eminent figure of the Church in Gaul and the Iberian Peninsula, Caesarius later became bishop of Arles, serving in that capacity for forty years at the beginning of the sixth century (502–542).

84. Rome, Biblioteca Apostolica Vaticana, ms. Palatinus Latinus 2.4.

85. Florence, Biblioteca Medicea-Laurenziana, ms. 63.1.

86. Symmachus, *Epistulae* 9.13.

87. Sidonius Apollinaris, *Epistulae* 8.3.1.

88. Three medieval codices of Horace bear the name of Vettius Agorius Mavortius, who was consul in 527 (R. J. Tarrant, "Horace," in *Texts and Transmission: A Survey of the Latin Classics*, ed. Leighton Durham Reynolds and Peter K. Marshall [Oxford: Clarendon, 1983], 185). In one of the two manuscript traditions of the works of Julius Caesar appear the names of the editors Iulius Celsus Constantinus, Flavius Licerius, and Firminus Lupicinus. See Virginia Brown, "Latin Manuscripts of Caesar's *Gallic War*," in *Paleografia Diplomatica et Archivistica*, ed. G. Battelli, Storia e Letteratura 139 (Rome: Edizioni di Storia e Letteratura, 1979), 122–123.

89. Florence, Biblioteca Medicea-Laurenziana, ms. 68.2.

90. Rome, Biblioteca Apostolica Vaticana, ms. Latinus 492.9.

91. It is not hard to find further examples, notable among which is a subscription from a manuscript of Martial, featuring Torquatus Gennadius, who is identified as the son of Gennadius the rhetor. He was recorded by Jerome in the *Chronicon* of 353 and received a laudatory poem by Claudian (*Carmina minora* 19). Some manuscripts of Juvenal record the name "Nicaeus," a student of Servius. Flavius Iulius Sabinus Tryphonianus is mentioned in the *subscriptiones* of a manuscript of Persius, copied in 402, between Barcelona and Toulouse. A figure by the name of "Paulus" is cited in a codex of Lucan, copied at Constantinople. The name "Iunius Laurentius" can be found in a manuscript of Pliny the Elder.

92. Jerome, *Praefatio in librum Job*.

93. *Patrologia Graeca* 59.186–187.

94. The same Felix, it should be added, is mentioned in a *subscriptio* as a teacher and supervisor to the aforementioned Agorius Mavortius.

95. Jerome, *De viris illustribus* 32.5.

96. C. H. Roberts and T. C. Skeat, *The Birth of the Codex* (Oxford: Oxford University Press, 1983), 50.

97. One scroll was necessary for the *Eclogues*, four for *Georgics*, and twelve for the *Aeneid*.

98. I.e., Florence, Biblioteca Medicea-Laurenziana, ms. 37.13.

99. Florence, Biblioteca Medicea-Laurenziana, ms. 68.1.

100. Florence, Biblioteca Medicea-Laurenziana, ms. 68.2.

101. J. E. G. Zetzel, "The Subscriptions in the Manuscripts of Livy and Fronto and the Meaning of *Emendatio*," *Classical Philology* 75 (1980): 38–59.

102. Markus Dubischar, "Preserved Knowledge: Summaries and Compilations," in

A Companion to Greek Literature, ed. Martin Hose (Chichester: Wiley-Blackwell, 2016), 427–440.

103. This story is preserved in the *Historia Augusta* 520–521. Cf. T. Keith Dix, "'Public Libraries' in Ancient Rome: Ideology and Reality," *Libraries and Culture* 29 (1994): 286–294.

104. Near Squillace in Calabria, its name comes from nurseries that the family owned in the area.

105. Hans Walther and Paul Gerhard Schmidt, *Proverbia sententiaeque Latinitatis medii ac recentioris aevi: Lateinische Sprichwörter und Sentenzen des Mittelalters und der frühen Neuzeit in alphabetischer Anordnung*, book 7, no. 37134 (Göttingen: Vandenhoeck and Ruprecht, 1982).

106. Ammianus Marcellinus, *Res gestae* 14.6.18 (our translation). R. Martorelli, "Riflessioni sulle attività produttive nell'età tardoantica e altomedievale: Esiste un artigianato ecclesiastico?" *Rivista di Archeologia Cristiana* 75 (1999): 571–596. See also Peter Brown, *Through the Eye of a Needle: Wealth, the Fall of Rome, and the Making of Christianity in the West* (Princeton: Princeton University Press, 2012), 497–498.

107. The English Benedictine Wynfrith (Saint Boniface, ca. 680–755) was a missionary to Germany: killed in Friesland, his body was removed to the imperial abbey of Fulda, which was founded by his student Sturmi (744); shortly afterward the Hersfeld monastery was founded (770). Already, a few decades before, the Visigoth Pirmin, perhaps an exile following the Arab conquest (711), founded, on Lake Constance, the monasteries of Reichenau (724) and Murbach (727).

108. Peter Brown, *The Rise of Western Christendom: Triumph and Diversity, AD 200–1000*, 2nd ed. (Malden, MA: Blackwell, 2003), 207–212.

109. L. D. Reynolds, *Texts and Transmission: A Survey of the Latin Classics* (Oxford: Clarendon, 1983), xvi–xvii.

110. B. Maurenbrecher, *C. Sallusti Crispi Historiarum Reliquiae* (Leipzig: Teubner, 1891), 5.

111. E.g., an open *a* resembling the letter *u* and a *t* whose crossbar curves down to the bottom of the vertical stroke. Cf. Bischoff (1990) 97–98.

112. Paul the Deacon's historical works are entitled *Historia romana* and *Historia Langobardorum*.

113. Einhard, *Vita Karoli Magni* 19.1: "*Liberos suos ita censuit instituendos, ut tam filii quam filiae primo liberalibus studiis, quibus et ipse operam dabat, erudirentur.*"

114. *Patrologia Latina* 101.853.

115. Berlin, Deutsche Staatsbibliothek, ms Diez B Sant. 66.

116. Einhard, *Vita Karoli Magni* 27. Another biblical name, "Bezalel," was also used by Einhard (*Gesta Abbatum*, 272).

117. Rosamond McKitterick, "Script and Book Production," in *Carolingian Culture: Emulation and Innovation* (Cambridge: Cambridge University Press, 1994), 221.

118. *Epistulae ad Lucilium*, *Apocolocyntosis Claudii*, *De beneficiis*, and *De clementia*.

119. Of the Flavian writers, Pliny the Elder along with Quintilian (*Institutio oratoria* and *Declamationes minores*) were favored. Statius (*Thebaid* and *Achilleid*) also was copied, as were Valerius Flaccus and Martial. Frontinus' *Stratagemata* and the *Corpus*

agrimensorum Romanorum. The last of these is preserved in the Herzog-August Bibliotek at Wolfenbüttel in a composite manuscript written in uncial script, of north Italian origin and dating to the late fifth/early sixth century: *Corpus Agrimensorum Romanorum (Guelf. 2° 36. 23)*. See the edition of H. Butzmann, *Corpus Agrimensorum Romanorum: Der Codex Arcenianus A der Herzog-August-Bibliotek zu Wolfenbüttel (cod. Guelf. 36. 23A)*, Codices Graeci et Latini 22 (Leiden: Brill, 1970).

120. One also finds many texts of late antiquity, including the *Anthologia Latina*; selections from Ausonius; Avianus; Claudian's *Carmina minora*; Servius, the Virgilian commentator; alongside Tiberius Donatus. The grammatical works of Charisius and Nonius, and Caelius Aurelian's *Medicinales responsiones* and *Tardae passiones* were also preserved.

121. The story is recorded in Bede, *Historia Abbatum 6*.

122. The dissemination of Isidorus' *Origines* is reconstructed by Bernhard Bischoff, "Die europäische Verbreitung der Werke Isidors von Sevilla," in *Isidoriana: Colección de estudios sobre Isidoro de Sevilla*, ed. Manuel Díaz y Díaz (León: Centro de Estudios San Isidoro, 1961), 317–344.

123. Also known as Ealhwine, Albinus, and Flaccus.

124. Leiden, Bibliotheek der Rijksuniversiteit, ms. Vossianus Latinus Folio 4, containing books 2–6 of Pliny, and the fragment of a manuscript of the historian Justin, housed in Weinheim.

125. Sometimes incorrectly attributed to Servius. Cf. Charles Murgia, *Prolegomena to Servius 5: The Manuscripts* (Berkeley: University of California Press, 1975), 71.

126. This center of manuscript production produced the codex that once incorporated the now-fragmentary ms. Vossianus Latinus Folio 111 of Leiden, Bibliotheek der Rijksuniversiteit, and the manuscript in Paris, Bibliothèque Nationale, ms. Latinus 8093, containing Ausonius, a collection of poetry merged then into the *Anthologia Latina*, along with other the texts of Visigothic origin.

127. Paris, Bibliothèque Nationale, ms. Latinus 10318.

128. The work of Pomponius Mela was used in his own *Cosmographia* by a certain Aethicus Ister, perhaps identifiable with the Irish Virgil, bishop of Salzburg.

129. The Codex Oblongus is now housed in Leiden, Bibliotheek der Rijksuniversiteit (ms. Vossianus Latinus Folio 30).

130. London, British Museum, ms. Harley 2736.

131. Rome, Biblioteca Apostolica Vaticana, ms. Latinus 4929.

132. In Berne, Burgerbibliothek ms. 366.

133. K. J. Conant, *Carolingian and Romanesque Architecture, 800–1200* (New Haven: Yale University Press, 1959), 66–67.

134. Remigius wrote commentaries on Holy Scripture, Latin grammarians, the *De consolatione philosophiae* of Boethius, and the work of Martianus Capella. His *De divisione naturae* was augmented in the twelfth century by the Platonic school of Chartres; as it involved theological controversies of that epoch, in 1210 the work was banned.

135. Paris, Bibliothèque Nationale, ms. Graecus 437.

136. Charles the Bald did so even though the abbot Hilduin had translated it into Latin in ca. 838 at the abbey church of Saint-Denis. Further, see the introductory essay

of Jean LeClercq, "Influence and Noninfluence of Dionysius in the Western Middle Ages," in *Pseudo-Dionysius: The Complete Works*, trans. C. Luibheid (New York: Paulist, 1987), 25–33.

137. Leiden, Bibliotheek der Rijksuniversiteit, ms. Vossianus Latinus Folio 94.

138. Rome, Biblioteca Apostolica Vaticana, ms. Latinus 3277.

139. Birger Munk Olsen, *I classici nel canone scolastico altomedievale* (Spoleto: Centro Italiano di Studi sull'Alto Medioevo, 1991).

140. *Eclogues, Georgics, Aeneid*.

141. *Ars poetica, Carmen saeculare, Epodes, Odes, Satires*.

142. *Metamorphoses*.

143. *Thebais*.

144. *De inventione, De officiis, Somnium Scipionis, De amicitia*, and the pseudo-Ciceronian *Rhetorica ad Herennium*.

145. Catilina, *De coniuratione Catalinae* and *Bellum Iugurthinum*.

146. *Epistula ad Lucilium*, including the apocryphal correspondence of Seneca and Saint Paul.

147. *Collectanea rerum memorabilium*.

148. Cassiodorus, *Institutiones* 1.15.7. Cassiodorus himself draws his list from that of the fourth-century grammarian Arusianus Messius, who had collected from these authors for his *Exempla elocutionum*. Other poets included in Cassiodorus' canon are those who were read and commented upon by late antique grammarians, among which Juvenal, whose literary value was acknowledged as early as Servius, stands out.

149. Most scholars date the publication of *De vulgari eloquentia* between 1302 and 1305; Dante began writing the *Comedia* in 1308. Cf. Steven Botterill, ed. and trans., *Dante: De Vulgari Eloquentia* (Cambridge: Cambridge University Press, 1991), xiii, 91 n. 17.

150. Beyond the commentary of Servius and other minor commentators, that of Tiberius Donatus on Virgil also survived, preserved in Luxeuil (Rome, Biblioteca Apostolica Vaticana, ms. Latinus 1512, of the eighth/ninth centuries). Other late antique commentaries that survive are those of Aelius Donatus on the comedies of Terence, the commentary by Lactantius Placidus on the *Thebaid* of Statius, and the work of anonymous commentators on Lucan.

151. Tarrant (1983).

152. This ensemble consisted of the *De senectute* (in later centuries often joined to the *De amicitia*), the *Tusculanae disputationes*, and a further group comprising the *De natura deorum, De divinatione, Timaeus, De fato, Paradoxa Stoicorum, Academica priora*, and the *De legibus*: this collection of eight works is preserved in the ms. Vossianus Latinus Folio 86, a ninth-century manuscript housed in Leiden, Bibliotheek der Rijksuniversiteit.

153. Rome, Biblioteca Apostolica Vaticana, ms. Palatinus Latinus 1547.

154. The two traditions are represented by the codices of the class A and the *Etruscus*. The latter of these dates to the eleventh century and was probably copied in Pomposa; it may have been a replica of a codex from Monte Cassino.

155. Munk Olsen (1991) 23–38.

156. Ludwig Traube, *Vorlesungen und Abhandlungen* II, ed. P. Lehmann (Munich: Beck, 1911), 113.

157. Paris, Bibliothèque Nationale, ms. Latinus 8071.

158. Biblioteca Medicea-Laurenziana, ms. 63.19, which was afterward copied by Petrarch into what is now London, British Library, ms. Harley 2493. Cf. Smith (2011) 153 n. 4.

159. Such as Codex Vetus (Palatinus Latinus 1615), Codex Decurtatus (Palatinus Latinus 1613), and the discovered Codex Ursinianus (Vaticanus Latinus 3870). Cf. W. M. Lindsay, *Introduction to Captivi of Plautus* (London: Methuen, 1900), 3.

160. Munich, Bayerische Staatsbibliothek, ms. CLM 18787.

161. *Pro lege Manilia*, *Pro Milone*, *Pro Plancio*, and *Pro Caecina*. Cf. *Revue d'histoire des textes* (Editions du Centre National de la Recherche Scientifique, 2008), 61.

162. This codex is known as Wolfenbüttel, Herzog-August Bibliotek, ms. Gudianus Latinus 335.

163. Cologny, Fondation Martin Bodmer, ms. Latinus 146.

164. Bamberg, Staatsbibliothek, ms. Class. 44.

165. Montpellier, Bibliothèque de la Faculté de Médecine, ms. H 126.

166. The name is suggestive of his relationship with the emperor: Pope Sylvester I had been the interlocutor of Constantine.

167. Florence, Biblioteca Medicea-Laurenziana, ms. 49.9.

168. Paris, Bibliothèque Nationale, ms. Latinus 7920.

169. Other works that emerge in this era, perhaps copies of lost Carolingian specimens, are the *Ilias Latina*, the *Astronomica* of Manilius, the *Medicamina faciei femineae* of Ovid, and the pseudo-Ovidian *Nux*.

170. Florence, Biblioteca Medicea-Laurenziana, ms. 51.10.

171. Milan, Biblioteca Ambrosiana, ms. C 90 inf.

172. *Apology*, *Metamorphosis*, and *Florida*.

173. In Tacitus' case, the works preserved are *Annales* 11–16 and *Historiae*. The manuscript is Florence, Biblioteca Medicea-Laurenziana, ms. 68.2.

174. Today's Naples, Biblioteca Nazionale, ms. Latinus IV.A.3.

175. A complete copy of the work had been used two centuries earlier by Paul the Deacon (Paulus Diaconus) for his *Epitome*.

176. Charles H. Haskins, *The Renaissance of the Twelfth Century* (Cambridge, MA: Harvard University Press, 1927).

177. Further, see Ernst R. Curtius, *European Literature and the Latin Middle Ages*, trans. Willard R. Trask (Princeton: Princeton University Press, 1953), 255 n. 23.

178. Richard H. Rouse and Mary A. Rouse, *Manuscripts and Their Makers: Commercial Book Producers in Medieval Paris, 1200–1500* (Turnhout: Harvey Miller, 2000), 1:263; Destrez (1935) 21.

179. G. Pollard, "The *Pecia* System in the Medieval Universities," in *Medieval Scribes: Manuscripts and Libraries. Essays Presented to N. R. Ker*, ed. M. Parks and A. Watson (London: Oxford University Press, 1978), 156.

180. Rouse and Rouse (2000) 263.

181. Rouse and Rouse (2000) 263.

182. Destrez (1935) 13. Further, R. Rouse, "The Book Trade in the University of Paris (ca. 1250-1350)," in *La production du livre universitaire au Moyen Âge: Exemplar et pecia*, ed. R. Rouse, L. G. Bataillion, and B. G. Guyot (Paris: Vrin, 1988), 41-114.

183. Pollard (1978) 156-157.

184. Pollard (1978) 152.

185. Rouse and Rouse (2000) 267.

186. Pollard (1978) 149.

187. Pollard (1978) 148-150.

188. Pollard (1978) 150. An exception is the University of Paris, which did include arts books among its *peciae* (151).

189. Lucien Febvre and Henri-Jean Martin, *The Coming of the Book: The Impact of Printing, 1450-1800* (London: NLB, 1976), 30-31.

190. As mentioned above, the seven liberal arts consisted of a lower division: the "trivium," comprised of grammar, logic, and rhetoric; as well as the upper division, the "quadrivium," comprising arithmetic, geometry, music, and astronomy. Together, these prepared for the highest study of philosophy, which included both of the modern divisions of philosophy and theology.

191. Rodney Stark, *The Victory of Reason: How Christianity Led to Freedom, Capitalism, and Western Success* (New York: Random House, 2005), viii-xi, 37.

192. Peter Abelard, Betty Radice, and Michael Clanchy, *The Letters of Abelard and Heloise* (London: Penguin, 2003).

193. Karl Sudhoff, "Konstantin der Afrikaner und die Medizinschule von Salerno," *Archiv für Geschichte der Medizin* 23 (1930): 113-198. Cf. also Charles S. F. Burnett and Danielle Jacquart, eds., *Constantine the African and ʿAlī Ibn Al-ʿAbbās Al-Maǧūsī: The Pantegni and Related Texts* (Leiden: Brill, 1994).

194. Michael R. McVaugh, "Niccolò da Reggio's Translations of Galen and Their Reception in France," *Early Science and Medicine* 11 (2006): 275-301. Also Caroline Bruzelius, *The Stones of Naples: Church Building in Angevin Italy (1266-1343)* (New Haven: Yale University Press, 2004), 2-4.

195. Lorenzo Minio-Paluello, "Boezio, Giacomo Veneto, Guglielmo di Moerbeke, Jacques Lefèvre d'Etaples e gli 'Elenchi Sophistici,'" *Rivista di Filosofia Neo-Scolastica* 44 (1952): 398-411.

196. Sylvain Goughuenheim, *Aristote au mont Saint-Michel: Les racines grecques de l'Europe chrétienne* (Paris: Seuil, 2008). Further, Bee Yun, "Does the History of Medieval Political Thought Need a Spatial Turn? The Murals of Longthorpe, the *Secretum secretorum*, and the Intercultural Transfer of Political Ideas in the High Middle Ages," in *Cultural Transfers in Dispute: Representations in Asia, Europe and the Arab World since the Middle Ages*, ed. Jörg Feuchter, Friedhelm Hoffmann, and Bee Yun (Frankfurt: Campus, 2011), 137-138.

197. Eckart Schütrumpf, *The Earliest Translations of Aristotle's Politics and the Creation of Political Terminology*, Morphomata Lectures Cologne 8 (Paderborn: Fink, 2014).

198. Traube (1911) 8.

199. Paris, Bibliothèque Nationale, lat. 8260.

200. S. J. Heyworth, *Cynthia: A Companion to the Text of Propertius* (Oxford: Oxford University Press, 2007), xv: "Codex Leidensis Vossianus lat.O.38 (desinit II i 63). ca. 1230-1260 pro Ricardo Furniualensi scriptus."

201. It is now in Wolfenbüttel, Herzog-August Bibliotek, ms. Gudianus Latinus 224.

202. Fernando Navarro Antolín, ed., *Lygdamus: Corpus Tibullianus III.1-6, Mnemosyne* Suppl. 154 (Leiden: Brill 1996), 33-35.

203. Monte Cassino, Biblioteca dell'Abbazia, ms. 361.

204. Ann E. Hanson and Monica H. Green, "Soranus of Ephesus: *Methodicorum princeps*," in *Aufstieg und Niedergang der römischen Welt*, II.37.2, ed. W. Haase and H. Temporini (Berlin: de Gruyter, 1994), 968-1075.

205. Edward Courtney, *The Fragmentary Latin Poets* (Oxford: Oxford University Press, 1993), 406.

206. Plutarch, *Sulla* 26.

207. Diodorus Siculus 16.3.8.

208. G. Cavallo, "Conservazione e perdita dei testi greci," in *Tradizione dei classici: Trasformazioni di cultura*, ed. A. Giardina (Bari: Laterza, 1986), 84.

209. Further on this period and its wake, see Tim Whitmarsh, *Beyond the Second Sophistic: Adventures in Greek Postclassicism* (Berkeley: University of California Press, 2013).

210. H. Strasburger, "Umblick im Trümmerfeld der griechischen Geschichtsschreibung," in *Historiographia Antiqua: Commentationes Lovanienses in honorem W. Peremans septuagenarii editae*, ed. C. Prins (Leuven: Leuven University Press, 1977), 10-15.

211. Our translation.

212. See Harry Y. Gamble, *Books and Readers in the Early Church: A History of Early Christian Texts* (New Haven: Yale University Press, 1995), 168.

213. See Mario Geymonat, *The Great Archimedes*, trans. and ed. R. Alden Smith (Waco, TX: Baylor University Press, 2010), 69-70.

214. Origen, *Patrologia Graeca* 10.1088a.

215. Stephen Gersh, *Middle Platonism and Neoplatonism: The Latin Tradition*, vol. 2 (South Bend, IN: University of Notre Dame Press, 1986), 767.

216. A. Pontani, "La filologia," in *Lo spazio letterario della Grecia antica*, vol. 2, ed. G. Cambiano, L. Canfora, D. Lanza (Rome: Salerno, 1995), 323.

217. Paul Maas, "Schicksale der antiken Literatur in Byzanz," in *Einleitung in die Altertumswissenschaft* I.2: *Textkritik*, ed. Alfred Gercke and Eduard Norden (Leipzig: Teubner, 1927), 2-5.

218. Cavallo (1986) 106-107.

219. Anthony Grafton and Megan Williams, *Christianity and the Transformation of the Book: Origen, Eusebius, and the Library of Caesarea* (Cambridge, MA: Harvard University Press, 2006), 88, 67-68, 116-117.

220. Maas (1927) 5.

221. Cavallo (1986) 166.

222. Jean Irigoin, "Accidents matériels et critiques des textes," *Revue d'Historie des Textes* 16 (1986): 12-14.

223. G. Cavallo, "I fondamenti culturali della trasmissione dei testi antichi a Bisanzio," in Cambiano, Canfora, and Lanza (1995) 265.

224. P. Lemerle, *Le premier humanisme byzantine: Notes et remarques sur enseignement et culture à Byzance des origines au Xe siècle*, Bibliothèque Byzantine, Études 6 (Paris: Presses Universitaires de France, 1971), 230-231.

225. The problem of the sources of Photius is complicated by the introduction of his *Bibliotheca*, which refers to an embassy in which Photius had participated. The delegation in question may have been that of 885-886, ordered by Empress Theodora to treat a prisoner exchange with the Arabs. It is thought that the embassy's destination was Baghdad, and that some of the books he cites were recovered on that occasion in that city, as Baghdad was a cultural center where, after the Arab occupation, Greek texts circulated. Further on Photius' influence, G. Ostrogorsky, *History of the Byzantine State* (New Brunswick, NJ: Rutgers University Press, 1969), 217-219.

226. Photius, *Bibliotheca* 129.

227. Florence, Biblioteca Medicea-Laurenziana, ms. Conventi Soppressi 627.

228. Robert Browning, *Studies on Byzantine History, Literature, and Education* (London: Variorum Reprints, 1977), 89.

229. These manuscripts are Oxford, Bodleian Library, ms. Clark 39, and Rome, Biblioteca Apostolica Vaticana, ms. Graecus I.

230. See Monica White, *Military Saints in Byzantium and Rus, 900-1200* (Cambridge: Cambridge University Press, 2013), 50-52. George T. Dennis, *The* Taktika *of Leo VI: Text, Translation, and Commentary* (Washington, DC: Dumbarton Oaks, 2010) offers an edition based closely on the Florentine manuscript (Florence, Biblioteca Medicea-Laurenziana, ms. 55.4).

231. Rome, Biblioteca Apostolica Vaticana, ms. Palatinus Graecus 23.

232. That manuscript, now Venice, Biblioteca Marciana, ms. Gr. 454, also contains in its margins a copious amount of scholia and commentary, derived from the Alexandrian tradition of Aristarchus.

233. This manuscript—London, British Museum, ms. Burney 86—is also known as the *Townleianus*, named after the English bibliophile who bought it in Rome in the eighteenth century.

234. That is the Codex Ravennas 429. The new plays were the *Acharnians* and the three dramas in which women's roles are prominent, the *Lysistrata*, *Thesmophoriazusae*, and *Ecclesiazusae*.

235. Venice, Biblioteca Marciana, ms. Gr. 474.

236. In a letter he recounts his personal admiration of the Parthenon, but laments how uncultured the Greeks he encountered were, a people quite different from that of the Athens of the past. The siege of Constantinople in 1204 compelled Choniates to take refuge in the island of Kos, where he died.

237. This manuscript is Milan, Biblioteca Ambrosiana, ms. C 126 inf., which contains his edition of the *Moralia*.

238. Milan, Biblioteca Ambrosiana, ms. E 157 sup.

239. Florence, Biblioteca Medicea-Laurenziana, ms. 32.16.

240. Venice, Biblioteca Marciana, ms. Gr. 481.

241. Florence, Biblioteca Medicea-Laurenziana, ms. 32.2.

242. Our translation of L. Canfora, "Le collezioni superstiti," in Cambiano, Canfora, Lanza (1995) 234.

CHAPTER 4

1. Jacob Burckhardt, *The Civilization of the Renaissance in Italy*, trans. S. G. C. Middlemore (London: G. Allen and Urwin, 1914), 175; E. Garin, "Interpretazioni del Rinascimento," in *Interpretazioni del Rianscimento*, ed. M. Ciliberto, vol. 2 (Rome: Edizioni di Storia e Letteratura, 2009), 3-14; J. Woolfson, "Burckhardt's Ambivalent Renaissance," in *Palgrave Advances in Renaissance Historiography*, ed. J. Woolfson (New York: Macmillan, 2005), 9-26.

2. R. Black, "Renaissance Humanism and Historiography Today," in Woolfson (2005) 97-117.

3. George Sarton, *Science and Learning in the Fourteen Century* (Baltimore, MD: Williams and Wilkins, 1947), 15.

4. James Franklin, "Science by Conceptual Analysis: The Genius of the Late Scholastics," *Studia Neoaristotelica* 9 (2012): 3-24.

5. Edward Grant, *The Foundations of Modern Science in the Middle Ages: Their Religious, Institutional, and Intellectual Contexts* (Cambridge: Cambridge University Press, 1996).

6. Pierre Maurice Marie Duhem, *Les origines de la statique* (Paris: A. Hermann, 1905), 38.

7. Ronald G. Witt, *In the Footsteps of the Ancients: The Origins of Humanism from Lovato to Bruni* (Leiden: Brill, 2000), 81.

8. William Philip Sisler, "An Edition and Translation of Lovato Lovati's 'Metrical Epistles,' with Parallel Passages from Ancient Authors" (PhD diss., Johns Hopkins University, 1977).

9. London, British Library, ms. Additional 19906.

10. It is interesting to note that Dante also highlighted this classical figure in *Inferno*, naming a circle of hell after him.

11. Nicholas Mann and B. Munk Olsen, eds., *Medieval and Renaissance Scholarship. Proceedings of the Second European Science Foundation Workshop on the Classical Tradition in the Middle Ages and the Renaissance*, Mittellateinische Studien und Texte 21 (Leiden: Brill, 1997), 6.

12. Gary R. Grund et al., *Humanist Tragedies* (Cambridge, MA: Harvard University Press, 2011), xxii.

13. Philip H. Wicksteed and Edmund G. Gardner, *Dante and Giovanni del Virgilio, including a Critical Edition of the Text of Dante's* Eclogae Latinae, *and of the Poetic Remains of Giovanni del Virgilio* (Westminster: Archibald Constable, 1902; rpt. London: Forgotten Books, 2013), 9.

14. Witt (2000) 130.

15. Grund (2011) xxi.

16. *Compendium moralium notabilium*, written in 1400.

17. Witt (2000) 81.

18. N. Sapegno, *Il Trecento* (Milan: Vallardi, 1960), 152 (our translation).

19. Petrarch, *Familiares* 2.1, 6.2.

20. Petrarch, *Seniles* 4.5.

21. Will Durant, *The Renaissance: A History of Civilization in Italy from 1304–1576 A.D.*, Story of Civilization, Part V (New York: Simon and Schuster, 1953), 14.

22. Petrarch, *Familiares* 10.3.

23. Victoria Kirkham and Armando Maggi, eds., *Petrarch: A Critical Guide to the Complete Works* (Chicago: University of Chicago Press, 2009), 403 n. 4.

24. Petrarch, *Familiares* 10.3 (our translation).

25. Cf. Julia Hairston and Walter Stevens, *The Body in Early Modern Italy* (Baltimore: Johns Hopkins University Press, 2010), 292.

26. Petrarch, *De sui ipsius et multorum ignorantia* 4.45: *si mirari autem Ciceronem, hoc est ciceronianum esse, ciceronianus sum* (our translation).

27. *Familiares* 3.186: *quippe cum certus michi uidear, quod Cicero ipse christianus fuisset, si uel Christum uidere, uel Christi doctrinam percipere potuisset*. Cf. Hairston and Stevens (2010) 292.

28. Milan, Biblioteca Ambrosiana, ms. S.P. 10/27.

29. This single codex, ms. Harley 2493 of the British Library in London, was identified as Petrarch's by Billanovich in 1951.

30. Paris, Bibliothèque Nationale, ms. Latinus 5690.

31. Geoffrey Chaucer, *The Canterbury Tales*, ed. John Urban Nicolson (N.p.: Courier Corporation, 2015), 213.

32. Florence, Biblioteca Medicea-Laurenziana, ms. Plutei 36.49 (vid. Dissertation 24), compiled in 1380 for Salutati, is the oldest such copy.

33. Rome, Biblioteca Apostolica Vaticana, ms. Palatinus Latinus 899.

34. The copy of this manuscript, made for Petrarch, is Paris, Bibliothèque Nationale, ms. Latinus 8500. Another of Ausonius' works is a manuscript in Leiden, Vossianus Latinus F. 111, originally prepared in Lyon by exiled Visigoths.

35. L. Furbetta, "Apostilles in Pomponius Mela between Petrarch and William Pastrengo," *Rassegna della Letteratura Italiana* 113 (2009): 573–574. Further, Giuseppe Frasso and Giuseppe Velli, *Petrarca e la Lombardia. Atti del convegno di studi, Milano, 22–23 Maggio 2003* (Padua: Antenore, 2005), 368; and Henry Hollway-Calthrop, *Petrarch: His Life and Times* (New York: Cooper Square, 1972), 223–224.

36. Vitruvius' *De architectura* is now housed in Oxford, Bodleian Library, ms. Auctarium F. 5.7.

37. F. Brunhölzl, "Zum Problem der Casineser Klassiküberlieferung," *Abhandlungen der Marburger Gelehrten Gesellschaft* 3 (1971): 114–143; see L. D. Reynolds and N. G. Wilson, *Scribes and Scholars: A Guide to the Transmission of Greek and Latin Literature*, 4th ed. (Oxford: Oxford University Press, 2013), 269.

38. Paul Oskar Kristeller, *Iter Italicum*, vol. 6: *Italy III and Alia Itinera IV* (Leiden: Brill, 1992), 516.

39. On Monte Cassino generally, see Herbert Bloch, *Monte Cassino in the Middle Ages* (Cambridge, MA: Harvard University Press, 1988).

40. Boccaccio made a copy of the Tacitus volume and bequeathed it to the Florentine monastery of Santo Spirito.

41. Florence, Biblioteca Medicea-Laurenziana, ms. 68.2.

42. Reynolds and Wilson (2013) 134.

43. Our translation.

44. Florence, Biblioteca Medicea-Laurenziana, ms. 2.9.2.

45. *Ad* Apuleius, *Metamorphoses* 10.21.

46. This description is no longer accepted, even though it was supported in the assessment of the text's seventeenth-century editor Elmenhurst. For the argument of refutation, see Ephraim Lytle, "Apuleius' *Metamorphoses* and the *Spurcum Additamentum* (10.21)," *Classical Philology* 98 (2003): 349–365.

47. Florence, Biblioteca Medicea-Laurenziana, ms. 33.31.

48. Florence, Biblioteca Medicea-Laurenziana, ms. 51.19.

49. Petrarch, *Familiares* 18.4.

50. Ambrosiano C 126 inf.

51. Ronald G. Witt, *The Two Latin Cultures and the Foundation of Renaissance Humanism in Medieval Italy* (Cambridge: Cambridge University Press), 380 n. 119.

52. Codex F, housed in the Biblioteca Ambrosiana in Milan.

53. Perhaps a codex housed in Paris, Bibliothèque Nationale, ms. Graecus 1807.

54. Petrarch, *Familiares* 18.2.

55. R. Weiss, *Medieval and Humanist Greek: Collected Essays* (Padua: Antenore, 1987), 170–171.

56. Steven Runciman, *The Great Church in Captivity: A Study of the Patriarchate of Constantinople from the Eve of the Turkish Conquest to the Greek War of Independence* (Cambridge: Cambridge University Press, 1985), 142. Christopher Kleinhenz, *Medieval Italy: An Encyclopedia* (London: Routledge, 2004), 97.

57. Miguel H. Bronchud, *The Secret Castle: The Key to Good and Evil* (N.p.: CreateSpace, 2007), 78.

58. A. F. Massera, *Giovanni Boccacci: Opere Latine Minori* (Bari: Laterza, 1928), 195 (our translation).

59. Having originated in Capodistria, which led to him being dubbed "Histrus" in the title of Bruni's dialogue, Pier Paolo Vergerio studied at Padua, Bologna, and Florence. He continued his activities in Padua and Rome and was later a pioneer of humanism in Hungary.

60. David Quint, "Humanism and Modernity: A Reconsideration of Bruni's Dialogues," *Renaissance Quarterly* 38 (1985): 426.

61. So Ernst Hans Gombrich and Rudolf Wittkower, *From the Revival of Letters to the Reform of the Arts: Niccolò Niccoli and Filippo Brunelleschi* (London: Phaidon, 1967), 82.

62. Milan, Biblioteca Ambrosiana, ms. R 26 sup.

63. Petrarch had recovered this manuscript in France, and it was likely a descendant of the copy of Tibullus that was part of the library of Charlemagne.

64. Florence, Biblioteca Medicea-Laurenziana, ms. 38.49.

65. Rome, Biblioteca Apostolica Vaticana, ms. Ottobonianus Latinus 1829.

66. Florence, Biblioteca Medicea-Laurenziana, ms. 49.9.

67. This manuscript (C 74a) is currently is stored in the Zentralbibliothek of Zurich.

68. See chapter 3 above, note 129 and relevant text.

69. A copy of which is now extant in Rome, Biblioteca Apostolica Vaticana, ms. Latinus 11458.

70. Poggio's transcription of the *Silvae* and the *Astronomica* of Manilius survives in Madrid, Biblioteca Nacional, ms. 3678.

71. The current Rome, Biblioteca Apostolica Vaticana, ms. Latinus 1873.

72. Poggio came upon the *Cena Trimalchionis* in a codex that later went missing but which fortunately reappeared in Trogir (modern Croatia) around 1650.

73. A subscription reveals that this manuscript had been transcribed by Peter the Deacon (ca. 1107–ca. 1153), librarian of Monte Cassino. See Robert H. Rodgers, ed., *Frontinus: De aquaeductu urbis Romae*, Cambridge Classical Texts and Commentaries 42 (Cambridge: Cambridge University Press, 2004), 37–44.

74. Johann Ramminger, "Humanists and the Vernacular: Creating the Terminology for a Bilingual Universe," *Renaessanceforum* 6 (2010): 7.

75. James Jerome Murphy, *Rhetoric in the Middle Ages: A History of Rhetorical Theory from Saint Augustine to the Renaissance* (Berkeley: University of California Press, 1981), 360.

76. Florence, Biblioteca Medicea-Laurenziana, ms. 39.38.

77. In the Middle Ages only eight of the twenty-one comedies of the corpus were known. The manuscript of the twelve comedies is currently housed as Vaticanus Latinus 3870.

78. Many years later, in the nineteenth century, the twenty-first (the mutilated *Vidularia*) would be retrieved from the Ambrosian Palimpsest (Milan, Biblioteca Ambrosiana, ms. C 90).

79. Durant (1953) 78 relates that "Ambrogio Traversari rescued Cornelius Nepos from oblivion in Padua (1434)."

80. Sir William Smith, *A Dictionary of Greek and Roman Biography and Mythology*, vol. 2: *Earinus–Nyx* (London: J. Murray, 1880), 1157; cf. Nicolaus Courtin, ed., *Cornelius Nepos De vitis excellentium imperatorum* (Whitefish, MT: Kessinger, 2010).

81. Sir John Edwin Sandys, *A History of Classical Scholarship*, vol. 2: *From the Revival of Learning to the End of the Eighteenth Century (In Italy, France, England, and the Netherlands)* (Cambridge: Cambridge University Press, 1908), 50.

82. Florence, Biblioteca Medicea-Laurenziana, ms. 73.1.

83. Brian Stock, *Listening for the Text: On the Uses of the Past* (Baltimore: John Hopkins University Press 1990), 161–162.

84. Laurentius Valla, *Elegantiae Linguae Latinae* (Venice: Johannes Baptista de Sessa, 1499) (our translation).

85. E. Garin, *Prosatori latini del Quattrocento* (Milan: Ricciardi, 1952), 623 (our translation).

86. A. Barbero, Guglielmo Cavallo, Claudio Leonardi, Enrico Menestò, and Piero

Boitani, "Età di mezzo e secoli bui," in *Lo spazio letterario del Medioevo*, ed. P. Boitanin, M. Mancini, and A. Varvaro (Rome: Salerno, 2003), 523. Jacques Le Goff, *The Medieval Imagination*, trans. Arthur Goldhammer (Chicago: University of Chicago Press, 1992), 18.

87. Cf. Theodor Ernst Mommsen, "Petrarch's Conception of the 'Dark Ages,'" *Speculum* 17 (1942): 226–242; taken from Craig W. Kallendorf, ed., *A Companion to the Classical Tradition* (Chichester: Wiley-Blackwell, 2007), 30.

88. Petrarch, *Africa* 9.456–457 (our translation).

89. Our translation. Cf. Garin (1952) 611.

90. Valla (1499) 13 (our translation). Cf. Garin (1952) 603.

91. Cf. his work entitled *Historiarum ab Inclinatione Romanorum imperii Decades*.

92. Salutati, *Epistulae* 7.2 (dated 1392).

93. Cf. B. Ullman, *Sicconis Polentoni Scriptorum illustrium latinae linguae libri XVIII* (Rome: American Academy, 1928), 125.

94. Bischoff (1990) 127.

95. Robert Black, "Humanism," in *The New Cambridge Medieval History*, ed. Christopher Allmand (Cambridge: Cambridge University Press, 2008), 253. Cf. B. L. Ullman, *The Origin and Development of Humanistic Script* (Rome: Edizioni di Storia e Letteratura, 1960), 12.

96. The idea of chronological assessment was taken up again in titles of works by the late nineteenth-/early twentieth-century scholars Felix Voigt (1888–1897: *Wiederbelebung*, i.e., "Revival") and Remigio Sabbadini (1905–1914: *Scoperte*, i.e., "Discoveries").

97. Ambrosianus F III sup. (our translation).

98. H. Harth, *Poggio Bracciolini: Lettere*, vol. 2 (Florence: Olschki, 1984), 155. Cf. Garin (1952) 243–245.

99. That is to say, *ab ipsis propriis Maronis exemplaribus*, or "from the very *exemplars* of Virgil*,*" meaning I saw Virgil's very own handwriting. Of course, this is impossible, as he could not have seen Virgil's own, exact copy of the *Aeneid*. This is just one example of a general failure to date manuscripts well. In any case, this manuscript was kept secure in Rome in the Basilica of St. Paul (see Craig Kallendorf, *A Bibliography of Venetian Editions of Virgil, 1470–1599* [Florence: Olschki, 1991], 22) before being acquired by Francesco I de' Medici and then moved to Florence, where it now resides.

100. Paris, Bibliothèque Nationale, ms. 1989. Kirkham and Maggi (2009) xx.

101. Petrarch, *Familiares* 18.3.

102. Petrarch, *Familiares* 13.1.

103. Historiated initials differ by the inclusion of a picture or scene beyond the adornment of decorated initials.

104. In other words, the design of what is now recognized as essentially a standard modern text.

105. Walter J. Ong, "System, Space, and Intellect in Renaissance Symbolism," *Bibliothèque d'Humanisme et Renaissance* 18 (1956): 228.

106. These forms would be appropriated to suit the revival of nationalism in the nineteenth century. Even today, the fortune of the humanistic script continues, evidenced clearly enough in the names of certain fonts, such as Italic and New Roman.

107. Eduard Norden, *Die antike Kunstprosa vom VI. Jahrhundert v. Chr. bis in die Zeit der Renaissance* (Leipzig: Teubner, 1923), 770.

108. Such a distinction in the Church, in which Latin was the official language of the Mass for the next six hundred years, would not be resolved until the Second Vatican Council (1962–1965), when the vernacular languages were finally made central for worship. With the recent return of the Latin Mass, the Catholic situation remains a bit unresolved.

109. Dante, *Inferno* 27.19-21; see Angelo Mazzocco, *Linguistic Theories in Dante and the Humanists: Studies of Language and Intellectual History in Late Medieval and Early Renaissance Italy* (Leiden: Brill, 1993), 257 ff.

110. Four years after this debate, the papacy was temporarily transferred to Florence (1439) because of a plague that broke out in Rome.

111. Mirko Tavoni, *Latino, grammatica, volgare: Storia di una questione umanistica* (Padua: Antenore, 1984), xv.

112. Cf. Tavoni (1984) 195–301.

113. Remigio Sabbadini, ed., *Epistolario di Guarino Veronese*, vol. 2 (Venice: Deputazione di Storia Patria, 1915), 503–511.

114. Mazzocco (1993) 51.

115. Isodorus of Seville, *Origines* 9.1.6–7.

116. Among other things, Alberti was the author of the *De Re Aedificatoria*, an architectural treatise modeled on Vitruvius.

117. Politian, *Epistulae* 8.16.

118. This work appeared in 1545.

119. David Greetham, "A History of Textual Scholarship," in *The Cambridge Companion to Textual Scholarship*, ed. Neil Fraistat and Julia Flanders (Cambridge: Cambridge University Press, 2013), 27.

120. David Nicholas, *The Transformation of Europe, 1300–1600* (New York: Bloomsbury Academic, 1999), 154.

121. David R. Carlson, "Greeks in England, 1400," in *Interstices: Studies in Late Middle English and Anglo-Latin Texts in Honour of A. G. Rigg*, ed. Richard Firth Green, Linne R. Mooney, and A. G. Rigg (Toronto: University of Toronto Press, 2004), 86.

122. See Leopoldo Tanfani, *Niccola Acciaiuoli: Studi storici, fatti principalmente sui documenti dell' archivio fiorentino* (Florence: Felice le Monnier, 1863), 32-33: "Those book donations by [Niccolò] Acciaiouli are not the only ones [in the fourteenth century] to the monastery. Previously there was the testament of a certain monk, Amico of Buonamico, stipulated by Corrado di Felice da Amalfi, on June 29, 1348, in which the monk left to the Certosa monastery all his manuscripts and two rugs. Everything he owned had been a gift of the monk Antonio da Siena, who brought the manuscripts and the rugs to Europe from Smyrna. These ancient manuscripts were written in Greek. Their arrival was an exceptional event for Western culture at that time. We can therefore consider Antonio da Siena one of the first humanists who traveled to the East and brought back classical Greek texts to be studied" (our translation). We here acknowledge the advice of Professor Piergiacomo Petrioli on this point.

123. Diego Bottoni, "I Decembrio e la traduzione della *Repubblica* di Platone: Dalle correzioni dell'autografo di Uberto alle integrazioni greche di Pier Candido," in *Vestigia: Studi in onore Giuseppe Billanovich*, ed. Rino Avesani (Rome: Edizioni di storia e letteratura, 1984), 75–76.

124. Other humanists who learned Greek staying in Constantinople were Aurispa Giovanni, Francesco Filelfo, and Giovanni Tortelli (see Nancy Bisaha, *Creating East and West: Renaissance Humanists and the Ottoman Turks* [Philadelphia: University of Pennsylvania Press, 2006], 103).

125. Ada Francesca Marcianò and Biagio Rossetti, *L'età di Biagio Rossetti: Rinascimenti di casa d'Este* (Ferrara: Gabriele Corbo, 1991), 477.

126. G. Abbamonte, "Considerazioni su alcune dediche di traduzioni latine di opere greche fatte da umanisti del Quattrocento," in *Pratiques latines de la dédicace: Permanence et mutations, de l'Antiquité à la Renaissance*, ed. J.-C. Juhle (Paris: Classiques Garnier, 2014), 523–559.

127. Other fifteenth-century translators were Aurispa, who worked on the *Dialogues* of Lucian; Pier Candido Decembrio, who translated Appian; and Valla, who worked on Herodotus, Thucydides, and Xenophon. Politian, too, made several translations, among which his works on Homer and Epictetus are especially noteworthy.

128. M. Cortesi, "Umanesimo greco," in *Lo spazio letterario del Medioevo: I. Il Medioevo latino*, vol. 3: *La ricezione del testo*, ed. G. Cavallo, C. Leonardi, and E. Menesto (Rome: Salerno, 1995), 503.

129. Among the Byzantine delegation were highly cultured individuals, such as Gemistus Plethon (1355–1452), a supporter of Platonism, who years earlier, when exiled from Constantinople because he was suspected of heresy and paganism, had founded a school at Mystras in the Peloponnese. His presence in Florence had considerable impact and caused the subsequent flowering of Neoplatonism with Marsilio Ficino (1433–1499). Among the illustrious admirers of Plethon, a group that included Cosimo de' Medici, who valued him among his counselors, Sigismondo Malatesta stands out. Malatesta, the lord of Rimini, in 1645 exhumed the remains of Plethon from Mystras and interred them in Rimini's Malatesta Temple (the cathedral church of Rimini officially named for Saint Francis).

130. After Pius II's death in 1464, Bessarion was on the verge of election to the pontificate.

131. John Monfasani, *George of Trebizond: A Biography and a Study of His Rhetoric and Logic* (Leiden: Brill, 1976) 70–113. Further, Karl Alexander, "Honor, Reputation, and Conflict: George of Trebizond and Humanist Acts of Self-Presentation" (PhD diss., University of Kentucky, 2013).

132. Andrea Moudarres, "Crusade and Conversion: Islam as Schism in Pius II and Nicholas of Cusa," *MLN* 128 (2013): 40–52; Monfasani (1976) 184–194.

133. Deno John Geanakoplos, *Constantinople and the West: Essays on the Late Byzantine (Palaeologan) and Italian Renaissances and the Byzantine and Roman Churches* (Madison: University of Wisconsin Press, 1989), 73–75.

134. Geanakoplos (1989) 24, 74.

135. Reynolds and Wilson (2013) 149-150.

136. Years ago, Hans Baron put forth the idea of "civic humanism"; see Hans Baron, *The Crisis of the Early Italian Renaissance: Civic Humanism and Republican Liberty in an Age of Classicism and Tyranny*, 2 vols. (Princeton: Princeton University Press, 1966). This concept has been revisited also by James Hankins, "The 'Baron Thesis' after Forty Years and Some Recent Studies of Leonardo Bruni," *Journal of the History of Ideas* 56 (1995): 309-338.

137. G. J. Dorleijn and Herman L. J. Vanstiphout, eds., *Cultural Repertoires: Structure, Function, and Dynamics*, Groningen Studies in Cultural Change 3 (Leuven: Peeters, 2003), 93-97; see Cicero, e.g., in *Pro Archia* 3; *Tusculanae disputationes* 1.1 and elsewhere; Aulus Gellius, *Noctes Atticae* 13.17.1-2: "But learned speakers generally have called *humanitas* what the Greeks call 'teaching' and what we call 'erudition' and 'instruction in the good arts.' Those who are especially learned are eager for and seek after these arts. For the care and discipline of this knowledge was granted only to humankind out of all living creatures. Therefore it was called *humanitas*" (our translation).

138. Augusto Campana, "The Origin of the Word 'Humanist,'" *Journal of the Warburg and Courtauld Institutes* 9 (1946): 60-73; see Paul Oskar Kristeller, *Eight Philosophers of the Italian Renaissance* (Stanford: Stanford University Press, 1964), 150, 170 n. 24; E. Peterson, "The Communication of the Dead: Notes on 'Studia Humanitatis' and the Nature of Humanist Philology," in *The Uses of Greek and Latin: Historical Essays*, ed. A. C. Dionisotti, Anthony Grafton, and Jill Kraye (London: Warburg Institute and University of London, 1988), 57-69.

139. Walter Rüegg, *Cicero und der Humanismus: Formale Untersuchungen über Petrarca und Erasmus* (Düsseldorf: Rhein, 1946), 65.

140. *De oratore* 1.17.

141. M. D. Reeve, "Classical Scholarship," in *The Cambridge Companion to Renaissance Humanism*, ed. J. Kraye (Cambridge: Cambridge University Press, 1996), 21-22.

142. François Rabelais and Henri Clouzot, *Gargantua and Pantagruel* (Paris: G. Crès, 1922), 3.XLVIII (our translation).

143. *Ars dictaminis*, set alongside *ars poetriae* and *ars praedicandi*, the arts of grammar and poetry and preaching, was the medieval invention of letter writing as an art. Previously letters were often merely meant to be read aloud, but the medieval period saw the development of robust and detailed rules of style for writing letters meant to be read not aloud but to oneself. Cf. Murphy (1981) 194-268.

144. Durant (1953) 250.

145. Thomas More, "Letter to Oxford University," trans. T. S. K. Scott-Craig, in *The Thought and Culture of the English Renaissance: An Anthology of Tudor Prose, 1481-1555*, ed. Elizabeth M. Nugent (The Hague: Martinus Nijhoff, 1969), 69.

146. Étienne Gilson, *Les idées et les lettres*, 2nd ed., Librairie Philosophique (Paris: Vrin, 1955), I, x-xi.

147. Cf. Vladimiro N. Zabughin, *Storia del rinascimento cristiano in Italia*, ed. Bruno Basile (Naples: La Scuola Pitagora, 2011) (original published in Milan: Treves, 1924); Giuseppe Toffanin, *La religione degli umanisti* (Bologna: Zanichelli 1950).

148. Cf. Charles Taylor, *A Secular Age* (Cambridge, MA: Belknap, 2007).

149. F. Novati, ed., *Epistolario di Coluccio Salutati*, vol. 3 (Rome: Forzani, 1896), 470-480.

150. Lucia Gualdo Rosa, "Leonardo Bruni, l'*Oratio in hypocritas* e i suoi difficili rapporti con Ambrogio Traversari," *Vita Monastica* 65 (1987): 89-111.

151. The fortune of this translation is documented by the more than two hundred extant codices.

152. Leonardo Bruni, "De studiis et litteris liber ad Baptistam de Malatestis" (trans. Craig W. Kallendorf), in *Humanist Educational Treatises*, ed. Craig W. Kallendorf (Cambridge, MA: Harvard University Press, 2002), 92-125. ¶26: *quia amores et flagitia in illis reperiuntur* (our translation).

153. Ibid., ¶26: *Penelopae erga Ulyxem fidelissimam castitatem.*

154. Ibid., ¶27: *amores paene insani.*

155. Ibid., ¶27: *at etiam Davidis amorem in Bersabeam et scelus in Uriam Salomonisque fratricidium et tam numerosum concubinarum gregem.*

156. Ibid., ¶27: *At illi forsan suo more vixerunt? Quasi vero honestas gravitasque morum non tunc eadem fuerit quae nunc est!*

157. This pamphlet is now lost.

158. Ludwig Friedrich August von Pastor, *Acta Inedita Historiam Pontificum Romanorum Praesertim saec. XV, XVI, XVII. Illustrantia. Edidit L. Pastor. Volumen I: A. 1376-1464* (Freiburg: Herder, 1904).

159. Documents reproduced in Pastor (1904) 741-748.

160. His *Laws*, publicly burned by the theologian George Scholarios, are lost.

161. Margaret Meserve and Marcello Simonetta, *Pius II: Commentaries*, vol. 2, I Tatti Renaissance Library 29 (Cambridge, MA: Harvard University Press, 2003), 327, 329.

162. Edward Muir, *The Culture Wars of the Late Renaissance* (Cambridge, MA: Harvard University Press, 2007), 63-67.

163. The *Agricola* was also preserved in a manuscript of Monte Cassino.

164. Cf. R. P. Robinson, *The Germania of Tacitus: A Critical Edition* (Middletown, CT: American Philological Association, 1935), xii, 8, and 341-356.

165. Cf. Giorgio Brugnoli and Carlo Santini, *L'Additamentum Aldinum di Silio Italico* (Rome: Accademia Nazionale dei Lincei, 1995).

166. Josef Delz, ed., *Sili Italici Punica*, Bibliotheca Scriptorum Graecorum et Romanorum Teubneriana (Stuttgart: Teubner, 1987), LXIV-LXIX.

167. During the fifteenth century, Florence was also an important seat for the production of hand-copied codices. The workshop was of Vespasiano da Bisticci (1421-1498) was particularly prominent, and it continued to operate for some time even after the introduction of printing.

168. Prominent examples included Alfonso V of Aragon's collection in Naples, the Visconti family's in Milan, the Este family's in Ferrara, Duke Federico of Montefeltro's in Urbino, and Cardinal Bessarion's endowment to the Biblioteca Marciana in Venice.

169. M. Clapham, "L'arte della stampa," in *Storia della tecnologia*, vol. 3, ed. Ch. Singer, E. J. Holmyard, A. R. Hall, and T. I. Williams (Torino: Boringhieri, 1963), 385.

170. One such text is an *incunabulum*.

171. Howard Jones, *Printing the Classical Text* (Utrecht: Hes and de Graaf, 2004), 25.

172. Texts discovered include those of Charisius, Sacerdos, and pseudo-Probus, as well as Rutilius Namatianus' *De reditu suo* and the so-called *Epigrammata Bobiensia*, a collection that includes the poems of Sulpicia, a poetess from the age of Domitian, sometimes cited by Martial.

173. Much of Parrasio's library, after some upheaval, is today housed in the national library of Naples. Those volumes that he found, including the work of the grammarian Charisius, were later given to the Austrians (in the seventeenth century), who housed them in the collection of the national library in Vienna. In 1918 they were returned to Italy as an aspect of compensation for the First World War. The grammatical texts were published in part shortly after their discovery and in part in the Viennese edition of 1837. The codex containing Rutilius and the *Epigrammata*, however, has since been lost. Fortunately, Rutilius was printed in 1520, but the *Epigrammata* is preserved only in a copy of a manuscript (an *apograph*) and came to light much later, in 1955.

174. Leiden, Bibliotheek der Rijksuniversiteit, ms. Vossianus Latinus Folio III.

175. Vienna, Österreichische Nationalbibliothek, ms. Lat. 277.

176. Florence, Biblioteca Medicea-Laurenziana, ms. 68.1.

177. Paris, Bibliothèque Nationale, ms. Latinus 10318 (Codex Salmasianus).

178. One part of it was included in the *Florilegium Gallicum* sent to Orléans in the twelfth century.

179. A fragment from Lorsch was discovered in 1921.

180. Ulrich von Wilamowitz-Moellendorff, *History of Classical Scholarship*, trans. Hugh Lloyd-Jones (Baltimore: Johns Hopkins University Press, 1982), 21.

181. In the texts transcribed or studied by Petrarch and also by other humanists of a later period, we find textual solutions that are even today of great interest, formulated on the basis of one criterion: the *usus scribendi* of the author under consideration.

182. Cf. Harth (1984) 284.

183. Only one copy remains, diligently prepared by Nicholas of Clémanges (Paris, Bibliothèque Nationale, ms. Latinus 14749). Cf. Torbjörn Jelerup, "The Renaissance, and the Rediscovery of Plato and the Greeks," *Fidelio* 12 (2003): 46–47.

184. Fortunately, as mentioned earlier, it reappeared two centuries later in Dalmatia (Paris, Bibliothèque Nationale, ms. Latinus 7989).

185. The twentieth-century scholar Giorgio Pasquali, citing this episode, noted that after its discovery "scholars throughout Italy rejoiced; every humanist, one might say, busied himself to procure one copy of the new script. It appears that that most precious codex, already seven years after the discovery, had disappeared again" (G. Pasquali, *Storia della tradizione e critica del testo* [Milan: Mondadori, 1974], 61).

186. For example, his footnotes to Quintilian's *Institutio oratoria*, found in Paris, Bibliothèque Nationale, ms. Latinus 7723, were published only in Lorenzo Valla and Lucia Cesarini Martinelli, *Le postille all'"Institutio Oratoria" di Quintiliano: Edizione critica*, ed. Lucia Cesarini Martinelli and Alessandro Perosa (Padua: Antenore, 1996).

187. Today's highly conservative approach, by contrast, exalts the distance of the manuscript from the modern reader. Today, many manuscript repositories and libraries will only allow a scholar access to the manuscript if there is a demonstrable need to see the original; otherwise, digital access alone is granted, often exclusively on-site. Fortu-

nately, these digital copies are of the highest quality and can afford the scholar a more careful inspection than might a blue light held over the manuscript proper.

188. Anthony Grafton, "On the Scholarship of Politian and Its Contexts," *Journal of the Warburg and Courtauld Institutes* 40 (1977): 150–152.

189. Wilamowitz-Moellendorff (1982) 28–29.

190. Only the first one hundred entries of the *Miscellany* were originally published, in 1489. The remainder were published in Angelo Poliziano, *Miscellaneorum centuria secunda: Di Angelo Poliziano; per cura di Vittore Branca e Manlio Pastore Stocchi. Editio minor*, ed. Vittore Branca and Manlio Pastore Stocchi (Florence: Olschki, 1972).

191. Both of these exemplars of an important first step in classical philology are now in Oxford's Bodleian Library.

CHAPTER 5

1. M. Pattison, *Isaac Casaubon, 1559–1614* (Geneva: Slatkine, 1970; originally published Oxford: Clarendon, 1892), 449. Further, Antony Grafton and Joanna Wineberg, *"I have always loved the Holy Tongue": Isaac Casaubon, the Jews, and a Forgotten Chapter in Renaissance Scholarship* (Cambridge, MA: Belknap, 2011), 9.

2. Isaac Newton, *Correspondence*, vol. 1 (Cambridge: Cambridge University Press, 1959), 416. Newton alluded to it in a letter to Robert Hooke in 1676, and it has since enjoyed signal fortune. In modern times, the verity of Bernard's aphorism has been explored by Umberto Eco in his foreword to Robert K. Merton, *On the Shoulders of Giants: A Shandean Postscript, the Post-Italianate Edition* (Chicago: University of Chicago Press, 1993). We see it occurring afresh in the title of a work edited by Stephen Hawking (*On the Shoulders of Giants*, 2002), in the motto of the free software movement, and, made popular yet again, in the name of the modern rock band "OTSOG."

3. David Woodward, *The History of Cartography*, vol. 3: *Cartography in the European Renaissance* (Chicago: University of Chicago Press, 2007), 285–364.

4. Marie Boas Hall, *The Scientific Renaissance, 1450–1630* (New York: Dover, 1994), 49.

5. At that time the most advanced discoveries of Hellenistic science were still unknown. Lucio Russo, *The Forgotten Revolution: How Science Was Born in 300 BC and Why It Had to Be Reborn*, trans. Silvio Levy (Berlin: Springer, 2004), however, considers these discoveries to have been precursors of the same scientific revolution.

6. Newton (1959) 416.

7. L. D. Reynolds and N. G. Wilson, *Scribes and Scholars: A Guide to the Transmission of Greek and Latin Literature*, 4th ed. (Oxford: Oxford University Press, 2013), 139.

8. For the development and spread of the printing press, see Febvre and Martin (1976) 29–71.

9. Anthony Levi, *Humanism in France at the End of the Middle Ages and in the Early Renaissance* (Manchester: Manchester University Press, 1970), 6–28.

10. Ignacio Enrique Navarrete, *Orphans of Petrarch: Poetry and Theory in the Spanish Renaissance* (Berkeley: University of California Press, 1994), 17–18.

11. Lwów (ancient Leopolis) is modern Lviv, Ukraine, about a hundred miles east of

Gregory's hometown, Sanok. Cf. Harold B. Segel, *Renaissance Culture in Poland: The Rise of Humanism, 1470–1543* (Ithaca, NY: Cornell University Press, 1989), 7.

12. Jerzy Axer, "Central Eastern Europe," in *A Companion to the Classical Tradition*, ed. C. W. Kallendorf (Chichester: Blackwell, 2007), 143.

13. Julian Raby, "A Sultan of Paradox: Mehmed the Conqueror as a Patron of the Arts," *Oxford Art Journal* 5 (1982): 6. On Mehmet II's political ascendancy and activity, see G. Ostrogorsky, *History of the Byzantine State* (New Brunswick, NJ: Rutgers University Press, 1969), 567–568.

14. Julian Raby, "Cyriacus of Ancona and the Ottoman Sultan Mehmed II," *Journal of the Warburg and Courtauld Institutes* 43 (1980): 242–246.

15. Franz Babinger, *Mehmed the Conqueror and His Time* (Princeton: Princeton University Press, 1992), 118.

16. Luca Landucci and Giuseppe Mazzotta, "The Entry of Charles VIII, King of France, into Florence," in *Images of Quattrocento Florence: Selected Writings in Literature, History, and Art*, ed. Stefano U. Baldassarri and Arielle Saiber (New Haven: Yale University Press, 2000), 115–122.

17. Carol Kidwell, *Pietro Bembo: Lover, Linguist, Cardinal* (Montreal: McGill-Queen's University Press, 2004), 162; L. M. Koff, "Dreaming the Dream of Scipio," in *Cicero Refused to Die: Ciceronian Influence through the Centuries*, ed. Nancy van Deusen (Leiden: Brill, 2013), 65.

18. Cf. Frank Tallett, "Soldiers in Western Europe, c. 1500–1790," in *Fighting for a Living: A Comparative Study of Military Labour, 1500–2000*, ed. Erik-Jan Zürcher (Amsterdam: Amsterdam University Press, 2013), 148.

19. Markjke Crab, "Josse Bade's *Familiaris Commentarius* on Valerius Maximus (1510): A School Commentary?," in *Transformations of the Classics via Early Modern Commentaries*, ed. Karl A. E. Enenkel (Leiden: Brill, 2013), 153–154.

20. Michael L. Rodkinson, trans., "Reuchlin, Pfefferkorn, and the Talmud in the Sixteenth and Seventeenth Centuries," in *The Babylonian Talmud: The History of the Talmud*, book 10, vol. I, ch. XIV (1918), 76.

21. David O. McNeil, *Guillaume Budé and Humanism in the Reign of Francis I*, Travaux d'Humanisme et Renaissance 142 (Geneva: Librairie Droz, 1971), 98.

22. R. Bracht Banham, "Utopian Laughter: Lucian and Thomas More," *Moreana* 86 (1985): 30–31.

23. Margaret L. King, *Renaissance Humanism: An Anthology of Sources* (Indianapolis: Hackett, 2014), 158.

24. Sir John Edwin Sandys, *A History of Classical Scholarship*, vol. 2: *From the Revival of Learning to the End of the Eighteenth Century (in Italy, France, England, and the Netherlands)* (Cambridge: Cambridge University Press, 1908), 214–215.

25. Peter Burke, *The European Renaissance: Centers and Peripheries* (Chichester: Wiley-Blackwell, 1998), 99.

26. Burke (1998) 97.

27. This pronunciation, later dubbed "Erasmian," remains in use to this day. The main change that Erasmus proposed concerned the pronunciation of the letter *H* (*eta*),

which in classical Greek has the value of *e* (etacism) but in Byzantine Greek was pronounced as *i* (iotacism). The Byzantine pronunciation, widespread among humanists before Erasmus, was defended by Reuchlin and thus named "Reuchlinian." Cf. N. F. Moore, *Ancient Mineralogy; or, An Inquiry Respecting Mineral Substances Mentioned by the Ancients* (New York: Carvill, 1834), 138.

28. Erika Rummel, *Erasmus as a Translator of the Classics*, Erasmus Studies 7 (Toronto: University of Toronto Press, 1985), 15-18.

29. For the opinion that the theses were never posted, see Erwin Iserloh, *The Theses Were Not Posted: Luther between Reform and Reformation*, trans. J. Wicks (Boston: Beacon, 1968).

30. *De Falso Credita et Ementita Constantini Donatione.* This debt of Luther to Valla and von Hutten was noted long ago by V. Chauffour-Kestner, *Ulrich von Hutten, Imperial Poet and Orator: The Great Knightly Reformer of the Sixteenth Century* (Edinburgh: T. and T. Clark, 1963), 56-61.

31. It is interesting to note that Calvin published a classical commentary in addition to his theological works. In 1532 he published his commentary on the *De clementia* of Seneca. Further, see Wulfert Greef, *The Writings of John Calvin: An Introductory Guide*, trans. L. Bierma (Louisville, KY: Westminster John Knox, 2008), 65.

32. L. Canfora, *Filologia e libertà* (Milan: Mondadori, 2008), 16.

33. Further, J. Lough, ed., *The "Encyclopédie" of Diderot and d'Alembert* (Cambridge: Cambridge University Press, 1954).

34. On Jesuit censorship generally, see G. Lewy, "The Struggle for Constitutional Government in the Early Years of the Society of Jesus," *Church History* 29 (1960): 156.

35. A. J. Boyle, "Seneca Inscriptus," in *Tragic Seneca: An Essay in the Theatrical Tradition*, ed. A. J. Boyle (London: Routledge, 1997), 142-143.

36. Elizabeth Armstrong, *Robert Estienne, Royal Printer: An Historical Study of the Elder Stephanus* (Cambridge: Cambridge University Press, 1954), 165-167.

37. Howard L. Goodman and Anthony Grafton, "Ricci, the Chinese, and the Toolkits of Textualists," *Asia Major* 3 (1990): 95-148.

38. Jane Kneller, "Imaginative Freedom and the German Enlightenment," *Journal of the History of Ideas* 51 (1990): 231.

39. In the lucky cases when the codices were preserved, we can verify the lack of careful preparation that characterized many of these first editions. An example of this is Erasmus's edition of Seneca. Erasmus correctly identified problems in the production process that he addressed in the preface of his 1529 second edition. Further, Reynolds and Wilson (2013) 163-164.

40. This codex is now housed in the Biblioteca Angelica in Rome.

41. Bussi was characteristic of a growing *haute couture*, which was generated in part under the cultural influence of Cardinal Nicholas of Cusa, a German theologian most famous for the mystical notion of "learned ignorance." Cf. H. Lawrence Bond, trans., *Nicholas of Cusa: Selected Spiritual Writings*, Classics of Western Spirituality 89 (Mahwah, NJ: Paulist, 2005), 20.

42. *Tum temporis invidia, tum litterarum neglegentis arrogantia, et librariorum inscitia;*

in Beriah Botfield, *Praefationes et Epistulae Editionibus Principibus Auctorum Veterum Praepositae* (Cambridge: Cambridge University Press, 1861), 141.

43. John Monfasani, "The First Call for Press Censorship: Niccolò Perotti, Giovanni Andrea Bussi, Antonio Moreto, and the Editing of Pliny's *Natural History*," *Renaissance Quarterly* 41 (1988): 1–31.

44. Urbinus Latinus 1146, now housed in the Biblioteca Apostolica Vaticana.

45. Though it eventually came into the possession of Cardinal Bessarion, the Codex Fuldense is currently housed in the library of the Academy of Medicine in New York.

46. E.g., Politian uses the term *"Longobardis . . . litteris"* in *Opera Omnia* 263. Cf. Silvia Rizzo, *Il lessico filologico degli umanisti*, Collana Sussidi Eruditi 26 (Rome: Edizioni di Storia e Letteratura, 1973), 124–125. Anthony Grafton, *Defenders of the Text: The Traditions of Scholarship in an Age of Science, 1450–1800* (Cambridge, MA: Harvard University Press, 1991), 59 and 264 n. 50.

47. Both kinds of emendation were performed frequently in the sixteenth through the eighteenth centuries, in ways vastly different from those that characterize philology in the nineteenth and twentieth centuries. Further, see John F. D'Amico, *Theory and Practice in Renaissance Textual Criticism: Beatus Rhenanus between Conjecture and History* (Berkeley: University of California Press, 1988), 9–11.

48. Seneca, *Apocolocyntosis* 12.3.15.

49. Anthony Grafton, *Joseph Scaliger: A Study on the History of Classical Scholarship* (Oxford: Clarendon, 1983, 1993), 2:79–82.

50. Cf. E. J. Kenney, *The Classical Text: Aspects of Editing in the Age of the Printed Book* (Chicago: University of Chicago Press, 1995), 26 (our translation).

51. *"Nobis et ratio et res ipsa centum codicibus potiores sunt"*; in Richard Bentley, ed., *Q. Horatius Flaccus* (Cambridge: Cambridge University Press, 1711), ad *Carmen* 3.27.15.

52. Patricia E. Easterling, "Notes on Notes: The Ancient Scholia on Sophocles," *Acta Universitatis Upsaliensis* 21 (2006): 21.

53. M. Lowry, "Girolamo Aleandro," in *Contemporaries of Erasmus: A Biographical Register of the Renaissance and Reformation*, vol. 1, ed. Peter G. Bietenholz (Toronto: University of Toronto Press, 1985), 28–34.

54. Reynolds and Wilson (2013) 167–168.

55. Reynolds and Wilson (2013) 174. On the French influence in classical philology generally, see Philip Ford, "France," in *A Companion to the Classical Tradition*, ed. C. W. Kallendorf (Chichester: Blackwell, 2007), 156–168.

56. R. Smith, *Virgil* (Chichester: Wiley-Blackwell, 2011), 26. Cf. also Wade Richardson, *Reading and Variant in Petronius: Studies in the French Humanists and Their Manuscript Sources* (Toronto: University of Toronto Press, 1993), 63–82.

57. Karl Enenkel and Henk Nellen, "Introduction," in *Neo-Latin Commentaries and Management of Knowledge in the Late Middle Ages and the Early Modern Period (1400–1700)*, ed. Karl Enenkel and Henk Nellen, Supplementa Humanistica Lovaniensia 33 (Leuven: Leuven University Press, 2013), 13.

58. Gilbert Tournoy, "Low Countries: The Seventeenth Century," in *A Companion to the Classical Tradition*, ed. C. W. Kallendorf (Chichester: Blackwell, 2007), 245–246.

59. The *Palatine Anthology* comprises the main part of the *Greek Anthology*, which also included parts of the *Anthology of Planudes*, among other materials.

60. A. J. Boyle, *Roman Tragedy* (New York: Routledge, 2006), 189.

61. The Reginenses manuscripts are now housed in the Vatican library (Biblioteca Apostolica Vaticana, 1945).

62. Sebastiano Timpanaro, *The Genesis of Lachmann's Method*, trans. Glenn W. Most (Chicago: University of Chicago Press, 2005), esp. 24–57.

63. Bruce G. Trigger, *A History of Archaeological Thought* (Cambridge: Cambridge University Press, 1989), 36.

64. Leonard Barkan, *Unearthing the Past: Archaeology and Aesthetics in the Making of Renaissance Culture* (New Haven: Yale University Press, 1999), 1–2.

65. Cf. also Reynolds and Wilson (2013) 169.

66. These different lines of research converged in nineteenth-century practices that offered a more refined basis for epigraphic research. In 1805 August Boeck started the publication of the *Corpus Inscriptionum Graecarum*. Then, in the second half of the century, Wilamowitz set about the enormous task of preparing the *Corpus Inscriptionum Latinarum* and *Inscriptiones Graecae* for publication. Cf. R. Pfeiffer, *History of Classical Scholarship, 1300–1850* (Oxford: Clarendon, 1976), 181.

67. M. D. Neuhofer, *In the Benedictine Tradition: The Origins and Early Development of Two College Libraries* (Lanham, MD: University Press of America, 1999), 19–20.

68. James J. John, "Latin Paleography," in *Medieval Studies: An Introduction*, 2nd ed., ed. James M. Powell (Syracuse, NY: Syracuse University Press, 1992), 3–4.

69. Carol C. Mattusch, "Programming Sculpture? Collection and Display in the Villa of the Papyri," in *The Villa of the Papyri at Herculaneum: Archaeology, Reception, and Digital Reconstruction*, ed. Mantha Zarmakoupi (Berlin: de Gruyter, 2010), 83–85.

70. The discoveries of Pompeii and Herculaneum became widely known in the second half of the nineteenth century, particularly as the Egyptian resorts of Arsinoe and Oxyrhynchus were studied more closely (see E. G. Turner, *Greek Papyri: An Introduction* [repr., Princeton: Princeton University Press, 2015], 81). In 1880 several thousands of papyri were purchased from a rug merchant in Cairo and moved to Vienna, a discovery that made the study of papyrus central to understanding antiquity. Universities, including those of Berlin, Heidelberg, and Oxford, acquired other substantial funds of papyrus. These papyri included hitherto unknown sections of Menander, Sappho, Callimachus, Poseidippus, and Philodemus, as well as Aristotle's *Athenaion politeia*, which was discovered in the nineteenth century. As recently as 2012, a fresh discovery of Sappho was obtained from cartonnage, the text of which was published by Simon Burris, Jeffrey Fish, and Dirk Obbink, "New Fragments of Book 1 of Sappho," *Zeitschrift für Papyrologie und Epigraphik* 189 (2014): 1–28. Another even more famous example is the so-called Artemidorus Papyrus, containing poems from the geographical work of Artemidorus of Ephesus (second century AD). The existence of this papyrus of unknown provenance was announced in 1998, and since 2004 it has been housed in Turin. Recent examination of the papyrus has raised suspicion that it is a counterfeit wrought by the nineteenth-century forger Constantine Simonides. Cf. L. Canfora,

The True History of the So-Called Artemidorus Papyrus (Bari: Edizioni di Pagina, 2007). Another voice in favor of it being a forgery is that of Richard Janko, "The Artemidorus Papyrus," *Classical Review* 59 (2009): 403–410.

71. Helmut Müller-Sievers and David E. Wellbery, *The Science of Literature: Essays on an Incalculable Difference* (Berlin: de Gruyter, 2015), 96.

72. James MacCaffrey, *The History of the Catholic Church in the Nineteenth Century (1789–1908)*, 2nd ed. (St. Louis: M. H. Gill and Son, 1910), 1:225.

73. The method used in the nineteenth century, based on chemical reagents, permitted the reading of calligraphy hitherto hidden from the naked eye. This method, however, also rendered the parchment permanently damaged. The procedure currently in use, based on ultraviolet rays, is safer for the preservation of the manuscript. Further on the use of ultraviolet technology, K. T. Knox, "Enhancement of Overwritten Text in the Archimedes Palimpsest," in *Society of Photo-Optical Instrumentation Engineers Conference Series* 6810 (2008), doi:10.1117/12.877135.

74. One of the most recent and significant discoveries of palimpsests is that of a codex containing the mathematical works of Archimedes. Copied in Constantinople at the beginning of the tenth century, this manuscript was probably stolen during the sack of 1204 and brought to Jerusalem, where in the thirteenth century it was used to transcribe a series of prayers. The Greek text of Archimedes was partially identified by the Danish philologist Johann Ludwig Heiberg at the beginning of the twentieth century. During the First World War, however, the codex was lost, only to reappear at an auction in 1998, and is currently owned by the Walters Art Museum in Baltimore. Noel W. Netz, *The Archimedes Codex: Revealing the Secrets of the World's Greatest Palimpsest* (London: Weidenfeld and Nicolson, 2007), 206–209.

75. Margaret W. Ferguson, *Dido's Daughters: Literacy, Gender, and Empire in Early Modern England and France* (Chicago: University of Chicago Press, 2003), 118.

76. Kaspar Schoppe, *Consultationes de Scholarum et Studiorum Ratione* (1631).

77. U. Foscolo, *Epistolario*, ed. P. Carli (Florence: Le Monnier, 1953), 3:67–68.

78. Cf. Robert Southard, *Droysen and the Prussian School of History* (Lexington: University of Kentucky Press, 1995), 12.

79. Michael Werner, "Philology in Germany: Textual or Cultural Scholarship?" in *Multiple Antiquities — Multiple Modernities: Ancient Histories in Nineteenth-Century European Cultures*, ed. Gabor Klaniczay, Michael Werner, and Otto Gecser (Frankfurt: Campus, 2011), 84–85.

80. Pfeiffer (1976) 173.

81. Dag Haug, "The Linguistic Thought of Friedrich August Wolf: A Reconsideration of the Relationship between Classical Philology and Linguistics in the Nineteenth Century," *Historiographia Linguistica* 32 (2005): 38.

82. Friedrich August Wolf, *Vorlesung über die Geschichte der römischen Litteratur* (Leipzig: August Lehnhold, 1832), 4.

83. Friedrich August Wolf, *Darstellung der Altertumswissenschaft nach Begriff, Umfang, Zweck und Wert* (Weinheim: Acta Humaniora, 1986), 85–86. Though originally published for the journal *Museum der Alterthums-Wissenschaft*, the most recent edition of the text comes in the form of a monograph.

84. Wolf (1986) 86.
85. Wolf (1986) 91.
86. Wolf (1986) 117.
87. Wolf (1986) 120.
88. Wolf (1986) 141. The framework outlined by Wolf ultimately prevailed in the educational model prevalent in Europe in the nineteenth and twentieth centuries: higher education and that of classically based Gymnasia included the study of Greek and Latin and the reading of the ancient literature; the scientific study of antiquity at the university level was reserved for those who specialized in this division of studies and for those who would become teachers of classical languages in the Gymnasia. For more on the "Darstellung," see Jay Bolter, "Friedrich August Wolf and the Scientific Study of Antiquity," *Greek, Roman, and Byzantine Studies* 21 (1980): 83–99.
89. Pfeiffer (1976) 169.
90. Wolf (1986) 19–20.
91. Wolf (1986) 22–23.
92. Wolf (1986) 19–20. Cf. Johann Joachim Winckelmann, *Geschichte der Kunst des Altertums* (Dresden, 1764). For a modern analysis see Miriam Leonard, *Socrates and the Jews: Hellenism and Hebraism from Moses Mendelssohn to Sigmund Freud* (Chicago: University of Chicago Press, 2012), 31–33. For a concise contextualization of the term "Altertumswissenschaft," see Ernst Robert Curtius, *European Literature and the Latin Middle Ages*, trans. Colin Burrow (repr., Princeton: Princeton University Press, 2013), 591–592.
93. Wolf (1986) 19.
94. Wolf (1986) 40–41.
95. A. Boeckh, *Enzyklopädie und Methodenlehre und der philologischen Wissenschaften* (Leipzig: Druck und Verlag von B. G. Teubner, 1877), 20–22.
96. Southard (1995) 12.
97. Cf. Michael Holquist, "World Literature and Philology," in *The Routledge Companion to World Literature*, ed. Theo D'haen, David Damrosch, and Djelal Kadir (London: Routledge, 2012), 150.
98. Sebastiano Timparano, *The Genesis of Lachmann's Method*, ed. and tran. Glenn W. Most (Chicago: University of Chicago Press, 2005), 84–89.
99. Timparano (2005) 76.
100. Giovanni Fiesoli has reprised discussion of Lachmann's contribution, considering in some detail both Lachmann's method and the legend surrounding it, a story that grew up the early years of the twentieth century. G. Fiesoli, *La genesi del lachmannismo* (Florence: Sismei, 2000), 359–463.
101. On the influence of positivism on philology during the nineteenth century, see Klaus Weimar, *Geschichte der deutschen Literaturwissenschaft bis zum Ende des 19. Jahrhunderts* (Munich: Wilhelm Fink, 1989).
102. Karl Lachmann, *Novum Testamentum Graece* (Berlin: G. Reimer, 1846).
103. *Recensere [. . .] sine interpretatione et possumus et debemus*; from the preface to Lachmann (1846) 1:v.
104. Timpanaro (2005) 45.

105. M. D. Reeve, "Shared Innovations, Dichotomies, and Evolution," in *Filologia classica e filologia romanza: Esperienze ecdotiche a confronto*, ed. Anna Ferrari (Spoleto: Centro Italiano di Studi sull'Alto Medioevo, 1999), 450.

106. Otto Ribbeck, *Friedrich Wilhelm Ritschl: Ein Beitrag zur Geschichte der Philologie* (Leipzig: Teubner, vol. 1, 1879; vol. 2, 1881).

107. A. Momigliano, "Jacob Bernays," *Mededeelingen der Koninklijke Nederlandse Akademie van Wetenschappen, Afdeeling Letterkunde* 32 (1969): 151–178.

108. Cf. Timpanaro (2005) 115 n. 3.

109. The term itself is derived from the Greek *stemma*, a wreath with which Greek families decorated and honored sculptured portraits of ancestors; the Romans associated the word with the concept of a "family tree."

110. Cf. R. Smith, *Virgil* (Chichester: Wiley-Blackwell, 2011), 150–151.

111. Elio Montanari, *La critica del testo secondo Paul Maas: Testo e commento*, Millennio Medievale 41: Strumenti e Studi 3 (Florence: Sismel, 2003) 402. Montanari calls it "a packaging of recensionist methodology" (our translation).

112. The original meaning of the term *contaminatio* derives from the process whereby Latin comedic dramatists, such as Plautus or Terence, combined into a single play scenes and stories drawn from more than one Greek comedy. Further on the term, George Duckworth's classic treatment *Nature of Roman Comedy: A Study in Popular Entertainment* (Princeton: Princeton University Press, 1952), 202–208.

113. Paul Maas, *Textkritik*, 4th ed. (Leipzig: Teubner, 1960), 31. An English translation based on the second edition of Maas's work is also available; see Paul Maas, *Textual Criticism*, trans. Barbara Flower (Oxford: Clarendon, 1958).

114. Such variations are not present in other codices that are dependent on the same subarchetype.

115. On *usus scribendi*, see Kerstin Güthert, *Herausbildung von Norm und Usus Scribendi im Bereich der Worttrennung am Zeilenende (1500–1800)* (Heidelberg: Winter, 2005).

116. Cf. Smith (2011) 151.

117. Cf. Mario Geymonat's *praefatio* to his *P. Vergili Maronis Opera* (Rome: Edizioni di Storia e Letteratura, 2008).

118. See Richard Tarrant, *Texts, Editors, and Readers* (Cambridge: Cambridge University Press, 2016), 126.

119. There are, it should be noted, critical editions prepared in modern languages; the use of Latin, however, continues to be employed by most editors of texts for the Bibliotheca Teubneriana and the Bibliotheca Oxoniensis.

CHAPTER 6

1. The title *Didaskaliai* is derived from the Greek verb *didaskein*, "to teach," a verb that was also used to indicate the production of a play. Further, cf. Blum (1991) 25–33.

2. James O'Donnell, review of Jon Solomon, *Accessing Antiquity: The Computeriza-*

tion of Classical Studies, in *Bryn Mawr Classical Review* 94.02.14 (http://bmcr.brynmawr.edu/1994/94.02.14.html).

3. David M. Schaps, *Handbook for Classical Research* (London: Routledge, 2011).

4. Manfred Schmidt, *Corpus Inscriptionum Latinarum* (Berlin: Berlin-Brandenburgische Akademie der Wissenschaften, 2001), 5–6.

5. Schmidt (2001) 6–8.

6. Theodor Mommsen, ed., *Inscriptionum Regni Neapolitani Latinae* (Leipzig: G. Wigand, 1852), XVI; see Schmidt (2001) 9.

7. John Bodel, *Epigraphic Evidence: Ancient History from Inscriptions* (New York: Routledge, 2001), 159; Schmidt (2001) 9.

8. Schmidt (2001) 9.

9. *CIL* III thus covers the provinces of Noricum, Raetia, Pannonia, Moesia, Dacia, Thrace, Greece, Crete, Asia Minor, Syria, Judaea, Arabia, Cyrenaica, and Egypt.

10. *CIL* IV also has a second supplement, divided into four fascicles.

11. These regions consist of Liguria (IX), Venetia and Histria (X), and Transpadana (XI).

12. *CIL* VII is a collection of inscriptions from Britain, and *CIL* VIII includes inscriptions from northern Africa (excluding Egypt), specifically the Roman provinces of Mauretania Tingitana, Mauretania Caesariensis, Mauretania Sitifensis, Numidia, and Africa Proconsularis. *CIL* IX, X, and XI all collect inscriptions from Italy: *CIL* IX is dedicated to the Augustan regions of Apulia and Calabria (II), Samnium and Sabine country (IV), and Picenum (V); *CIL* X covers the regions of Latium, Bruttium, and Campania (I) and Sicily and Sardinia (III); *CIL* XI contains inscriptions from Umbria (VI), Etruria (VII), and Aemilia (VIII). The next volume, *CIL* XII, returns to the provinces with inscriptions from Narbonese Gaul (in the western Alps). More provinces are represented in *CIL* XIII, including the three parts of Gaul (Aquitania, Lugdunensis, and Belgica) as well as the German provinces. *CIL* XIV, the last of the regionally organized volumes, holds inscriptions from Latium, including Ostia and Portus. The four remaining volumes are organized according to types of inscriptions: *CIL* XV gathers *instrumentum domesticum* from Rome, *CIL* XVI is a collection of military diplomas, *CIL* XVII is exclusively milestones, and *CIL* XVIII, which is still a work in progress, is devoted to inscribed poetry (*carmina*). On the reliability and accuracy of various volumes, see Bodel (2001) 162.

13. This webpage is accessible from http://cil.bbaw.de/cil_en/index_en.html.

14. The effort to digitize the enormous collection of *CIL* began in the 1990s with searchable CD-ROMs, which have now been superseded by web-based projects that seek to create searchable databases of the inscriptions found not only in *CIL* but in other collections as well. One of particular note is the Epigraphic Database Heidelberg. Still in progress, its database comprises transcriptions, photos, and bibliography of numerous volumes of *CIL*, as well as inscriptions from other collections. This webpage can be accessed from http://edh-www.adw.uni-heidelberg.de/home.

15. Hermann Dessau, *Inscriptiones Latinae Selectae* (Berlin: Weidmann, 1902), IV.

16. Dessau (1902) IV: "*Seligere volui, et in pauca volumina redactas proponere, ex ingenti inscriptionum Latinarum numero eas quae res scitu maxime dignas continerent.*"

17. Of its five volumes, the fourth and fifth are wholly devoted to indices and appendices, and volume 5 in particular is noteworthy for its extensive compendium of epigraphic abbreviations.

18. Woodhead (1992) 97–98.

19. Cf. Bodel (2001) 166–167, and Woodhead (1992) 98–99, for a fuller list of collections of Greek inscriptions from outside Europe.

20. Woodhead (1992) 98.

21. A third edition of volume 1 (*IG* I³), which contains Attic inscriptions down to the fourth century, was published in 1998. A third edition of volumes 2 and 3 (*IG* II/III³) is underway. These volumes encompass Attic inscriptions from the fourth century on, and, because of the monumental nature of the task, the work has been divided among a group of scholars into seven parts, each of which covers a particular type of inscription. To date, only fascicles 2, 4, and 5 of part 1 (*Leges et Decreta*, laws and decrees) and fascicle 1 of part 4 (*Dedicationes*, dedications) have been published. Additional fascicles for *IG* IV² and XII appeared in 2016.

22. This collection can be accessed at http://epigraphy.packhum.org.

23. Bodel (2001) 157.

24. Each volume of *AE* is organized along the same geographic principles as *CIL*, with inscriptions from areas outside *CIL*'s coverage linked to the province to which each is geographically closest. *AE* is a bit more difficult to access electronically than *CIL* or *IG*. Some inscriptions from *AE* have been incorporated into the Epigraphic Database Heidelberg, yet to view the volumes themselves, currently one must have access to JSTOR, an online repository of scholarly publications from across many disciplines. Despite the difficulty of accessing *AE* outside of institutional holdings, it is an invaluable tool for locating current scholarship on inscriptions.

25. H. W. Pleket and R. S. Stroud, eds., "Preface," *Supplementum Epigraphicum Graecum* 37 (1990): v.

26. Charlotte Tupman, "Contextual Epigraphy and XML: Digital Publication and Its Application to the Study of Inscribed Funerary Monuments," in *Digital Research in the Study of Classical Antiquity*, ed. Gabriel Bodard and Simon Mahony (Burlington, VT: Ashgate, 2010), 78–79.

27. Because of the universality of XML, encoded transcriptions and scholarly notes are broadly available to anyone with access to the Internet.

28. See http://usepigraphy.brown.edu/projects/usep/collections/.

29. The online papyrological sources noted here are chosen for the breadth of material they cover, though it should be noted that images of many papyri can be found at websites established by the institutions that own collections of papyri.

30. Richard Janko, "The Herculaneum Library: Some Recent Developments," *Estudios Clásicos* 121 (2002): 25.

31. Janko (2002) 26–27. Piaggio's catalogue was only recently discovered. See D. L. Blank and F. Longo, "An Inventory of the Herculaneum Papyri from Piaggio's Time," *Cronache Ercolanesi* 30 (2000): 131–147.

32. Janko (2002) 28.

33. I.e., the International Center for the Study of Herculanean Papyri.

34. See http://www.herculaneum.ox.ac.uk/?q=books.

35. Also helpful is George Houston's appendix, which contains a list of known or probable works preserved on the Herculaneum papyri. The list does not account for all of the papyri but only those that contain identifiable works. Houston collects the authors, titles, number of books in the work, and the number found at the villa, and the relevant *P.Herc.* numbers; see George Houston, *Inside Roman Libraries* (Chapel Hill: University of North Carolina Press, 2014), 280–286.

36. Hélène Cuvigny, "The Finds of Papyri: The Archaeology of Papyri," in *The Oxford Handbook of Papyrology*, ed. Roger Bagnall (Oxford: Oxford University Press, 2009), 34.

37. *P.Oxy.* I 1898, v.

38. Many volumes of the Graeco-Roman Memoirs (GRM) series are devoted to Oxyrhynchus papyri, but not all are exclusively so; volumes 60–64 of GRM, for instance, collect papyri from Tebtunis. The GRM series, and specifically its volumes of *Oxyrhynchus Papyri*, are the most famous published collections of papyri; unfortunately, other important collections are more obscure and can be difficult to locate (as, for example, is the case for the Herculaneum papyri). Even the volumes of *P.Oxy.* can challenge the reader; each is accompanied by its own set of indices, but there is no print version of indices for the collection as a whole, which now totals eighty tomes; see GRM 101.

39. See http://www.papyrology.ox.ac.uk/POxy/.

40. Peter Van Minnen, "The Future of Papyrology," in *The Oxford Handbook of Papyrology*, ed. Roger Bagnall (Oxford: Oxford University Press, 2009), 644–645. These are not exclusively Latin and Greek papyri, of course.

41. Van Minnen (2009) 645, estimates the number of published papyri since 1895 at 72,500.

42. For APIS, see http://www.columbia.edu/cu/lweb/projects/digital/apis/index.html; for the DDbDP, see now http://papyri.info/; for Trismegistos, see http://www.trismegistos.org/index.html.

43. Another valuable resource is Papyri.info. One of its most ground-breaking features is a contribution to the problem of the backlog of unpublished papyri. In addition to the "papyrological navigator" (PN), which allows the user to search the site's papyrological archives, the site also includes a "papyrological editor" (PE), which allows users to contribute new texts to the site and propose emendations to existing entries. The proposed changes are voted on by a panel of papyrologists before being added to the DDbDP. The ability of any scholar worldwide to augment and emend the papyrological database through a peer-reviewed process is a striking example of how dramatically technology is changing the fields of ancient studies, not only for experts but for scholars and students new to the fields as well. The potential of the Papyri.info's papyrological editor also underscores the need for students and new scholars to familiarize themselves with at least some basic aspects of coding. See http://www.papyri.info/docs/about.

44. See the transcription of Joshua Sosin's 2010 talk "Digital Papyrology," presented at the 26th Congress of the International Association of Papyrologists: http://www.stoa.org/archives/1263.

45. Another database of papyri to be noted is CEDOPAL (Centre de Documentation de Papyrologie Littéraire), which maintains a website that has collected references to literary papyri (http://cipl93.philo.ulg.ac.be/Cedopal/MP3/dbsearch_en.aspx). CEDOPAL's database is based on *The Greek and Latin Literary Texts from Greco-Roman Egypt* (also known as Mertens-Pack), and its creators are in the process of incorporating all the data from the upcoming third edition. This site allows for a number of search criteria, but most notably makes it possible for the user to gather easily all the papyrological evidence available in Mertens-Pack for a particular author; the database extends beyond papyri covered by the papyri.info sources and includes material published in a variety of places. Modern publication information for each papyrus is conveniently included in the metadata for each entry, for those who want more details on a particular papyrus, but the abbreviations for the modern resources can be difficult to locate on the website. CEDOPAL abbreviations are available at http://web.philo.ulg.ac.be/cedopal/eng/liste-des-abreviations/. Those interested in paleographical issues related to papyri should consult the PapPal website (http://www.pappal.info/). Using online collections, PapPal aggregates images of papyri of known dates (from the third century BC to the eighth century AD) specifically for paleographical analysis. Each images is accompanied by a set of metadata listing the date, provenance, material, and links to other papyrological websites with data on the papyrus (such the DDbDP or HGV). A variety of search terms are permitted, including title of the work in which the papyrus was published, provenance, date, and PPbDP and HGV number. A keyword search is also possible, although it is currently enabled only for German terms (the rest of the website is in English).

46. http://www.ancientlives.org.

47. Volumes 3 and 4 were edited by Stefan Radt. The remaining volumes also received updates; the most recent, volume 5 on Euripides, came out in 2004.

48. Volume 1 is on the minor tragedians, while volume 2 covers adespota. Volumes 3–5 are dedicated to the fragments of Aeschylus, Sophocles, and Euripides, respectively.

49. For more on this, see R. Kassel and C. Austin, *Poetae Comici Graeci* 1:vii.

50. With the exceptions of volume 1 (Doric comedy) and volume 8 (*adespota*), the remaining volumes are devoted to known fragments of Attic comedy. Most volumes contain comedies from multiple poets (e.g., volume 2 is encompasses poets from Agathenor to Aristonymus), but volumes 3 (Aristophanes) and 6 (Menander) attend to individual authors.

51. Hermann Diels, *Die Fragmente der Vorsokratiker* (Berlin: Weidmann, 1906), v.

52. Mortimer Chambers, "Felix Jacoby (biografie_jacoby)," *Die Fragmente der Griechischen Historiker Part I–III*, general editor Felix Jacoby, BrillOnline (2016): http://referenceworks.brillonline.com/entries/die-fragmente-der-griechischen-historiker-i-iii/felix-jacoby-biografiejacoby-abiografie_jacoby.

53. Mortimer Chambers, "The organization of the FGrHist (organisation_fgrhist)," *Die Fragmente der Griechischen Historiker Part I–III*, general editor Felix Jacoby, BrillOnline(2016): http://referenceworks.brillonline.com/entries/die-fragmente-der-griechischen-historiker-i-iii/the-organization-of-the-fgrhist-organisationfgrhist-aorganisation_fgrhist.

54. Mortimer Chambers, "The Continuation of the FGrHist (continuation_fgrhist)," *Die Fragmente der Griechischen Historiker Part I–III*, general editor Felix Jacoby, BrillOnline, (2016): http://referenceworks.brillonline.com/entries/die-fragmente-der-griechischen-historiker-i-iii/the-continuation-of-the-fgrhist-continuationfgrhist-acontinuation_fgrhist.

55. See http://referenceworks.brillonline.com/cluster/Jacoby%20Online.

56. Theodore Brunner, "Classics and the Computer: The History of a Relationship," in *Accessing Antiquity: The Computerization of Classical Databases*, ed. Jon Solomon (Tucson: University of Arizona Press, 1993), 11–15.

57. Brunner (1993) 27.

58. Rob Latousek, "Fifty Years of Classical Computing: A Progress Report," *CALICO* 18 (2001): 213.

59. The term "N-gram" simply refers to an undefined sequence of letters or words. In the case of the *TLG*, the N-gram is a series of words with significant overlap between texts.

60. See the Perseus website at http://www.perseus.tufts.edu/hopper/.

61. Greg Crane, "Classics and the Computer: An End of the History," in *A Companion to Digital Humanities*, ed. Susan Schreibman, Ray Siemens, and John Unsworth (Malden, MA: Blackwell, 2004), 48.

62. Crane (2004) 48–49.

63. In addition to the short definition given, there are further links to standard lexica, such as the LSJ (on which, see below) and Autenrieth's *A Homeric Dictionary*. Citations of classical sources found in the lexica also link to their texts. Beyond grammatical and lexical data, the site contains a rich database of images and information on material culture. Thus, a search for "Odysseus" yields many results from texts, but also forty-four images and references to thirty-nine artifacts (vases, sculpture, coins, etc.). There is some overlap between the information to be found on the Perseus Project website and the *TLG*.

64. Henry Liddell, Robert Scott, and Henry Stuart Jones, *A Greek–English Lexicon* (Oxford: Clarendon Press, 1996), iii–iv. For the rich history of the LSJ, see the preface to the 1925 edition.

65. Anthony Corbeill, "'Going Forward': A Diachronic Analysis of the *Thesaurus Linguae Latinae*," *American Journal of Philology* 128 (2007): 469; *Thesaurus Linguae Latinae: Praemonenda de Rationibus et Usu Operis* (Leipzig: Teubner, 1990), 26.

66. *TLL: Praemonenda* (1990) 26.

67. Corbeill (2007) 469.

68. Corbeill (2007) 469.

69. The information in the headings of the *TLL* can be extensive. As it is presented in a relatively compact form, subheadings must be noted carefully. All of the data provided are evidenced by examples from literature or material culture, sometimes with a brief citation from the passage cited. Definitions of the word follow the general information of the heading. These definitions are most often given in the form of Latin synonyms, sometimes augmented with a comparable word in Greek.

70. Olivier Reverdin, "Préface," in *Lexicon Iconographicum Mythologiae Classicae* I 1 (Zurich: Artemis, 1981), vii.

71. Many of the images can be viewed in part 2 of the volume; those images in the catalogue with corresponding plates in part 2 are marked with an asterisk, and entries accompanied by a bold dot can be viewed as a drawing in part 1 of the volume.

72. See http://www.iconiclimc.ch/.

73. See http://www.limc-france.fr/.

74. Vassilis Lambrinoudakis, "Preface," in *Thesaurus Cultus et Rituum Antiquorum*, ed. Vassilis Lambrinoudakis and Jean Ch. Balty (Los Angeles: J. Paul Getty Museum, 2004), ix.

75. John Boardman, "Introduction," in Lambrinoudakis and Balty, *ThesCRA*, xi.

76. Volumes 1–5 form a set that covers the activities of cult practice in volumes 1–3 (processions, sacrifices, purification, prayer, etc.) and static elements in volumes 4–5 (locations, ritual instruments, etc). Volume 6 concerns stages and circumstances of life, volume 7 treats festivals and games, and volume 8 includes articles on private and public space, polarities in religious life, and religious interrelations between ancient civilizations, as well as an addendum to volume 6 on death and burial and a supplement on plants and animals.

77. Lisa Carson, "CD-ROM Technology and *L'Année Philologique*," *Classical World* 91 (1998): 553.

78. J. Marouzeau, "Avant-propos," *Dix années de bibliographie classique* (Paris: Société d'Édition "Les Belles Lettres," 1957), v.

79. Carson (1998) 554.

80. The American office is currently housed at Duke University and is under the direction of Lisa Carson.

81. Hans-Friedrich Mueller, "An Inside View of *L'Année Philologique*: American Office," *BMCR* 94.03.25 (http://bmcr.brynmawr.edu/1994/94.03.25.html). The transition to computers was lengthy, and by 1994 the *APh* offices had still not fully incorporated computers into the production of volumes; see Carson (1998) 555.

82. Carson (1998) 555. Progress on the DCB continued into 2001, by which time the database had incorporated all of *APh*'s most recent volumes and lacked only volumes 1–37, the compilation of which was estimated at five years (APA newsletter April 2001 [24.2], 3). In the same year a proposal was announced to combine the DCB's database with online data recently added by *APh*; this project was put into place by 2002, which led to the current online version (available from http://www.annee-philologique.com/).

83. A few of the search features are particularly noteworthy for their ability to handle the vast amount of scholarship available in multiple languages. First, entries returned by searches can be narrowed according to discipline and subdiscipline by choosing the "subjects and disciplines" option in an advanced search, which opens a menu of disciplines to choose from. Also noteworthy is that search terms followed by a tilde (~) will return variations of the term, including its forms in other languages (e.g., a search for "Theocritus~" will return results for "Theocritus," "Teocrito," "Theokrit," etc.). *APh*, in both its online and its print forms, is a venerable and vital resource for all students and scholars of Greco-Roman civilization.

84. See http://projects.chass.utoronto.ca/amphoras/tocs.html.

85. Walther Sontheimer and Konrat Ziegler, "Vorwort," in *Der kleine Pauly* (Stuttgart: Alfred Druckenmüller, 1964), v.

86. Hubert Cancik and Helmuth Schneider, "Vorwort," in *Der neue Pauly* (Stuttgart: Metzler, 1996), v.

87. The overall structure is twelve sections, each of which is divided into multiple volumes. Many volumes are further divided into fascicles.

88. See the Digital Scriptorium website at http://vm136.lib.berkeley.edu/BANC/digitalscriptorium/.

BIBLIOGRAPHY

1990. *Thesaurus Linguae Latinae: Praemonenda de Rationibus et Usu Operis.* Leipzig: Teubner.
2008. *Revue d'histoire des textes.* Editions du Centre National de la Recherche Scientifique.
Abbamonte, G. 2014. "Considerazioni su alcune dediche di traduzioni latine di opere greche fatte da umanisti del Quattrocento." In *Pratiques latines de la dédicace: Permanence et mutations, de l'Antiquité à la Renaissance,* ed. J.-C. Juhle, 523-559. Paris: Classiques Garnier.
Abelard, Peter, Betty Radice, and Michael Clanchy. 2003. *The Letters of Abelard and Heloise.* London: Penguin.
Affleck, Michael. 2013. "Priests, Patrons, and Playwrights: Libraries in Rome before 168 BC." In *Ancient Libraries,* ed. Jason König, Katerina Oikonomopoulou, and Greg Woolf, 124-136. Cambridge: Cambridge University Press.
Alexander, Karl. 2013. "Honor, Reputation, and Conflict: George of Trebizond and Humanist Acts of Self-Presentation." PhD diss., University of Kentucky.
Anderson, R. D., P. J. Parsons, and R. G. M. Nisbet. 1979. "Elegiacs by Gallus from Qaṣr Ibrîm." *Journal of Roman Studies* 69:125-155.
Armstrong, Elizabeth. 1954. *Robert Estienne, Royal Printer: An Historical Study of the Elder Stephanus.* Cambridge: Cambridge University Press.
Assmann, Jan. 2009. *The Price of Monotheism.* Trans. Robert Savage. Stanford: Stanford University Press.
Atkinson, J. B., and D. Sices, ed. and trans. 1996. *Machiavelli and His Friends: Their Personal Correspondence.* De Kalb: Northern Illinois University Press.
Avrin, Leila. 1991. *Scribes, Script, and Books: The Book Arts from Antiquity to the Renaissance.* Chicago: American Library Association.
Axer, Jerzy. 2007. "Central Eastern Europe." In *A Companion to the Classical Tradition,* ed. C. W. Kallendorf, 132-155. Chichester: Blackwell.
Babinger, Franz. 1992. *Mehmed the Conqueror and His Time.* Princeton: Princeton University Press.
Balogh, Josef. 1927. "*Voces paginarum.*" *Philologus* 82:84-240.

Banham, R. Bracht. 1985. "Utopian Laughter: Lucian and Thomas More." *Moreana* 86: 23–43.

Barbero, A., Guglielmo Cavallo, Claudio Leonardi, Enrico Menestò, and Piero Boitani. 2003. "Età di mezzo e secoli bui." In *Lo spazio letterario del Medioevo*, ed. P. Boitanin, M. Mancini, and A. Varvaro, 505–526. Rome: Salerno.

Barkan, Leonard. 1999. *Unearthing the Past: Archaeology and Aesthetics in the Making of Renaissance Culture*. New Haven: Yale University Press.

Barker, Graeme, and Tom Rassmussen. 2000. *The Etruscans*. Oxford: Blackwell.

Baron, Hans. 1966. *The Crisis of the Early Italian Renaissance: Civic Humanism and Republican Liberty in an Age of Classicism and Tyranny*. 2 vols. Princeton: Princeton University Press.

Bellamy, Rufus. 1989. "Bellerophon's Tablet." *Classical Journal* 84:289–307.

Bentley, Richard, ed. 1711. *Q. Horatius Flaccus*. Cambridge: Cambridge University Press.

Bérard, Francoise, Denis Feissel, Pierre Petitmengin, Denis Rousset, and Michel Séve, eds. 2000. *Guide de l'épigraphiste*. Paris: Rue d'Ulm.

Berti, Monica, and Virgilio Costa. 2010. *La biblioteca di Alessandria: Storia di un paradiso perduto*. Rome: Tored.

Best, Edward E., Jr. 1965. "Attitudes toward Literacy Reflected in Petronius." *Classical Journal* 61:72–76.

Biles, Zachary. 2011. *Aristophanes and the Poetics of Competition*. Cambridge: Cambridge University Press.

Birt, Theodor. 1882. *Das antike Buchwesen in seinem Verhältniss zur Litteratur*. Berlin: W. Herz.

Bisaha, Nancy. 2006. *Creating East and West: Renaissance Humanists and the Ottoman Turks*. Philadelphia: University of Pennsylvania Press.

Bischoff, Bernhard. 1961. "Die europäische Verbreitung der Werke Isidors von Sevilla." In *Isidoriana: Colección de estudios sobre Isidoro de Sevilla*, ed. Manuel Díaz y Díaz, 317–344. León: Centro de Estudios San Isidoro.

———. 1990. *Latin Palaeography: Antiquity and the Middle Ages*. Trans. Dáibhí Ó Cróinín and David Ganz. Cambridge: Cambridge University Press.

Black, R. 2005. "Renaissance Humanism and Historiography Today." In *Palgrave Advances in Renaissance Historiography*, ed. Jonathan Woolfson, 97–117. New York: Macmillan.

Black, Robert. 2008. "Humanism." In *The New Cambridge Medieval History*, ed. Christopher Allmand, 243–277. Cambridge: Cambridge University Press.

Blank, D. L., and F. Longo. 2000. "An Inventory of the Herculaneum Papyri from Piaggio's Time." *Cronache Ercolanesi* 30:131–147.

Blank, David. 1983. "Remarks on Nicanor, the Stoics, and the Ancient Theory of Punctuation." *Glotta* 61:48–67.

Bloch, Herbert. 1988. *Monte Cassino in the Middle Ages*. Cambridge, MA: Harvard University Press.

Blum, Rudolf. 1991. *Kallimachos: The Alexandrian Library and the Origins of Bibliography*. Trans. Hans Wellisch. Madison: University of Wisconsin Press.

Boatwright, Mary T. 2013. "Hadrian and the Agrippa Inscription of the Pantheon." In

Hadrian: Art, Politics, and Economy, ed. T. Opper, 19–30. British Museum Research Publications 175. Cambridge: Cambridge University Press.

Bodel, John. 1994. *Graveyards and Groves: A Study of the Lex Lucerina*. Cambridge, MA: Harvard University Press.

———. 2001. "Epigraph and the Ancient Historian." In *Epigraphic Evidence: Ancient History from Inscriptions*, ed. John Bodel, 1–56. New York: Routledge.

Boeckh, A. 1987. *La filologia come scienza storica*. Naples: Guida.

Bolter, Jay. 1980. "Friedrich August Wolf and the Scientific Study of Antiquity." *Greek, Roman, and Byzantine Studies* 21:83–99.

Bond, H., trans. 2005. *Nicholas of Cusa: Selected Spiritual Writings*. Classics of Western Spirituality 89. Mahwah, NJ: Paulist.

Bonghi Jovino, M. 1986. *Gli etruschi di Tarquinia*. Modena: Panini.

Botfield, Beriah. 1861. *Praefationes et epistolae editionibus principibus auctorum veterum praepositae*. Cambridge: Cambridge University Press.

Botterill, Steven, ed. and trans. 1991. *Dante: De Vulgari Eloquentia*. Cambridge: Cambridge University Press.

Bottoni, Diego. 1984. "I Decembrio e la traduzione della *Repubblica* di Platone: Dalle correzioni dell'autografo di Uberto alle integrazioni greche di Pier Candido." In *Vestigia. Studi in onore Giuseppe Billanovich*, ed. Rino Avesani, 75–92. Rome: Edizioni di Storia e Letteratura.

Bowman, A. K., and J. D. Thomas. 1983. *Vindolanda: The Latin Writing Tablets*. Britannia Monograph Series. London: Society for the Promotion of Roman Studies.

Boyle, A. J. 1997. "Seneca Inscriptus." In *Tragic Seneca: An Essay in the Theatrical Tradition*, ed. A. J. Boyle, 141–166. London: Routledge.

———. 2006. *Roman Tragedy*. New York: Routledge.

Briggs, Ward, Jr. 1983. "Housman and Polar Errors." *American Journal of Philology* 104:268–277.

Bronchud, Miguel H. 2007. *The Secret Castle: The Key to Good and Evil*. N.p.: CreateSpace.

Brown, Peter. 1974. *The World of Late Antiquity: From Marcus Aurelius to Muhammad*. London: Thames and Hudson.

———. 1995. *The Rise of Western Christendom: Triumph and Diversity, AD 200–1000*. 2nd ed. Malden, MA: Blackwell.

———. 2012. *Through the Eye of a Needle: Wealth, the Fall of Rome, and the Making of Christianity in the West*. Princeton: Princeton University Press.

Brown, Virginia. 1979. "Latin Manuscripts of Caesar's *Gallic War*." In *Paleografia Diplomatica et Archivistica*, ed. G. Battelli. Storia e Letteratura 139. Rome: Edizioni di Storia e Letteratura.

Browning, Robert. 1977. *Studies on Byzantine History, Literature, and Education*. London: Variorum Reprints.

Brugnoli, Giorgio, and Carlo Santini. 1995. *L'additamentum Aldinum di Silio Italico*. Rome: Accademia Nazionale dei Lincei.

Brunhölzl, F. 1971. "Zum Problem der Casineser Klassiküberlieferung." *Abhandlungen der Marburger Gelehrten Gesellschaft* 3:114–143.

Bruni, Leonardo. 2002. "De studiis et litteris liber ad Baptistam de Malatestis" (trans. Craig W. Kallendorf). In *Humanist Educational Treatises*, ed. C. W. Kallendorf, 94–125. Cambridge, MA: Harvard University Press.

Brunner, Theodore. 1993. "Classics and the Computer: The History of a Relationship." In *Accessing Antiquity: The Computerization of Classical Databases*, ed. Jon Solomon, 10–33. Tucson: University of Arizona Press.

Bruzelius, Caroline. 2004. *The Stones of Naples: Church Building in Angevin Italy (1266–1343)*. New Haven: Yale University Press.

Bülow-Jacobson, Adam. 2009. "Writing Materials in the Ancient World." In *The Oxford Handbook of Papyrology*, ed. Roger Bagnall, 3–29. Oxford: Oxford University Press.

Burckhardt, Jacob. 1914. *The Civilization of the Renaissance in Italy*. Trans. S. G. C. Middlemore. London: G. Allen and Urwin. [First German ed. 1860.]

Buringh, Eltjo. 2011. *Medieval Manuscript Production in the Latin West: Exploration with a Global Database*. Leiden: Brill.

Burke, Peter. 1998. *The European Renaissance: Centers and Peripheries*. Chichester: Wiley-Blackwell.

Burnett, Charles S. F., and Danielle Jacqua, eds. 1994. *Constantine the African and ʿAlī Ibn Al-ʿAbbās Al-Maǧūsī: The* Pantegni *and Related Texts*. Leiden: Brill.

Burris, Simon, Jeffrey Fish, and Dirk Obbink. 2014. "New Fragments of Book 1 of Sappho." *Zeitschrift für Papyrologie und Epigraphik* 189:1–28.

Butzmann, H. 1970. *Corpus Agrimensorum Romanorum: Der Codex Arcenianus A der Herzog-August-Bibliotek zu Wolfenbüttel (Cod. Guelf. 36. 23A)*. Codices Graeci et Latini 22. Leiden: Brill.

Calhoun, George Miller. 1919. "Oral and Written Pleading in Athenian Courts." *Transactions and Proceedings of the American Philological Association* 50:177–193.

Cambiano, G., L. Canfora, and D. Lanza. 1995. *Lo spazio letterario della Grecia antica*. Vol. 2. Rome: Salerno.

Cameron, Alan. 2011. *The Last Pagans of Rome*. Oxford: Oxford University Press.

Campana, Augusto. 1946. "The Origin of the Word 'Humanist.'" *Journal of the Warburg and Courtauld Institutes* 9:60–73.

Canfora, L. 1995. "Le collezioni superstiti." In *Lo spazio letterario della Grecia antica*, vol. 2, ed. G. Cambiano, L. Canfora, and D. Lanza, 95–245. Rome: Salerno.

———. 2007. *The True History of the So-Called Artemidorus Papyrus*. Bari: Edizioni di Pagina.

———. 2008a. *Filologia e libertà*. Milan: Mondadori.

———. 2008b. *Il papiro di Artemidoro*. Bari: Laterza.

———. 2011. *La meravigliosa storia del falso Artemidoro*. Palermo: Sellerio.

Capps, Edward, Jr. 1927. "The Style of Consular Diptychs." *Art Bulletin* 10:60–101.

Cardauns, Burkhart. 2001. *Marcus Terentius Varro: Einführung in sein Werk*. Heidelberg: Winter.

Cardini, Franco. 2011. *Cristiani perseguitati e persecutori*. Rome: Salerno.

Carlson, David R. 2004. "Greeks in England, 1400." In *Interstices: Studies in Late Middle English and Anglo-Latin Texts in Honour of A. G. Rigg*, ed. Richard Firth Green, Linne R. Mooney, and A. G. Rigg, 74–98. Toronto: University of Toronto Press.

Carson, Lisa. 1998. "CD-ROM Technology and *L'Année Philologique*." *Classical World* 91:553–564.

Cavallo, G. 1986. "Conservazione e perdita dei testi greci." In *Tradizione dei classici: Trasformazioni di cultura*, ed. A. Giardina, 83–172. Bari: Laterza.

———. 1995. "I fondamenti culturali della trasmissione dei testi antichi a Bisanzio." In *Lo spazio letterario della Grecia antica*, vol. 2., ed. G. Cambiano, L. Canfora, and D. Lanza, 265–306. Rome: Salerno.

Cavallo, Gugliemo, and Herwig Maehler, eds. 2008. *Hellenistic Bookhands*. Berlin: de Gruyter.

Chaucer, Geoffrey. 2015. *The Canterbury Tales*. Ed. John Urban Nicolson. N.p.: Courier Corporation.

Chauffour-Kestner, V. 1963. *Ulrich von Hutten, Imperial Poet and Orator: The Great Knightly Reformer of the Sixteenth Century*. Edinburgh: T. and T. Clark.

Citroni, Mario. 2009. "Poetry in Augustan Rome." In *A Companion to Ovid*, ed. Peter E. Knox, 8–25. Chichester: Wiley-Blackwell.

Clapham, M. 1963. "L'arte della stampa." In *Storia della tecnologia*, vol. 3, ed. Ch. Singer, E. J. Holmyard, A. R. Hall, and T. I. Williams, 385–419. Turin: Boringhieri.

Clarysse, Willy. 1993. "Egyptian Scribes Writing Greek." *Chronique d'Égypte* 68:186–201.

Clemens, Raymond, and Timothy Graham. 2007. *Introduction to Manuscript Studies*. Ithaca, NY: Cornell University Press.

Cohen-Mushlin, Aliza. 2010. "A School for Scribes." In *Teaching Writing, Learning to Write. Proceedings of the XVIth Colloquium of the Comité International de Paléographie Latine*, ed. Pamela Robinson, 61–90. London: King's College London.

Coleman-Norten, P. R. 1950. "Cicero's Contribution to the Text of the Twelve Tables." *Classical Journal* 46:51–60.

Conant, K. J. 1959. *Carolingian and Romanesque Architecture, 800–1200*. New Haven: Yale University Press.

Cook, R. M., and A. G. Woodhead. 1959. "The Diffusion of the Greek Alphabet." *American Journal of Archaeology* 63:175–178.

Cooley, Alison. 2012. *The Cambridge Manual of Latin Epigraphy*. Cambridge: Cambridge University Press.

Coqueugniot, Gaëlle. 2013. "Where Was the Royal Library of Pergamum? An Institution Found and Lost Again." In *Ancient Libraries*, ed. Jason König, Katerina Oikonomopoulou, and Greg Woolf, 109–123. Cambridge: Cambridge University Press.

Corbeill, Anthony. 2007. "'Going Forward': A Diachronic Analysis of the *Thesaurus Linguae Latinae*." *American Journal of Philology* 128:469–496.

Cortesi, M. 1995. "Umanesimo greco." In *Lo spazio letterario del Medioevo* I: *Il Medioevo latino*, vol. 3: *La ricezione del testo*, ed. G. Cavallo, C. Leonardi, and E. Menesto, 457–507. Rome: Salerno.

Costa, Gabriele. 2000. *Sulla preistoria della tradizione italica*. Florence: Olschki.

Courtin, Nicolaus, ed. 2010. *Cornelius Nepos De Vitis Excellentium Imperatorum*. Whitefish, MT: Kessinger.

Courtney, E. 2009. "Housman's Manilius." In *A. E. Housman: Classical Scholar*, ed. D. J. Butterfield and Christopher Stray, 29–44. London: Duckworth.

Courtney, Edward. 1935. *The Fragmentary Latin Poets*. Oxford: Oxford University Press.

Crab, Markjke. 2013. "Josse Bade's *Familiaris Commentarius* on Valerius Maximus (1510): A School Commentary?" In *Transformations of the Classics via Early Modern Commentaries*, ed. Karl A. E. Enenkel, 153–166. Leiden: Brill.

Crane, Greg. 2004. "Classics and the Computer: An End of the History." In *A Companion to Digital Humanities*, ed. Susan Schreibman, Ray Siemens, and John Unsworth, 46–55. Malden, MA: Blackwell.

Cribiore, Raffaella. 2001. *Gymnastics of the Mind: Greek Education in Hellenistic and Roman Egypt*. Princeton: Princeton University Press.

Curchin, Leonard A. 1995. "Literacy in the Roman Provinces: Qualitative and Quantitative Data from Central Spain." *American Journal of Philology* 116:461–476.

Curtius, Ernst Robert. 1953. *European Literature and the Latin Middle Ages*. Trans. Willard R. Trask. Princeton: Princeton University Press.

Cuvigny, Hélène. 2009. "The Finds of Papyri: The Archaeology of Papyri." In *The Oxford Handbook of Papyrology*, ed. Roger Bagnall, 30–58. Oxford: Oxford University Press.

Cyrus, Cynthia. 2009. *The Scribes for Women's Convents in Late Medieval Germany*. Toronto: University of Toronto Press.

Dalton, O. M. 1961. *Byzantine Art and Archaeology*. New York: Dover.

D'Amico, John. 1988. *Theory and Practice in Renaissance Textual Criticism: Beatus Rhenanus between Conjecture and History*. Berkeley: University of California Press.

Delz, Josef, ed. 1987. *Sili Italici. Punica*. Bibliotheca Scriptorum Graecorum et Romanorum Teubneriana. Stuttgart: Teubner.

Dennis, George T. 2010. *The Taktika of Leo VI: Text, Translation, and Commentary*. Washington, DC: Dumbarton Oaks.

Derolez, Albert. 1973. "Codicologie ou archéologie du livre?" *Scriptorium* 27:47–49.

De Roover, Florence Edler. 1957. "The Scriptorium." In *The Medieval Library*, ed. James Westfall Thompson, 594–612. New York: Hafner.

Dessau, Herman, ed. 1902. *Inscriptiones Latinae Selectae*. 5 vols. Berlin: Weidmann.

Destrez, Jean. 1935. *La pecia dans les manuscrits universitaires du XIIIe et du XIVe siècles*. Paris: Jacques Vautrain.

de Vos, M. 1983. s.v. "Horti Maecenatis. 'Auditorium.'" *Lexicon Topographicum Urbis Romae* 3:74–75.

Dickey, Eleanor. 2007. *Ancient Greek Scholarship*. Oxford: Oxford University Press.

———. 2011. "The Sources of Our Knowledge of Ancient Scholarship." In *From Scholars to Scholia: Chapters in the History of Ancient Greek Scholarship*, ed. Franco Montanari and Lara Pagani, 459–514. Berlin: de Gruyter.

Diels, Hermann. 1906. *Die Fragmente der Vorsokratiker*. 2nd ed. Berlin: Weidmann.

Diringer, David. 1982. *The Book before Printing: Ancient, Medieval, and Oriental*. New York: Dover.

Dix, T. Keith. 1994. "'Public Libraries' in Ancient Rome: Ideology and Reality." *Libraries and Culture* 29:282–296.

Dorleijn, G. J., and Herman L. J. Vanstiphout, eds. 2003. *Cultural Repertoires: Structure, Function, and Dynamics*. Groningen Studies in Cultural Change 3. Leuven: Peeters.

Duckworth, George. 1952. *Nature of Roman Comedy: A Study in Popular Entertainment*. Princeton: Princeton University Press.

Duhem, Pierre Maurice Marie. 1905. *Les origines de la statique*. Paris: A. Hermann.

Durant, Will. 1953. *The Renaissance: A History of Civilization in Italy from 1304–1576 A.D.* Story of Civilization V. New York: Simon and Schuster.

Easterling, Patricia E. 2006. "Notes on Notes: The Ancient Scholia on Sophocles." *Acta Universitatis Upsaliensis* 21:21–36.

Edmondson, Jonathan. 2014. "Inscribing Roman Texts: *Officinae*, Layout, and Carving Techniques." In *The Oxford Handbook of Roman Epigraphy*, ed. Christer Bruun and Jonathan Edmondson, 111–130. Oxford: Oxford University Press.

Edmunds, Lowell. 1992. *From a Sabine Jar: Reading Horace* Odes *1.9*. Chapel Hill: University of North Carolina Press.

Eliot, T. S. 1921. "Tradition and the Individual Talent." In *The Sacred Wood: Essays on Poetry and Criticism*. New York: Knopf.

Elliott, Jackie. 2013. *Ennius and the Architecture of the* Annales. Cambridge: Cambridge University Press.

Enenkel, Karl, and Henk Nellen. 2013. "Introduction." In *Neo-Latin Commentaries and Management of Knowledge in the Late Middle Ages and the Early Modern Period (1400–1700)*, ed. Karl Enenkel and Henk Nellen, 1–75. Supplementa Humanistica Lovaniensia 33. Leuven: Leuven University Press.

Fairclough, H. Rushton. 1922. "The Poems of the Appendix Vergiliana." *Transactions and Proceedings of the American Philological Association* 53:5–34.

Farrell, Joseph. 2001. *Latin Language and Latin Culture: From Ancient to Modern Times*. Roman Literature and Its Contexts. Cambridge: Cambridge University Press.

Febvre, Lucien, and Henri-Jean Martin. 1976. *The Coming of the Book: The Impact of Printing, 1450–1800*. London: NLB.

Ferguson, Margaret W. 2003. *Dido's Daughters: Literacy, Gender, and Empire in Early Modern England and France*. Chicago: University of Chicago Press.

Feuchter, Jörg, Friedhelm Hoffmann, and B. Yun, eds. 2011. *Cultural Transfers in Dispute: Representations in Asia, Europe, and the Arab World since the Middle Ages*. Frankfurt: Campus.

Fiesoli, G. 2000. *La genesi del lachmannismo*. Florence: Sismei.

Filoramo, Giovanni. 2011. *La croce e il potere: I cristiani da martiri a persecutori*. Bari: Laterza.

Fisher, Jay. 2014. *The* Annals *of Quintus Ennius and the Italic Tradition*. Baltimore: Johns Hopkins University Press.

Forbes, Clarence A. 1936. "Books for the Burning." *Transactions and Proceedings of the American Philological Association* 67:114–125.

Ford, Philip. 2007. "France." In *A Companion to the Classical Tradition*, ed. C. W. Kallendorf, 156–168. Chichester: Blackwell.

Foscolo, U. 1953. *Epistolario*, Vol. 3. Ed. P. Carli. Florence: Le Monnier.

Franklin, James. 2012. "Science by Conceptual Analysis: The Genius of the Late Scholastics." *Studia Neoaristotelica* 9:3-24.

Fraschetti, Augusto. 1995. "Trent'anni dopo 'Il conflitto fra paganesimo e cristianesimo nel secolo IV.'" In *Pagani e cristiani da Giuliano l'Apostata al sacco di Roma. Atti del Convegno Internazionale di Studi: Rende, 12-13 novembre 1993 / a cura di Franca Ela Consolin*, ed. F. E. Consolino, 5-14. Soveria Mannelli: Rubbettino.

Fraser, P. M. 1985. *Ptolemaic Alexandria*. Oxford: Oxford University Press.

Frasso, Giuseppe, and Giuseppe Velli. 2005. *Petrarca e la Lombardia. Atti del convegno di studi, Milano, 22-23 Maggio 2003*. Padua: Antenore.

Frost, Gary. 2013. "Material Quality of Medieval Bookbindings." In *Scraped, Stroked, and Bound: Materially Engaged Readings of Medieval Manuscripts*, ed. Jonathan Wilcox, 129-134. Turnhout: Brepols.

Funaioli, Gino. 1969. *Grammaticae Romanae Fragmenta*. Stuttgart: Teubner.

Furbetta, L. 2009. "Apostilles in Pomponius Mela between Petrarch and William Pastrengo." *Rassegna della Letteratura Italiana* 113:573-574.

Gamble, Harry Y. 1995. *Books and Readers in the Early Church: A History of Early Christian Texts*. New Haven: Yale University Press.

Garin, E. 1952. *Prosatori latini del Quattrocento*. Milan: Ricciardi.

———. 2009. "Interpretazioni del Rinascimento." In *Interpretazioni del Rinascimento*, ed. M. Ciliberto, 2:3-14. Rome: Edizioni di Storia e Letteratura.

Gavrilov, A. K. 1997. "Techniques of Reading in Classical Antiquity." *Classical Quarterly* 47:56-73.

Geanakoplos, Deno John. 1989. *Constantinople and the West: Essays on the Late Byzantine (Palaeologan) and Italian Renaissances and the Byzantine and Roman Churches*. Madison: University of Wisconsin Press.

Gersh, Stephen. 1986. *Middle Platonism and Neoplatonism: The Latin Tradition*. Vol. 2. South Bend, IN: University of Notre Dame Press.

Geymonat, Mario. 2008. "Praefatio." In *P. Vergili Maronis Opera*. Rome: Edizioni di Storia e Letteratura.

———. 2010. *The Great Archimedes*. Trans. and ed. R. A. Smith. Waco, TX: Baylor University Press.

Gilson, Étienne. 1955. *Les idées et les lettres*. 2nd ed. Librairie Philosophique. Paris: Vrin.

Goldberg, Sander. 2005. *Constructing Literature in the Roman Republic: Poetry and Its Reception*. Cambridge: Cambridge University Press.

Gombrich, Ernst Hans, and Rudolf Wittkower. 1967. *From the Revival of Letters to the Reform of the Arts: Niccolò Niccoli and Filippo Brunelleschi*. London: Phaidon.

Goodman, Howard L., and Anthony Grafton. 1990. "Ricci, the Chinese, and the Toolkits of Textualists." *Asia Major* 3:95-148.

Gordan, Phyllis, and Walter Goodhard. 1974. *Two Renaissance Book Hunters: The Letters of Poggius Bracciolini to Nicolaus de Niccolis*. New York: Columbia University Press.

Goughuenheim, Sylvain. 2008. *Aristote au mont Saint-Michel: Les racines grecques de l'Europe chrétienne*. Paris: Seuil.

———. 2009. *Regards sur le Moyen Âge: Quarante histoires médiévales*. Milan: Rizzoli.

Grafton, Anthony. 1977. "On the Scholarship of Politian and Its Contexts." *Journal of the Warburg and Courtauld Institutes* 40:150–188.

———. 1983, 1993. *Joseph Scaliger: A Study on the History of Classical Scholarship*. 2 vols. Oxford: Clarendon.

———. 1991. *Defenders of the Text: The Traditions of Scholarship in an Age of Science, 1450–1800*. Cambridge, MA: Harvard University Press.

Grafton, Anthony, and Megan Williams. 2008. *Christianity and the Transformation of the Book: Origen, Eusebius, and the Library of Caesarea*. Cambridge, MA: Harvard University Press.

Grant, Edward. 1996. *The Foundations of Modern Science in the Middle Ages: Their Religious, Institutional, and Intellectual Contexts*. Cambridge: Cambridge University Press.

Greef, Wulfert. 2008. *The Writings of John Calvin: An Introductory Guide*. Trans. L. Bierma. Louisville, KY: Westminster John Knox.

Greenwood, David Neal. 2014. "The *Alethes Logos* of Celsus and the Historicity of Christ." *Anglican Theological Review* 96:705–713.

Greetham, David. 2013a. "A History of Textual Scholarship." In *The Cambridge Companion to Textual Scholarship*, ed. Neil Fraistat and Julia Flanders, 16–41. Cambridge: Cambridge University Press.

———. 2013b. *Textual Scholarship: An Introduction*. Garland Reference Library of the Humanities 1417. London: Routledge.

Grenfell, Bernard P., and Arthur S. Hunt. 1898. *The Oxyrhynchus Papyri*. Part 4. London: Egypt Exploration Fund.

Gruen, Erich. 1984. *The Hellenistic World and the Coming of Rome*. 2 vols. Berkeley: University of California Press.

Gruijs, Albert. 1972. "Codicology or the Archaeology of the Book? A False Dilemma." *Quaerendo* 2:87–108.

Grund, Gary R., et al. 2011. *Humanist Tragedies*. Cambridge, MA: Harvard University Press.

Gualdo Rosa, Lucia. 1987. "Leonardo Bruni, l'*Oratio in hypocritas* e i suoi difficili rapport con Ambrogio Traversari." *Vita Monastica* 65:89–111.

Gullick, Michael. 1992. "How Fast Did Scribes Write? Evidence from Romanesque Manuscripts." In *Making the Medieval Book: Techniques of Production*, ed. Linda Brownrigg, 39–58. Los Altos Hills, CA: Anderson-Lovelace.

Güthert, Kerstin. 2005. *Herausbildung von Norm und Usus Scribendi im Bereich der Worttrennung am Zeilenende (1500–1800)*. Heidelberg: Winter.

Hairston, Julia, and Walter Stevens. 2010. *The Body in Early Modern Italy*. Baltimore: Johns Hopkins University Press.

Hall, Marie Boas. 1994. *The Scientific Renaissance, 1450–1630*. New York: Dover.

Hammond, C. P. 1978. "A Product of a Fifth-Century Scriptorium Preserving Conventions Used by Rufinus of Aquileia." *Journal of Theological Studies* 29:366–391.

———. 1984. "Products of Fifth-Century Scriptoria Preserving Conventions Used by Rufinus of Aquileia: Script." *Journal of Theological Studies* 35:347–393.

Hankins, James. 1995. "The 'Baron Thesis' after Forty Years and Some Recent Studies of Leonardo Bruni." *Journal of the History of Ideas* 56:309–338.

Hansen, Esther. 1971. *The Attalids of Pergamon*. Ithaca, NY: Cornell University Press.

Hanson, Ann E., and Monica H. Green. 1994. "Soranus of Ephesus: *Methodicorum princeps*." In *Aufstieg und Niedergang der römischen Welt*, II.37.2, ed. W. Haase and H. Temporini, 968–1075. Berlin: de Gruyter.

Harder, Annette. 2012. *Callimachus: Aetia*. Vol. 1. Oxford: Oxford University Press.

Hardwick, Lorna, and Christopher Stray, eds. 2008. *A Companion to Classical Receptions*. Oxford: Blackwell.

Harris, William V. 1989. *Ancient Literacy*. Cambridge, MA: Harvard University Press.

Harth, Helene. 1984. *Poggio Bracciolini: Lettere*. Vol. 1. Florence: Olschki.

Harvey, F. D. 1966. "Literacy in the Athenian Democracy." *Revue des Études Grecques* 79:585–635.

Haskins, Charles H. 1927. *The Renaissance of the Twelfth Century*. Cambridge, MA: Harvard University Press.

Häuber, Ruth Christine. 1990. "Zur Topographie der Horti Maecenatis und der Horti Lamiani auf dem Esquilin in Rom." *Kölner Jahrbuch* 23:11–107.

Haug, Dag. 2005. "The Linguistic Thought of Friedrich August Wolf: A Reconsideration of the Relationship between Classical Philology and Linguistics in the Nineteenth Century." *Historiographia Linguistica* 32:35–60.

Hawking, Stephen. 2002. *On the Shoulders of Giants*. London: Penguin.

Healey, John. 1990. *The Early Alphabet*. Berkeley: University of California Press.

Hedrick, Charles W., Jr. 2000. *History and Silence: Purge and Rehabilitation of Memory in Late Antiquity*. Austin: University of Texas Press.

Hexter, Ralph, and David Townsend. 2012. *The Oxford Handbook of Medieval Latin Literature*. New York: Oxford University Press.

Heyworth, S. J. 2007. *Cynthia: A Companion to the Text of Propertius*. Oxford: Oxford University Press.

Hickey, Todd M. 2009. "Writing Histories from the Papyri." In *The Oxford Handbook of Papyrology*, ed. Roger Bagnall, 495–520. Oxford: Oxford University Press.

Highet, Gilbert. 2015. *The Classical Tradition: Greek and Roman Influences on Western Literature*. Oxford: Oxford University Press.

Hollway-Calthrop, Henry. 1972. *Petrarch: His Life and Times*. New York: Cooper Square.

Holquist, Michael. 2012. "World Literature and Philology." In *The Routledge Companion to World Literature*, ed. Theo D'haen, David Damrosch, and Djelal Kadir, 147–157. London: Routledge.

Holzberg, Niklas. 2004. "Impersonating Young Virgil: The Author of the *Catalepton* and His *Libellus*." *Materiali e Discussion per l'Analisi dei Testi Classici* 52:29–40.

Horn, Walter, and Ernest Born. 1986. "The Medieval Monastery as a Setting for the Production of Manuscripts." *Journal of the Walters Art Gallery* 44:16–47.

Houston, George. 2014. *Inside Roman Libraries*. Chapel Hill: University of North Carolina Press.

Humphrey, J. H., ed. 1991. *Literacy in the Ancient World*. Ann Arbor: University of Michigan Press.

Huot, Sylvia. 1988. "'Ci parle l'aucteur': The Rubrication of Voice and Authorship in *Roman de la Rose* Manuscripts." *Substance* 17:42–48.

Immerwahr, Henry. 2006. "Nonsense Inscriptions and Literacy." *Kadmos* 45:136–172.

Inge, W. R. 1900. "The Permanent Influence of Neoplatonism upon Christianity." *American Journal of Theology* 4:328–344.

Irigoin, Jean. 1986. "Accidents matériels et critiques des textes." *Revue d'Historie des Textes* 16:1–36.

Iserloh, Erwin. 1968. *The Theses Were Not Posted: Luther between Reform and Reformation*. Trans. J. Wicks. Boston: Beacon.

Isserlin, B. S. J. 1982. "The Earliest Alphabetic Writing." In *The Cambridge Ancient History*, vol. 3, part 3, ed. John Boardman, I. E. S. Edwards, N. G. L. Hammond, and E. Sollberger, 794–818. Cambridge: Cambridge University Press.

Janko, Richard. 2002a. "The Derveni Papyrus: An Interim Text." *Zeitschrift für Papyrologie und Epigraphik* 141:1–62.

———. 2002b. "The Herculaneum Library: Some Recent Developments." *Estudios Clásicos* 121:25–41.

———. 2009. "The Artemidorus Papyrus." *Classical Review* 59:403–410.

Jauss, Hans Robert. 1982. *Toward an Aesthetic of Reception*. Theory and History of Literature 2. Minneapolis: University of Minnesota Press.

Jeffery, L. H. 1982. "Greek Alphabetic Writing." In *The Cambridge Ancient History*, vol. 3, part 3, ed. John Boardman, I. E. S. Edwards, N. G. L. Hammond, and E. Sollberger, 819–833. Cambridge: Cambridge University Press.

———. 1990. *The Local Scripts of Archaic Greece*. Oxford: Clarendon.

Jelerup, Torbjörn. 2003. "The Renaissance, and the Rediscovery of Plato and the Greeks." *Fidelio* 12:36–55.

Johansson, Mikael. 2001. "The Inscription from Troizen: A Decree of Themistocles?" *Zeitschrift für Papyrologie und Epigraphik* 137:71–72.

John, James J. 1992. "Latin Paleography." In *Medieval Studies: An Introduction*, 2nd. ed., ed. James M. Powell, 1–68. 2nd ed. Syracuse, NY: Syracuse University Press.

Johnson, William A. 1993. "Pliny the Elder and Standardized Roll Heights in the Manufacture of Papyrus." *Classical Philology* 88:46–50.

———. 1994. "The Function of the Paragraphus in Greek Literary Prose Texts." *Zeitschrift für Papyrologie und Epigraphik* 100:65–68.

———. 2000. "Toward a Sociology of Reading in Classical Antiquity." *American Journal of Philology* 121:593–627.

———. 2004. *Bookrolls and Scribes in Oxyrhynchus*. Toronto: University of Toronto Press.

———. 2009. "The Ancient Book." In *The Oxford Handbook of Papyrology*, ed. Roger Bagnall, 256–281. Oxford: Oxford University Press.

Jones, Howard. 2004. *Printing the Classical Text*. Utrecht: Hes and de Graaf.

Jones, Leslie. 1946. "Pricking Manuscripts: The Instruments and Their Significance." *Speculum* 21:389–403.

———. 1961. "Ancient Prickings in Eighth Century Manuscripts." *Scriptorium* 15:14–22.

Kallendorf, Craig. 1991. *A Bibliography of Venetian Editions of Virgil, 1470–1599.* Florence: Olschki.

———, ed. 2007. *A Companion to the Classical Tradition.* Chichester: Wiley-Blackwell.

———. 2013. "Virgil in the Renaissance Classroom: From Toscanella's *Osservationi . . . sopra l'opere di Virgilio* to the *Exercitationes rhetoricae*." In *The Classics in the Medieval and Renaissance Classroom: The Role of Ancient Texts in the Arts Curriculum as Revealed by Surviving Manuscripts and Early Printed Books,* ed. J. F. Ruys, J. Ward, and M. Heyworth, 309–328. Disputatio 20. Turnhout: Brepols.

———. 2015. *The Protean Virgil.* Oxford: Oxford University Press.

———. Forthcoming. "Commentaries, Censorship, and Printed Books: Neo-Latin in a Transnational World." In *Acta Conventus Neo-Latini Vindobonensis. Proceedings of the Sixteenth International Congress of Neo-Latin Studies,* ed. Astrid Steiner-Weber and Franz Römer. Leiden: Brill.

Kaster, Robert A. 1988. *Guardians of Language: The Grammarian and Society in Late Antiquity.* Transformation of the Classical Heritage 11. Berkeley: University of California Press.

———, ed. and trans. 1995. *De grammaticis et rhetoribus.* Oxford: Oxford University Press.

Keenan, James. 2009. "The History of the Discipline." In *The Oxford Handbook of Papyrology,* ed. Roger Bagnall, 59–78. Oxford: Oxford University Press.

Kenney, E. J. 1995. *The Classical Text: Aspects of Editing in the Age of the Printed Book.* Chicago: University of Chicago Press.

Kent, Roland Grubb. 1936. "On the Text of Varro, *de Lingua Latina*." *Transactions and Proceedings of the American Philological Association* 67:64–82.

Keppie, Lawrence. 1991. *Understanding Roman Inscriptions.* Baltimore: Johns Hopkins University Press.

Keulen, Wytse Hette. 2009. *Gellius the Satirist: Roman Cultural Authority in* Attic Nights. Mnemosyne Supplements. Leiden: Brill.

Kidwell, Carol. 2004. *Pietro Bembo: Lover, Linguist, Cardinal.* Montreal: McGill-Queen's University Press.

King, Margaret L. 2014. *Renaissance Humanism: An Anthology of Sources.* Indianapolis: Hackett.

Kirchhoff, A. 1877. *Studien zur Geschichte des griechischen Alphabets.* Berlin: Ferd. Dümmler's Verlags-Buchhandlung, Harrwitz und Gossmann.

Kirchner, Joachim. 1950. *Germanistische Handschriftenpraxis: Ein Lehrbuch für die Studierenden der deutschen Philologie.* Munich: Beck.

Kirkham, Victoria, and Armando Maggi, eds. 2009. *Petrarch: A Critical Guide to the Complete Works.* Chicago: University of Chicago Press.

Kleinhenz, Christopher. 2004. *Medieval Italy: An Encyclopedia.* London: Routledge.

Kneller, Jane. 1990. "Imaginative Freedom and the German Enlightenment." *Journal of the History of Ideas* 51:231.

Knox, Bernard. 1968. "Silent Reading in Antiquity." *Greek, Roman, and Byzantine Studies* 9:421–435.

Knox, K. T. 2008. "Enhancement of Overwritten Text in the Archimedes Palimp-

sest." In *Society of Photo-Optical Instrumentation Engineers Conference Series* 6810. doi:10.1117/12.877135.
Koff, L. M. 2013. "Dreaming the Dream of Scipio." In *Cicero Refused to Die: Ciceronian Influence through the Centuries*, ed. Nancy van Deusen, 65-83. Presenting the Past 4. Leiden: Brill.
Kristeller, Paul Oskar. 1992. *Iter Italicum*, vol. 6: *Italy III and Alia Itinera IV*. Leiden: Brill.
Lachmann, Carolus. 1846. *Novum Testamentum Graece*. Berlin: G. Reimer.
Lambot, C. 1939. "Letter inédite de S. Augustin relative au *De civitate dei*." *Revue bénédictine* 51:109-121.
Landucci, Luca, and Giuseppe Mazzotta. 2000. "The Entry of Charles VIII, King of France, into Florence." In *Images of Quattrocento Florence: Selected Writings in Literature, History, and Art*, ed. Stefano U. Baldassarri and Arielle Saiber, 115-122. New Haven: Yale University Press.
Langdon, Merle K. 2005. "A New Greek Abecedarium." *Kadmos* 44:175-182.
Lapini, Walter. 2015. "Philological Observations and Approaches to Language in the Philosophical Context." In *Brill's Companion to Ancient Scholarship*, ed. Franco Montanari, Stephanos Matthaios, and Antonios Rengakos, 1012-1056. Leiden: Brill.
Latousek, Rob. 2001. "Fifty Years of Classical Computing: A Progress Report." *CALICO* 18:211-222.
LeClercq, Jean. 1987. "Influence and Noninfluence of Dionysius in the Western Middle Ages" (trans. C. Luibheid). In *Pseudo-Dionysius: The Complete Works*, 25-33. New York: Paulist.
Le Goff, Jacques. 1992. *The Medieval Imagination*. Trans. Arthur Goldhammer. Chicago: University of Chicago Press.
———. 2008. *Intellectuals in the Middle Ages*. Trans. Teresa Lavender Fagan. Cambridge: Blackwell.
Lemerle, P. 1971. *Le premier humanisme byzantine: Notes et remarques sur enseignement et culture à Byzance des origines au Xe siècle*. Bibliothèque Byzantine, Études 6. Paris: Presses Universitaires de France.
Leonard, Miriam. 2012. *Socrates and the Jews: Hellenism and Hebraism from Moses Mendelssohn to Sigmund Freud*. Chicago: University of Chicago Press.
Levi, Anthony. 1970. *Humanism in France at the End of the Middle Ages and in the Early Renaissance*. Manchester: Manchester University Press.
Lewy, G. 1960. "The Struggle for Constitutional Government in the Early Years of the Society of Jesus." *Church History* 29:141-160.
Lindsay, W. M., ed. 1900. "Introduction." *Captivi of Plautus*, 1-102. London: Methuen.
Löffler, Karl. 1929. *Einführung in die Handschriftenkunde*. Leipzig: K. W. Hiersemann.
Lough, J., ed. 1954. *The 'Encyclopédie' of Diderot and d'Alembert*. Cambridge: Cambridge University Press.
Lowry, M. 1985. "Girolamo Aleandro." In *Contemporaries of Erasmus: A Biographical Register of the Renaissance and Reformation*, vol. 1, ed. Peter G. Bietenholz, 28-34. Toronto: University of Toronto Press.

Lytle, Ephraim. 2003. "Apuleius' *Metamorphoses* and the *Spurcum Additamentum* (10.21)." *Classical Philology* 98:349–365.

Maas, Paul. 1927. "Schicksale der antiken Literatur in Byzanz." In *Einleitung in die Altertumswissenschaft* I.2: *Textkritik*, ed. Alfred Gercke and Eduard Norden, 1–17. Leipzig: Teubner.

———. 1958. *Textual Criticism*. Trans. Barbara Flower. Oxford: Clarendon.

———. 1960. *Textkritik*. Leipzig: Teubner.

———. 1962. *Greek Metre*. Oxford: Clarendon.

MacCaffrey, James. 1910. *The History of the Catholic Church in the Nineteenth Century (1789–1908)*. 2nd ed. 2 vols. St. Louis: M. H. Gill and Son.

MacGregor, James. 1891. *The Apology of Christian Religion: Historically Regarded with Reference to Supernatural Revelation and Redemption*. Edinburgh: T. and T. Clark.

Mallon, Jean. 1952. *Paléographie romaine*. Madrid: Consejo Superior de Investigaciones Científicas, Instituto Antonio de Nebrija de Filología.

Manguel, Alberto. 1996. *A History of Reading*. New York: Viking.

Mann, Nicholas, and B. Munk Olsen, eds. 1997. *Medieval and Renaissance Scholarship. Proceedings of the Second European Science Foundation Workshop on the Classical Tradition in the Middle Ages and the Renaissance*. Mittellateinische Studien und Texte 21. Leiden: Brill.

Marcianò, Ada Francesca, and Biagio Rossetti. 1991. *L'età di Biagio Rossetti: Rinascimenti di casa d'Este*. Ferrara: Gabriele Corbo.

Marrou, H. I. 1949. "La technique de l'edition a l'epoque patristique." *Vigiliae Christianae* 3:208–224.

Martindale, Charles. 1993. *Redeeming the Text: Latin Poetry and the Hermeneutics of Reception*. Cambridge: Cambridge University Press.

———. 2006. "Introduction: Thinking through Reception." In *Classics and the Uses of Reception*, ed. Charles Martindale and Richard F. Thomas, 1–13. Chichester: Wiley-Blackwell.

Martorelli, R. 1999. "Riflessioni sulle attività produttive nell'età tardoantica e altomedievale: Esiste un artigianato ecclesiastico?" *Rivista di Archeologia Cristiana* 75:571–596.

Masai, François. 1950. "Paléographie et codicologie." *Scriptorium* 4:279–293.

Massera, A. F. 1928. *Giovanni Boccacci: Opere latine minori*. Bari: Laterza.

Matthaios, Stephanos. 2015. "Philology and Grammar in the Imperial Era and Late Antiquity: An Historical and Systematic Outline." In *Brill's Companion to Ancient Scholarship*, ed. Franco Montanari, Stephanos Matthaios, and Antonios Rengakos, 184–296. Leiden: Brill.

Mattusch, Carol C. 2010. "Programming Sculpture? Collection and Display in the Villa of the Papyri." In *The Villa of the Papyri at Herculaneum: Archaeology, Reception, and Digital Reconstruction*, ed. Mantha Zarmakoupi, 79–88. Berlin: de Gruyter.

Maurenbrecher, B. 1891. *C. Sallusti Crispi Historiarum Reliquiae*. Leipzig: Teubner.

Mayerson, Philip. 1997. "The Role of Flax in Roman and Fatimid Egypt." *Journal of Near Eastern Studies* 56:201–207.

Mazzocco, Angelo. 1993. *Linguistic Theories in Dante and the Humanists: Studies of Lan-*

guage and Intellectual History in Late Medieval and Early Renaissance Italy. Leiden: Brill.

McGann, Jerome. 1983. *A Critique of Modern Textual Criticism*. Chicago: University of Chicago Press.

McGushin, P. 1972. Review of P. K. Marshall, *A. Gellii, Noctes Atticae. Mnemosyne* 25: 95–98.

McKitterick, Rosamond. 1994. "Script and Book Production." In *Carolingian Culture: Emulation and Innovation*, 241–247. Cambridge: Cambridge University Press.

McNeil, David O. 1971. *Guillaume Budé and Humanism in the Reign of Francis I*. Travaux d'humanisme et Renaissance 142. Geneva: Librairie Droz.

McVaugh, Michael R. 2006. "Niccolò da Reggio's Translations of Galen and Their Reception in France." *Early Science and Medicine* 11:275–301.

Merton, Robert. 1993. *On the Shoulders of Giants: A Shandean Postscript, the Post-Italianate Edition*. Chicago: University of Chicago Press.

Meserve, Margaret, and Marcello Simonetta. 2003. *Pius II: Commentaries*. Vol. 2. I Tatti Renaissance Library 29. Cambridge, MA: Harvard University Press.

Metcalf, William E. 2012. *The Oxford Handbook of Greek and Roman Coinage*. Oxford: Oxford University Press.

Meyer, Jesse. 2013. "Parchment Production: A Brief Account." In *Scraped, Stroked, and Bound: Materially Engaged Readings of Medieval Manuscripts*, ed. Jonathan Wilcox, 93–96. Turnhout: Brepols.

Minio-Paluello, Lorenzo. 1952. "Boezio, Giacomo Veneto, Guglielmo di Moerbeke, Jacques Lefèvre d'Etaples e gli 'Elenchi Sophistici.'" *Rivista di Filosofia Neo-Scolastica* 44:398–411.

Momigliano, A. 1950. "Ancient History and the Antiquarian." *Journal of the Warburg and Courtauld Institutes* 13:285–315.

———. 1963. *The Conflict between Paganism and Christianity in the Fourth Century*. Oxford: Clarendon.

———. 1969. "Jacob Bernays." *Mededeelingen der Koninklijke Nederlandse Akademie van Wetenschappen, Afdeeling Letterkunde* 32:151–178.

Mommsen, Theodor, ed. 1852. *Inscriptiones Regni Neapolitani Latinae*. Leipzig: G. Wigand.

Mommsen, Theodor Ernst. 1942. "Petrarch's Conception of the 'Dark Ages.'" *Speculum* 17:226–242.

Monfasani, John. 1976. *George of Trebizond: A Biography and a Study of His Rhetoric and Logic*. Columbia Studies in the Classical Tradition 1. Leiden: Brill.

———. 1988. "The First Call for Press Censorship: Niccolò Perotti, Giovanni Andrea Bussi, Antonio Moreto, and the Editing of Pliny's *Natural History*." *Renaissance Quarterly* 41:1–31.

Montana, Fausto. 2015. "Hellenistic Scholarship." In *Brill's Companion to Ancient Scholarship*, ed. Franco Montanari, Stephanos Matthaios, and Antonios Rengakos, 60–183. Leiden: Brill.

Montanari, Elio. 2003. *La critica del testo secondo Paul Maas: Testo e commento*. Millennio Medievale 41: Strumenti e Studi 3. Florence: Sismel.

Montanari, Franco. 2011. "Correcting a Copy, Editing a Text: Alexandrian Ekdosis and Papyri." In *From Scholars to Scholia: Chapters in the History of Ancient Greek Scholarship*, ed. Franco Montanari and Lara Pagani, 1–16. Berlin: de Gruyter.

Montevecchi, Orsolina. 2008. *La papirologia*. Milan: Vita e Pensiero.

Moore, N. F. 1834. *Ancient Mineralogy: An Inquiry Respecting Mineral Substances Mentioned by the Ancients*. New York: Carvill.

More, Thomas. 1969. "Letter to Oxford University" (trans. T. S. K. Scott-Craig). In *The Thought and Culture of the English Renaissance: An Anthology of Tudor Prose, 1481–1555*, ed. Elizabeth M. Nugent, 66–71. The Hague: Martinus Nijhoff.

Morgan, T. J. 1999. "Literate Education in Classical Athens." *Classical Quarterly* 49: 46–61.

Morison, Stanley. 1972. *Politics and Script: Aspects of Authority and Freedom in the Development of Graeco-Latin Script from Sixth Century B.C. to Twentieth Century A.D.* Lyell Lectures. Oxford: Oxford University Press.

Moss, Candida R. 2012. *Ancient Christian Martyrdom: Diverse Practices, Theologies, and Traditions*. New Haven: Yale University Press.

Moudarres, Andrea. 2013. "Crusade and Conversion: Islam as Schism in Pius II and Nicholas of Cusa." *MLN* 128: 40–52.

Muir, Edward. 2007. *The Culture Wars of the Late Renaissance*. Cambridge, MA: Harvard University Press.

Müller-Sievers, Helmut, and David E. Wellbery. 2015. *The Science of Literature: Essays on an Incalculable Difference*. Berlin: de Gruyter.

Munk Olsen, Birger. 1991. *I classici nel canone scolastico altomedievale*. Spoleto: Centro Italiano di Studi sull'Alto Medioevo.

Murgia, Charles. 1975. *Prolegomena to Servius 5: The Manuscripts*. Berkeley: University of California Press.

Murphy, James Jerome. 1981. *Rhetoric in the Middle Ages: A History of Rhetorical Theory from Saint Augustine to the Renaissance*. Berkeley: University of California Press.

Navarrete, Ignacio Enrique. 1994. *Orphans of Petrarch: Poetry and Theory in the Spanish Renaissance*. Berkeley: University of California Press.

Navarro Antolín, Fernando, ed. 1996. *Lygdamus: Corpus Tibullianum III.1–6. Mnemosyne* Suppl. 154. Leiden: Brill.

Nesselrath, Heinz-Günther. 2014. "Language and (in-)Authenticity: The Case of the (Ps.-)Lucianic *Onos*." In *Fakes and Forgers of Classical Literature*, ed. Javier Martinez, 195–206. Leiden: Brill.

Netz, Noel W. 2007. *The Archimedes Codex: Revealing the Secrets of the World's Greatest Palimpsest*. London: Weidenfeld and Nicolson.

Neuhofer, M. D. 1999. *In the Benedictine Tradition: The Origins and Early Development of Two College Libraries*. Lanham, MD: University Press of America.

Newton, Isaac. 1959. *Correspondence*. Vol. 1. Ed. Herbert Westren Turnbull. Cambridge: Cambridge University Press.

Nicholas, David. 1999. *The Transformation of Europe, 1300–1600*. New York: Bloomsbury Academic.

Nicholls, Matthew. 2010. "Parchment Codices in a New Text of Galen." *Greece & Rome* 57:378–386.

———. 2011. "Galen and Libraries in the *Peri alupias*." *Journal of Roman Studies* 101: 123–142.

Norden, Eduard. 1923. *Die antike Kunstprosa vom VI. Jahrhundert v. Chr. bis in die Zeit der Renaissance*. Leipzig: Teubner.

North, John A. 1998. "The Books of the *Pontifices*." In *La mémoire perdue: Recherches sur l'administration romaine*, ed. C. Moatti, 45–63. Rome: Collection de l'École française de Rome.

Novati, F., ed. 1896. *Epistolario di Coluccio Salutati*. Vol. 3. Rome: Forzani.

O'Donnell, James. 1979. *Cassiodorus*. Berkeley: University of California Press.

Olsen, Munk. 1991. *I classici nel canone scolastico altomedievale*. Spoleto: Centro Italiano di Studi sull'Alto Medioevo.

Ong, Walter J. 1956. "System, Space, and Intellect in Renaissance Symbolism." *Bibliothèque d'Humanisme et Renaissance* 18:222–239.

Ostrogorsky, G. 1969. *History of the Byzantine State*. New Brunswick, NJ: Rutgers University Press.

Pack, Roger. 1980. "Scribal Errors in an Autograph Manuscript." *American Journal of Philology* 101:459–461.

Papaioannou, Sophia. 2013. *Michael Psellos: Rhetoric and Authorship in Byzantium*. Cambridge: Cambridge University Press.

Parkes, M. B. 1993. *Pause and Effect: Punctuation in the West*. Berkeley: University of California Press.

———. 2008. *Their Hands Before Our Eyes: A Closer Look at Scribes*. Burlington, VT: Ashgate.

Pasquali, Giorgio. 1974. *Storia della tradizione e critica del testo*. Milan: Mondadori.

Pastor, Ludwig Friedrich August von. 1904. *Acta Inedita Historiam Pontificum Romanorum Praesertim saec. XV, XVI, XVII. Illustrantia. Edidit L. Pastor. Volumen I: A. 1376–1464*. Freiburg: Herder.

Pattison, M. 1970. *Isaac Casaubon, 1559–1614*. Geneva: Slatkine. [Originally published: Oxford: Clarendon, 1892.]

Pecere, Oronzo. 1986. "La tradizione dei testi latini tra IV e V secolo attraverso i libri sottoscritti." In *Società romana e impero tardoantico*, vol. 4: *Tradizione dei classici, trasformazioni della cultura*, ed. Andrea Giardina, 19–81. Bari: Laterza.

Peirano, Irene. 2012. *Rhetoric of the Roman Fake: Latin Pseudoepigraphia in Context*. Cambridge: Cambridge University Press.

Peterson, E. 1988. "The Communication of the Dead: Notes on 'Studia Humanitatis' and the Nature of Humanist Philology." In *The Uses of Greek and Latin: Historical Essays*, ed. A. C. Dionisotti, Anthony Grafton, and Jill Kraye, 57–69. London: Warburg Institute and University of London.

Pettitmengin, Pierre, and Bernard Flusin. 1984. "Le livre antique et la dicteé: Nouvelles recherches." In *Mémorial André-Jean Festugiére: Antiquité païnne et chrétienne*, ed. A. J. Festugière, Enzo Lucchesi, and H. D. Saffrey, 247–262. Geneva: Cramer.

Pfeiffer, Rudolf. 1976. *History of Classical Scholarship, 1300–1850*. Oxford: Clarendon.

Piccaluga, Giulia. 1994. "La specifità dei libri lintei romani." *Scrittura e Civiltà* 18:5-22.

Platner, S. B., and T. Ashby. 1929. *A Topographical Dictionary of Ancient Rome*. Oxford: Oxford University Press.

Pleket, H. W., and R. S. Stroud, eds. 1990. "Preface." *Supplementum Epigraphicum Graecum* 37:v-vii.

Poliziano, Angelo. 1972. *Miscellaneorum centuria secunda: Di Angelo Poliziano; per cura di Vittore Branca e Manlio Pastore Stocchi. Editio minor.* Ed. Vittore Branca and Manlio Pastore Stocchi. Florence: Olschki.

Pollard, Graham. 1978. "The *Pecia* System in the Medieval Universities." In *Medieval Scribes, Manuscripts, and Libraries. Essays Presented to N. R. Ker*, ed. M. B. Parkes and Andrew G. Watson, 145-161. Oxford: Oxford University Press.

Potani, A. 1995. "La filologia." In *Lo spazio letterario della Grecia antica*, vol. 2, ed. G. Cambiano, L. Canfora, and D. Lanza, 307-351. Rome: Salerno.

Powell, Barry. 1991. *Homer and the Origin of the Greek Alphabet*. Cambridge: Cambridge University Press.

———. 1997. "Homer and Writing." In *A New Companion to Homer*, ed. Ian Morris and Barry Powell, 3-32. Leiden: Brill.

Probert, Philomen. 2006. *Ancient Greek Accentuation: Synchronic Patterns, Frequency Effects, and Prehistory*. Oxford: Oxford University Press.

Quint, David. 1985. "Humanism and Modernity: A Reconsideration of Bruni's Dialogues." *Renaissance Quarterly* 38:423-445.

Rabelais, François, and Henri Clouzot. 1922. *Gargantua and Pantagruel*. Paris: G. Crès.

Raby, Julian. 1980. "Cyriacus of Ancona and the Ottoman Sultan Mehmed II." *Journal of the Warburg and Courtauld Institutes* 43:242-246.

———. 1982. "A Sultan of Paradox: Mehmed the Conqueror as a Patron of the Arts." *Oxford Art Journal* 5:3-8.

Ramminger, Johann. 2010. "Humanists and the Vernacular: Creating the Terminology for a Bilingual Universe." *Renaessanceforum* 6:1-22.

Rees, Valery. 2013. "Ciceronian Echoes in Marsilio Ficino." In *Cicero Refused to Die: Ciceronian Influence through the Centuries*, ed. Nancy van Deusen, 141-162. Leiden: Brill.

Reeve, M. D. 1978. "The Tradition of *Consolatio ad Liviam*." *Revue d'Histoire des Textes* 6:79-98.

———. 1996. "Classical Scholarship." In *The Cambridge Companion to Renaissance Humanism*, ed. J. Kraye, 20-46. Cambridge: Cambridge University Press.

———. 1999. "Shared Innovations, Dichotomies, and Evolution." In *Filologia classica e filologia romanza: Esperienze ecdotiche a confronto*, ed. Anna Ferrari, 445-505. Spoleto: Centro Italiano di Studi sull'Alto Medioevo.

Reverdin, Olivier. 1981. "Préface." In *Lexicon Iconographicum Mythologiae Classicae* I 1. Zurich: Artemis.

Reynolds, L. D., ed. 1983. *Texts and Transmission: A Survey of the Latin Classics*. Oxford: Clarendon.

Reynolds, L. D., and N. G. Wilson. 2013. *Scribes and Scholars: A Guide to the Transmission of Greek and Latin Literature*. 4th ed. Oxford: Oxford University Press.

Ribbeck, Otto. 1879, 1881. *Friedrich Wilhelm Ritschl: Ein Beitrag zur Geschichte der Philologie*. 2 vols. Leipzig: Teubner.

Richardson, Larry, Jr. 1988. *Pompeii: An Architectural History*. Baltimore: Johns Hopkins University Press.

———. 1992. *A New Topographical Dictionary of Ancient Rome*. Baltimore: Johns Hopkins University Press.

Richardson, Wade. 1993. *Reading and Variant in Petronius: Studies in the French Humanists and Their Manuscript Sources*. Toronto: University of Toronto Press.

Rico, F. 1998. *Il sogno dell'umanesimo: Da Petrarca a Erasmo*. Turin: Einaudi.

Rizzo, Silvia. 1973. *Il lessico filologico degli umanisti*. Collana Sussidi Eruditi 26. Rome: Edizioni di Storia e Letteratura.

Rizzo, Silviana. 1983. "L'Auditorium di Mecenate." In *Archeologia in Roma capitale, 1870–1911*, 231–247. Venice: Marsilio.

Robb, Kevin. 1994. *Literacy and Paideia in Ancient Greece*. Oxford: Oxford University Press.

Roberts, C. H., and T. C. Skeat. 1983. *The Birth of the Codex*. Oxford: Oxford University Press.

Robinson, R. P. 1935. *The* Germania *of Tacitus: A Critical Edition*. Middletown, CT: American Philological Association.

Rodgers, Robert H., ed. 2004. *Frontinus: De aquaeductu urbis Romae*. Cambridge Classical Texts and Commentaries 42. Cambridge: Cambridge University Press.

Rodkinson, Michael L., trans. 1918. "Reuchlin, Pfefferkorn, and the Talmud in the Sixteenth and Seventeenth Centuries." In *The Babylonian Talmud: The History of the Talmud*, book 10, vol. I, ch. XIV, 76–99. Boston: Talmud Society.

Rouse, Mary, and Richard Rouse. 1991. *Authentic Witnesses: Approaches to Medieval Texts and Manuscripts*. Notre Dame, IN: University of Notre Dame Press.

Rouse, Richard. 1988. "The Book Trade at the University of Paris (ca. 1250–1350)." In *La production du livre universitaire au Moyen Age: Exemplar et pecia*, ed. R. Rouse, L. G. Bataillion, and B. G. Guyot, 41–114. Paris: Vrin.

Rouse, Richard H., and Mary A. Rouse. 2000. *Manuscripts and Their Makers: Commercial Book Producers in Medieval Paris, 1200–1500*. Vol. 1. Turnhout: Harvey Miller.

Rowe, Gregory. 2014. "The Roman State: Laws, Lawmaking, and Legal Documents." In *The Oxford Handbook of Roman Epigraphy*, ed. Christopher Bruun and Jonathan Edmondson, 299–318. Oxford: Oxford University Press.

Rüegg, Walter. 1946. *Cicero und der Humanismus: Formale Untersuchungen über Petrarca und Erasmus*. Düsseldorf: Rhein.

Rummel, Erika. 1985. *Erasmus as a Translator of the Classics*. Erasmus Studies 7. Toronto: University of Toronto Press.

Runciman, Steven. 1985. *The Great Church in Captivity: A Study of the Patriarchate of Constantinople from the Eve of the Turkish Conquest to the Greek War of Independence*. Cambridge: Cambridge University Press.

Russo, Lucio. 2004. *The Forgotten Revolution: How Science Was Born in 300 BC and Why It Had to Be Reborn*. Trans. Silvio Levy. Berlin: Springer.

Salomies, Olli. 2001. "Names and Identities: Onomastics and Prosopography." In *Epi-

graphic Evidence: Ancient History from Inscriptions, ed. John Bodel, 73–94. New York: Routledge.

Salway, Benet. 2014. "Late Antiquity." In *The Oxford Handbook of Roman Epigraphy*, ed. Christer Bruun and Jonathan Edmondson, 364–396. Oxford: Oxford University Press.

Sandys, John Edwin. 1908. *A History of Classical Scholarship*, vol. 2: *From the Revival of Learning to the End of the Eighteenth Century (in Italy, France, England, and the Netherlands)*. Cambridge: Cambridge University Press.

———. 1927. *Latin Epigraphy*. 2nd ed. Chicago: Ares.

Sapegno, N. 1960. *Il Trecento*. Milan: Vallardi.

Sarton, George. 1947. *Science and Learning in the Fourteen Century*. Baltimore: Williams and Wilkins.

Schmidt, Manfred. 2001. *Corpus Inscriptionum Latinarum*. Berlin: Berlin-Brandenburgische Akademie der Wissenschaften.

Schoppe, Kasper. 1631. *Consultationes de Scholarum et Studiorum Ratione*.

Schubert, Paul. 2009. "Editing a Papyrus." In *The Oxford Handbook of Papyrology*, ed. Roger Bagnall, 197–215. Oxford: Oxford University Press.

Schütrumpf, Eckart. 2014. *The Earliest Translations of Aristotle's* Politics *and the Creation of Political Terminology*. Morphomata Lectures Cologne 8. Paderborn: Fink.

Segel, Harold B. 1989. *Renaissance Culture in Poland: The Rise of Humanism, 1470–1543*. Ithaca, NY: Cornell University Press.

Shackleton-Bailey, D. R. 1965. *Cicero's Letters to Atticus*. Cambridge: Cambridge University Press.

Shanzer, Danuta. 2006. "Latin Literature, Christianity, and Obscenity in the Later Roman West." In *Medieval Obscenities*, ed. Nicola McDonald, 179–202. Woodbridge: Boydell and Brewer.

Sisler, William Philip. 1977. "An Edition and Translation of Lovato Lovati's 'Metrical Epistles,' with Parallel Passages from Ancient Authors." PhD diss., Johns Hopkins University.

Skeat, T. C. 1956. "The Use of Dictation in Ancient Book-Production." *Proceedings of the British Academy* 42:179–208.

———. 1995. "Was Papyrus Regarded as 'Cheap' or 'Expensive' in the Ancient World?" *Aegyptus* 75:75–93.

Smith, R. Alden. 2005. *The Primacy of Vision in Virgil's* Aeneid. Austin: University of Texas Press.

———. 2011. *Virgil*. Chichester: Wiley-Blackwell.

Smith, William. 1880. *A Dictionary of Greek and Roman Biography and Mythology*, vol. 2: *Earinus–Nyx*. London: J. Murray.

Southard, Robert. 1995. *Droysen and the Prussian School of History*. Lexington: University of Kentucky Press.

Stark, Rodney. 2005. *The Victory of Reason: How Christianity Led to Freedom, Capitalism, and Western Success*. New York: Random House.

Steinmann, Martin. 2010. "Lesen und Schreiben in den Klöstern des frühen Mittelalters." In *Teaching Writing, Learning to Write. Proceedings of the XVIth Colloquium*

of the Comité International de Paléographie Latine, ed. Pamela Robinson, 25–36. London: King's College London.
Stephen, Gwendolen. 1959. "The Coronis." *Scriptorium* 13:3–14.
Stock, Brian. 1990. *Listening for the Text: On the Uses of the Past.* Baltimore: John Hopkins University Press.
Strasburger, H. 1977. "Umblick im Trümmerfeld der griechischen Geschichtsschreibung." In *Historiographia Antiqua. Commentationes Lovanienses in honorem W. Peremans septuagenarii editae*, ed. C. Prins, 3–52. Leuven: Leuven University Press.
Sudhoff, Karl. 1930. "Konstantin der Afrikaner und die Medizinschule von Salerno." *Archiv für Geschichte der Medizin* 23:113–198.
Szirmai, Janos. 1992. "Carolingian Bindings in the Abbey Library of St. Gall." In *Making the Medieval Book: Techniques of Production*, ed. Linda Brownrigg, 157–197. Los Altos Hills, CA: Anderson-Lovelace.
———. 1999. *The Archaeology of Medieval Bookbinding.* Brookfield, VT: Ashgate.
Tallett, Frank. 2013. "Soldiers in Western Europe, c. 1500–1790." In *Fighting for a Living: A Comparative Study of Military Labour, 1500–2000*, ed. Erik-Jan Zürcher, 135–168. Amsterdam: Amsterdam University Press.
Tanfani, Leopoldo. 1863. *Niccola Acciaiuoli: Studi storici, fatti principalmente sui documenti dell' archivio fiorentino.* Florence: Felice le Monnier.
Tarrant, R. J. 1983. "Horace." In *Texts and Transmission: A Survey of the Latin Classics*, ed. L. D. Reynolds and Peter K. Marshall, 182–186. Oxford: Clarendon.
———. 2016. *Texts, Editors, and Readers: Methods and Problems in Latin Textual Criticism.* Cambridge: Cambridge University Press.
Tavoni, Mirko. 1984. *Latino, grammatica, volgare: Storia di una questione umanistica.* Padua: Antenore.
Taylor, Charles. 2007. *A Secular Age.* Cambridge, MA: Belknap.
Teixidor, Jaxier. 1977. *The Pagan God: Popular Religion in the Greco-Roman Near East.* Princeton: Princeton University Press.
Thein, Alexander G. 2011. "Auditorium Maecenatis." *Digital Augustan Rome*, no. 313. http://digitalaugustanrome.org.
Thomas, Rosalind. 2009. "Writing, Reading, Public and Private 'Literacies': Functional Literacy and Democratic Literacy in Greece." In *Ancient Literacies: The Culture of Reading in Greece and Rome*, ed. William Johnson and Holt Parker, 13–45. Oxford: Oxford University Press.
Thompson, Edward Maunde. 1912. *An Introduction to Greek and Latin Paleography.* Oxford: Clarendon.
Thompson, James. 1957. *The Medieval Library.* New York: Hafner.
Thorndike, Lynn. 1937. "Copyists' Final Jingles in Mediaeval Manuscripts." *Speculum* 12:268.
———. 1956. "More Copyists' Final Jingles." *Speculum* 31:321–328.
Timpanaro, Sebastiano. 2005. *The Genesis of Lachmann's Method.* Trans. Glenn W. Most. Chicago: University of Chicago Press.
Tixeront, Joseph. 1920. *A Handbook of Patrology.* St. Louis: Herder.
Toffanin, Giuseppe. 1950. *La religione degli umanisti.* Bologna: Zanichelli.

Tournoy, Gilbert. 2007. "Low Countries: The Seventeenth Century." In *A Companion to the Classical Tradition*, ed. C. W. Kallendorf, 237–251. Chichester: Blackwell.
Traube, Ludwig. 1911. *Vorlesungen und Abhandlungen*. Vol. 2. Ed. P. Lehmann. Munich: Beck.
Trigger, Bruce G. 1989. *A History of Archaeological Thought*. Cambridge: Cambridge University Press.
Trout, Dennis E. 1999. *Paulinus of Nola*. Berkeley: University of California Press.
Tupman, Charlotte. 2010. "Contextual Epigraphy and XML: Digital Publication and Its Application to the Study of Inscribed Funerary Monuments." In *Digital Research in the Study of Classical Antiquity*, ed. Gabriel Bodard and Simon Mahony, 73–86. Burlington, VT: Ashgate.
Turner, E. G. 1952. *Athenian Books in the Fifth and Fourth Centuries B.C.* London: H. K. Lewis.
———. 1977. *The Typology of the Early Codex*. Eugene, OR: Wipf and Stock.
———. 2015. *Greek Papyri: An Introduction*. Princeton Legacy Library. Princeton: Princeton University Press.
Ullman, B. L. 1911. "The Manuscripts of Propertius." *Classical Philology* 6:282–301.
———. 1927. "The Etruscan Origins of the Roman Alphabet." *Classical Philology* 22: 372–377.
———. 1928. *Sicconis Polentoni Scriptorum illustrium latinae linguae libri XVIII*. Rome: American Academy.
———. 1960. *The Origin and Development of Humanistic Script*. Rome: Edizioni di Storia e Letteratura.
Valla, Laurentius. 1499. *Elegantiae Linguae Latinae*. Venice: Johannes Baptista de Sessa.
Valla, Lorenzo, and Lucia Cesarini Martinelli. 1996. *Le postille all' "Institutio Oratoria" di Quintiliano: Edizione critica*. Ed. Lucia Cesarini Martinelli and Alessandro Perosa. Padua: Antenore.
van der Meer, L. B. 2007. *Liber Linteus Zagrabiensis: The Linen Book of Zagreb. A Comment on the Longest Etruscan Text*. Monographs on Antiquity 4. Leuven: Peeters.
Van Groningen, B. A. 1940. *Short Manual of Greek Palaeography*. Leiden: A. W. Sijthoff.
Van Hoesen, Henry. 1915. *Roman Cursive Writing*. Princeton: Princeton University Press.
van Minnen, Peter. 2009. "The Future of Papyrology." In *The Oxford Handbook of Papyrology*, ed. Roger Bagnall, 644–660. Oxford: Oxford University Press.
Veyne, Paul. 2010. *When Our World Became Christian (312–394)*. Trans. Janet Lloyd. Cambridge: Polity.
Vogt, Joseph. 1973. "Alphabet für Freie und Sklaven: Zum sozialen Aspekt des antiken Elementarunterrichts." *Rheinisches Museum* 116:129–142.
Walther, Hans, and Paul Gerhard Schmidt. 1982. *Proverbia sententiaeque Latinitatis medii ac recentioris aevi: Lateinische Sprichwörter und Sentenzen des Mittelalters und der frühen Neuzeit in alphabetischer Anordnung*. Göttingen: Vandenhoeck and Ruprecht.
Ward-Perkins, Bryan. 2006. *The Fall of Rome and the End of Civilization*. Oxford: Oxford University Press.
Watkins, Calvert. 1976. "Observations on the 'Nestor's Cup' Inscription." *Harvard Studies in Classical Philology* 80:25–40.

———. 1995. "Greece in Italy outside Rome." *Harvard Studies in Classical Philology* 97:35-50.
Weimar, Klaus. 1989. *Geschichte der deutschen Literaturwissenschaft bis zum Ende des 19. Jahrhunderts.* Munich: Wilhelm Fink.
Weiss, R. 1987. *Medieval and Humanist Greek: Collected Essays.* Padua: Antenore.
Werner, Michael. 2011. "Philology in Germany: Textual or Cultural Scholarship?" In *Multiple Antiquities—Multiple Modernities: Ancient Histories in Nineteenth-Century European Cultures,* ed. Gábor Klaniczay, Michael Werner, and Ottó Gecser, 89-110. Frankfurt: Campus.
West, M. L. 1973. *Textual Criticism and Editorial Technique Applicable to Greek and Latin Texts.* Stuttgart: Teubner.
———. 1982. *Greek Metre.* Oxford: Clarendon.
White, Monica. 2013. *Military Saints in Byzantium and Rus, 900-1200.* Cambridge: Cambridge University Press.
Whitmarsh, Tim. 2013. *Beyond the Second Sophistic: Adventures in Greek Postclassicism.* Berkeley: University of California Press.
Wicksteed, Philip H., and Edmund G. Gardner. 2013. *Dante and Giovanni del Virgilio, Including a Critical Edition of the Text of Dante's* Eclogae Latinae, *and of the Poetic Remains of Giovanni del Virgilio.* London: Forgotten Books. [Originally published: Westminster: Archibald Constable, 1902.]
Wilamowitz-Moellendorff, Ulrich von. 1903. *Timotheos: Die Perser.* Leipzig: J. C Hinrichs'sche Buchhandlung.
———. 1907. *Einleitung in die griechische Tragödie.* Berlin: Weidman.
———. 1927. *Geschichte der Philologie.* Leipzig: Teubner.
Willis, James. 1972. *Latin Textual Criticism.* Urbana: University of Illinois Press.
Wilson, N. G. 1967. "A Chapter in the History of Scholia." *Classical Quarterly* 17:244-256.
———. 1983. *Scholars of Byzantium.* Baltimore: Johns Hopkins University Press.
Wilson-Okamura, David Scott. 2010. *Virgil in the Renaissance.* Cambridge: Cambridge University Press.
Winckelmann, Johann Joachim. 1764. *Geschichte der Kunst des Altertums.* Dresden.
Wingo, Otha. 1972. *Latin Punctuation in the Classical Age.* The Hague: Mouton.
Winsbury, Rex. 2009. *The Roman Book: Books, Publishing, and Performance in Classical Rome.* London: Duckworth.
Witt, Ronald G. 2000. *In the Footsteps of the Ancients: The Origins of Humanism from Lovato to Bruni.* Leiden: Brill.
———. 2005. *Sulle tracce degli antichi: Padova, Firenze e le origini dell'umanesimo.* Rome: Donzelli.
———. 2012. *The Two Latin Cultures and the Foundation of Renaissance Humanism in Medieval Italy.* Cambridge: Cambridge University Press.
Wolf, Friedrich August. 1832. *Vorlesung über die Geschichte der römischen Litteratur.* Leipzig: August Lehnhold.
———. 1986. *Darstellung der Altertumswissenschaft nach Begriff, Umfang, Zweck und Wert.* Weinheim: Acta Humaniora. [Orig. published Berlin, 1807.]

Woodhead, A. G. 1992. *The Study of Greek Inscriptions*. Norman: University of Oklahoma Press.

Woodward, David. 2007. *The History of Cartography*, vol. 3: *Cartography in the European Renaissance*. Chicago: University of Chicago Press.

Woolf, Greg. 2009. "Literacy or Literacies in Rome?" In *Ancient Literacies: The Culture of Reading in Greece and Rome*, ed. William Johnson and Holt Parker, 46-68. Oxford: Oxford University Press.

Woolfson, Jonathan. 2005. "Burckhardt's Ambivalent Renaissance." In *Palgrave Advances in Renaissance Historiography*, ed. Jonathan Woolfson, 9-26. New York: Macmillan.

Young, Douglas. 1965. "Some Types of Scribal Error in Manuscripts of Pindar." *Greek, Roman, and Byzantine Studies* 6:247-273.

Yun, Bee. 2011. "Does the History of Medieval Political Thought Need a Spatial Turn? The Murals of Longthorpe, the *Secretum secretorum*, and the Intercultural Transfer of Political Ideas in the High Middle Ages." In *Cultural Transfers in Dispute: Representations in Asia, Europe, and the Arab World since the Middle Ages*, ed. Jörg Feuchter, Friedhelm Hoffmann, and Bee Yun, 135-148. Frankfurt: Campus.

Zabughin, Vladimiro N. 2011. *Storia del rinascimento cristiano in Italia*. Ed. Bruno Basile. Naples: La Scuola Pitagora. [Originally published: Milan: Treves, 1924.]

Zanker, Paul. 1988. *Pompeii: Public and Private Life*. Trans. D. L. Schneider. Cambridge, MA: Harvard University Press.

Zetzel, J. E. G. 1973. "*Emendavi ad Tironem*: Some Notes on Scholarship in the Second Century A.D." *Harvard Studies in Classical Philology* 77:225-243.

———. 1980. "The Subscriptions in the Manuscripts of Livy and Fronto and the Meaning of *Emendatio*." *Classical Philology* 75:38-59.

———. 1981. *Latin Textual Criticism in Antiquity*. New York: Arno.

INDEX

Italicized page number indicates illustration.

abbreviations, epigraphic, 21
abecedarium, 8
Accessing Antiquity: The Computerization of Classical Studies (Solomon), 222
Acta martyrum scillitanorum, 264n31
Act of Supremacy, 194
Advanced Papyrological Information System (APIS), 229
Aelian: *Tactica*, 175
Aelius Stilo, 81
Aemilius Paullus, 39, 42, 59
Africa, 110
Agrippa, M. Vispasianus, 20
Alberti, Leon Battista, 170
Alcuin of York, 117–119, 122–123
Aldhelm of Malmesbury, 110
Aleandro, Girolamo: *Greek to Latin Dictionary*, 201
Alexander Aetolus, 54
Alexandria, 16
Alfonzo V of Aragon, 193
alphabets: Greek, 6–9; in Italy, 16; Latin, 9, 16, 243n23; Phoenician, 6, 7. *See also* scripts; writing
Altertumswissenschaft, 207–211, 213
Ambrose, Saint: and Altar of Victory, 98; use of Cicero by, 95
American Journal of Philology, 214
Americans, Native, 197
Americas, discovery of, 191, 192
Ammianus Marcellinus: codex of, 160; on libraries, 111
Ampère, Jean-Jacques, *Histoire littéraire de la France avant le douzième siècle*, 116
ampersand, 64
Anacreon, 54
analogy, and *anomaly*, 55
Anastasius Bibliothecarius, 109, 124
Ancient Lives Project, 229–230
Ancilla to the Pre-Socratic Philosphers (Freeman), 231
Anglican Church. *See* Church, Anglican
Annales maximi, 17
anomaly, and *analogy*, 55
anonymous works, 78–79
Antenor, 151, 275n10
Anthologia Palatina, 145, 202, 289n59
anthologies, 135
antigraph, 73, 122, 190
Antimachus, 52
Antiochus IV, 33
Antiochus of Ascalon, 59
antiquarianism: in ancient Rome, 59; in Carolingian period, 125; in Early Modern period, 203; in late antiquity, 106–107; in Renaissance, 203

325

anti-Semitism, 212
Apellicon of Teos, 39
Apicius: *De re coquinaria*, 121, 199; manuscript tradition, 199
Apollinaris, Sidonius, 99, 100
Apollo Belvedere, 203
Apollonius Dyscolus, 56–57
Apollo Temenites (statue), 40
apologists, Christian, 94
apparatus criticus, 24, 219–220
Appendix Vergiliana, 80, 102; in Carolingian period, 121
Apronianus Asterius, 101
Apuleius: in Carolingian period, 121; in late antiquity, 100; manuscript tradition, 156; "spurcum addimentum," 156
Arabic (language), 133–134
archetype (of manuscript), 214, 216–217, 219
Archilochus, 14
Archimedes, 141
architecture, Gothic, 136
Arch of Septimius Severus, 21
Arethas of Caesarea, 144, 145
Ariosto, Ludovico: *Orlando Furioso*, 197
Aristarchus, 53, 54
Aristonicus, 56
Aristophanes: *Clouds*, 14–15, 84; commentaries on, 54; *Frogs*, 40–41
Aristophanes of Byzantium, 53–54; charges of plagiarism, 81
Aristotle: and ancient preservation, 139; as book collector, 37, 42; and censorship, 134; *Didaskaliai*, 221; and Latin translations, 108, 109, 133, 134; 13th-c. rediscovery of, 132, 134
Arles, convent of, 68
Arval brotherhood, 17
asceticism, religious, 97
Asconius, 160
Asinius Pollio, Gaius, 39
Asterius, Turcius Rufius Apronianus, 99
Ataktoi glossai (Philetas), 52
Athenaeus: *Deipnosophistae*, 174

Athenian Academy, 142
Athens, ancient: education in, 15–16; fall of (86 BC), 39; literacy in, 14–15; papyrus use in, 26
Atrium libertatis, 39
Atticism, revival of, 140, 273n209
Atumano, Simon, 157
Auctus, Servius: *Epistolae ad familiares*, 122
auditorium Maecenatis, 42, 252n247
Augustine: on Christian use of pagan texts, 95–96; on Claudian, 98; on copying *City of God*, 84; *De civitate Dei*, 95; *De doctrina christiana*, 95; on silent reading, 50; use of Cicero by, 95; view of history of, 162
Augustus, Octavian: and literary censorship, 88; and punishment of *equites*, 50
Aulus Gellius: in Carolingian period, 122; manuscript tradition, 136; on Nigidius Figulus, 60; *Noctes Atticae*, 89, 136; on Plautine plays, 87; and search for ancient books, 89–90
Aurelianus, Caelius, 108
Aurispa, Giovanni, 161, 173–175
Ausonius, Decimus Magnus, 97, 155, 187
Austin, Colin, 230
authors, canonical: of Carolingian period, 126; in late antiquity, 126, 270n148
Avignon, 153, 157

Bacchanalia, 88, 223–224
Bacon, Roger, 134
Badius, Josse (Ascensius), 194
bankers, Athenian, 15, 245n56
Barlaam of Seminara, 157
Barzizza, Guiniforte, 193
Basil, Saint: *Ad adulescentes*, 173, 178, 182
Basilica of Santa Maria in Trastevere, 124
Battle of Artemesium, 23
Battle of Marathon, 15
Bellerophon, 8, 242n17
Bellini, Gentile, 193

Bembo, Pietro, 193
Benedict of Nursia, Saint: *Rules*, 112
Bengtson, Hermann, 239
Bentley, Richard, 200, 202–203, 212
Bergamo, Gasparinus de, 160–161
Bernard of Chartres, 191
Bernays, Jacob, 214
Bessarion: *In Calumniatorem Platonis*, 176
Bible, 70; Gutenberg, 167, *168*, 186; vulgate, 93, 108, 123
Bibliotheca Scriptorum Classicorum, 237
bifolia, 60
Biondo, Flavio, 160, 163; *Roma Instaurata*, 203
Birth of the Codex (Roberts and Skeat), 32
Bobbio, monastery of, 139, 165, 187
Boccaccio, Giovanni, 155–156, 157
Bodel, John, 226
Boeckh, August, 20, 289n66; *Corpus Inscriptionum Graecarum*, 223; *Enzyklopädie und Methodenlehre der philologischen Wissenschaften*, 212
Boeotia, 7
Boethius, 108–109
Boileau, Nicolas: *Satires*, 197
Boivin, Jean, 205
Boniface, Saint, 268n107
bookkeeping, 15, 31
book production: in Carolingian period, 122; in "Dark Ages," 144; late antique, 91, 96; in Middle Ages, 67–72, 70–76, 91, 111, 130–131; in Renaissance, 166–167. *See also* printing press
books, classical: aesthetics of, 49; burning of, 89, 142; in Byzantine period, 140–148; and Christian culture, 97; circulation of, 44, 122, 138–139; collecting of, 37, 89–90; cost of, 41, 85; and Greek etymology (*biblos*), 27; and Latin etymology (*liber*), 9; legibility of, 49; length of, 29; linen in, 10; as papyrus roll, 37; reproduction of, 31, 42, 70; textual integrity of, 101–102; Ulpian on, 36–37; as war spoils, 39.

See also bookkeeping; book production; books, late antique; booksellers; bookshops
books, late antique: arrangement of, 103, 104; authenticity of, 102–103; epitomes of, 105–106; rare, 106–107
books, medieval: assembly of, 72; copying of, 70; *pecia* system, 131
books, renaissance: *incunabula*, 167; *mise-en-page*, 166
booksellers, 43
bookshops, 44
Books of Numa, 27
book trade: in Athens, 26, 40–41; in Roman Italy, 27, 42–45
boustrophedon writing, 6, 7, *7*, 242n4
British Isles, 110, 122
Brown, Peter, 90
Bruni, Leonardo: and defense of pagan texts, 181–182; as Greek translator, 173, 175, 178
— Works: *De Interpretatione Recta*, 175; *De studiis et litteris*, 182; *Dialogi ad Petrum Histrum*, 158–159; *Dialogues for Pier Paulo Vergerio*, 162
brush, as writing instrument, 64
Bucolicon Olibrii, 139
Budé, Guillaume: *Commentarii Linguae Graecae*, 194; *De Asse et Partibus*, 194
Buonaccorsi, Filippo, 183, 193
Burckhardt, Jacob, 149
Burgundio of Pisa, 133
Bussi, Giovanni Andrea, 198
Buttmann, Philip, 208
Byzantine period, 110, 140–148

Cadmus, 7
Caelius Aurelianus, 136; *Medicinales responsiones*, 139
Caesar, Julius, 17; assassins of, 177; misattributed speeches of, 79; plans of, for public library, 59; and silent reading, 50; spurious works of, 103
Caesarius of Arles, 99–100, 267n83

INDEX / *327*

calamus (pen), 63–64
Caligula, Gaius Julius, 89
Callimachus: *Aetia*, 262n238; *Pinakes*, 38, 54
Calvin, John, 196, 287n31
Calvinism, 202
Cameron, Alan: *The Last Pagans of Rome*, 96
Camões, Luís Vaz de, 197; *Os Lusíadas*, 197
Canfora, Luciano, 147
Capitoline Hill, 151
Carmen de bello Actiaco, 79, 91, 92
Carolingian renaissance, 116–120
Carrion, Luis, 125
Carthage, destruction of, 39
Casa, Giovanni della: *Il Galateo, overo de' costumi*, 179
Casaubon, Isaac, 201–202
Cassiodorus, 68; and canonical authors, 126; *Ecclesiastical History*, 109; and foundation of Vivarium library, 111–112
Cassius Dio, 104
Cassius Severus, 88
Castiglione, Baldassare: *Cortegiano*, 179, 193
catacombs, Roman, 203
catharsis, 198
Catholic Church. See Church, Catholic
Catholic University of Leuven, 202
Cato the Elder, 87; *de agricultura*, 16, 103
Cavallo, Guglielmo, 140
Cefala, Constantine, 145
Celsus: *Alethes logos*, 94; *De medicina*, 121, 161, 189
cenobite monasticism, 67
censorship: in archaic Athens, 88; in Imperial period, 88–89; medieval, 77; religious, 132, 134, 196–197; in Roman regal period, 88
Centre de Documentation de Papyrologie Littéraire (CEDOPAL), 296n45

Centro Internazionale per lo Studio dei Papiri Ercolanesi, 227
certame coronario, 170
Charlemagne, 117–118
Charles IV of Habsburg, 152
Charles V, duke of Bourbon, 193
Charles VIII, 193
Charles the Bald, 124, 269n136
Charta Borgiana, 205
Chartres, cathedral school of, 132
China, 197
chirography, 62–63, 204
Christ, Jesus, year of birth, 109
"christian humanism," 181
Christianity: and book production, 66–67; in Eastern Roman Empire, 142; emergence of, in Roman world, 93–94, 96; and paganism, 94–95, 96, 99; and Platonic philosophy, 94, 95; and preference for codex, 36; and preservation of classics, 142–142; and technical studies, 132; and textual integrity; 101–102; and uncial writing, 92, 265n42
Christina, Queen of Sweden, 202
Chrysoloras, Emmanuel, 172–173, 174, 185
Church, Anglican, 194
Church, Catholic: and censorship, 3–4, 132, 134, 196–197; conflict of, with Eastern Church, 144, 175, 281n129; as cultural watchdog, 152; doctrine of, 134; and "Index of Prohibited Books," 195, 196; and Renaissance humanists, 182–183; rift of, with Anglican, 194–195
Church, Orthodox, 142, 144
Cicero, 42, 59; ancient publication of, 82–83; attribution of *Rhetorica ad Herennium* to, 80; on bookshops, 43; in Carolingian period, 120–121; as Christian, 153; and codices, 124; educational vision of, 177; in late Middle

Ages, 126–127, 135; manuscript tradition, 127, 129–130, 153, 155, 160, 205; in Ottonian renaissance, 129; and palimpsest, 205; personal changes of, to editions, 83; and printing press, 186; in Renaissance, 177–178
— Works: *Brutus*, 160; *De officiis*, 186; *De re publica*, 127: *Epistulae ad Atticum*, 153, 159; *Epistolae ad familiares*, 159; *In Clodium et Curionem*, 82; *Pro Archia*, 153, 177–178; *Pro Cluentio*, 130, 153, 155, 156; *Pro Murena*, 160; *Pro Roscio Amerino*, 160; *Somnium Scipionis*, 100, 127, 129, 152
Ciceronian model, 171
"civic humanism," 177
"classical tradition," 1–2, 85
Classical Tradition, The (Highet), 1
Classics (discipline): invention of, 207–211, 291n88; modern changes in, 2
Claudian, 98, 129; *De raptu Proserpinae*, 136
Claudius (emperor), 9
Claudius Quadrigarius: *Annales* of, 89
Clodius, Servius, 81
Cluny, abbey of, 160
codex: and book production, 96; construction of, 60–62; Christian preference for, 36; dating, 164, 204; digitization of, 232–233, 284n187; forgeries of, 184–185; and *kolleisis*, 36; as literary edition, 35–37; luxury editions of, 101; vs. papyrus roll, 35; and printing press, 187, 198, 189; single- and multi-quire, 61; as single author volume, 102; as status symbol, 99, 101; as tablet, 11, 33–34, 243n34. *See also* codicology
codices: Ambrosian palimpsest, 92, 104; Ambrosian Quintilian, 165; Augusteus, 92; Bambergenis, 129; Bembinus, 103; *Blandinius vetustissimus*, 102; *Einsidlensis*, 222; Etruscus, 103, 130, 150–151, 202; Farnesianus, 130; Fuldense, 199; Harleian, 124; Hersfeldense manuscript, 184; Mediceus (*prior* and *alter*), 91, 99, 101, 104, 105, 128, 165, 279n99; Monacensis, 129; Nag Hammadi, 61; New York Academy of Medicine, 136; *Oblongus*, 123, 160, 125; Palatine Tradition, 103; Palatinus, 91; *Priapea*, 156; *Quadratus*, 125; Romanus, 91; Sangallensis, 92; Troyes, Bibliothèque Municipale, 152, 155; Venetus A Manuscript (VMK), 56, 57
codicology, 62–63
coin production, 10
cola, 54
Collége de France, 194
College of the Sorbonne, 132
Colonna, Cardinal, 151
colophons, 66, 67, 98–99; legal value of, 102; medieval copies of, 100–101
Columbus, Christopher, 191–192
Columella: *De rustica*, 121
Comicorum Graecorum Fragmenta, 230
commentaries, 2–3; and transition to scholia, 56
communication, oral, 48–49
Comnena, Anna: *Alexiad*, 146
composition, oral, 14
Consolatio ad Liviam (att. Ovid), 79
consonants, 9
Constantine VII Porphyrogenitus, 145
Constantine the African, 133
Constantinople: fall of (1453), 148, 176, 193; sack of (1204), 146; as source for Greek manuscripts, 173, 176, 185
Constantius II, 140–141
consular diptychs, 13
contaminatio, 217, 292n112
conventions, editorial, 53–54
Coptic language, 25
Corneille, Pierre, 198
Cornelius Gallus: *Amores*, 88, 263n15
Cornelius Nepos, 135–136, 161

Corpus Inscriptionum Graecarum, 20, 223, 224, 289n66
Corpus Inscriptionum Latinarum, 20, 222–224, 289n66, 292n12
corrector of texts (*diorthotes*), 52
corruption, textual: by printing press, 198–199, 287n39; typology of, 74. *See also* criticism, textual; emendations, textual; errors, scribal
Corsini, Pietro (Cardinal), 157
Corvey, Wibald, 135
Council of Constance, 192
Council of Florence, 191
Council of Trent, 196
Counter-Reformation, 197
Crane, Greg, 232
Crates of Mallus, 55, 57, 86
Cremonini, Cesare, 197
Cremutius Cordus, 89
Crete, 6
Crinito, Pietro: *De Poetis Latinis Libri Quinque*, 206
Crisis of the Third Century, 91
Crispus Sallustius, 100
criticism, textual, 4, 52, 55, 214. *See also* emendations, textual
Cronache Ercolanesi, 227
crux desperationis, 219
culture, humanistic: and Cicero's *pro Archia*, 177–178; education in, 177–179; and *lingua franca*, 192; municipal context of, 177; and rediscovery of science, 191–192; in universities, 193. *See also* humanism, Renaissance; Renaissance, the
cursive. *See under* scripts
Cyprion of Toulon, 99
Cyriacus of Ancona, 193, 203, 223

dactylic hexameter, 59
Dain, Alphonse, 63
Damascus I (pope), 108
d'Angelo da Scarperia, Jacopo, 172, 173, 174

Daniel, Pierre: "Servius Danielis," 201
Dante: *De vulgari eloquentia*, 126, 169; *Divine Comedy*, 169; *Inferno*, 275n10
"Dark Ages," 143–144, 162
Database of Classical Bibliography, 238
dating: of inscriptions, 22–23; of manuscripts, 164, 204, 213
Dea Dia festival, 17
De bellis Macedonicis, 79
Decameron, 156
Decembrio, Uberto, 173
dedications: honorary, 19; monumental, 19–20; sacred (*tituli sacri*), 19. *See also* inscriptions
Dell'Acqua, Antonio, 177
Demetrius of Phalerum, 37–38
Demotic language, 25
Derveni Papyrus, 51
Dessau, Hermann, 224
di Capelli, Pasquino, 159
Didymus, 56
Diels, Hermann, 230
Diet of Worms, 201
digamma (Greek), 9, 202–203
Digital Scriptorium (DS) database, 240
digitization, 232–233, 284n187
diglossia debate, 169
Diocletian, price edict on books, 43
Diodorus Siculus: *Bibliotheca*, 104; on Theopompus, 139–140
Dionysius Exiguus, 109
Dionysius Thrax, 55, 57
di Rienzo, Cola, 151
dis manibus, 22
Dix années de bibliographique classique, 237
documents, legal, 15, 244n5
documents, public, 10, 12
Dolet, Étienne, 3
Dominici, Giovanni: *Lucula Noctis*, 181
"Donation of Constantine," 79, 182–183
Donatus: *Ars Maior*, 124; commentary on Terence, 103, 129, 161; *Vita Virgilii*, 102

dreams, Christian, 96, 100, 152
Dryden, John: *Discourse concerning the Origins and Progress of Satire*, 197
Duke Databank of Documentary Papyri (DDbDP), 229
Dungal, 117, 123

Early Modern period, 191, 197, 200, 203–204, 206
earth, circumference of, 191
editions (of manuscripts), 52–53, 83; arranged by author, 102–103; arranged by genre, 103–104; critical, 219–220; *editio princeps*, 198; emendations to, 199–200; multi-volume, 103–104
education: in Athens, 15–16; during Carolingian period, 117, 128; in Eastern Roman Empire, 141–142; in Germany, 208–211; in Hellenistic Egypt, 16; Jesuit, 197; in late antiquity, 90; in medieval monasteries, 67; during Middle Ages, 110–111, 128, 179, 282n143; during Renaissance, 177–179. See also quadrivium; trivium
Egypt: libraries of, 37–38; literacy in, 16; papyri of, 227, 228; as source of linen, 35; as source of papyrus, 25, 35; stylus tablets in, 12; war of, with Antiochus IV, 33
Egypt Exploration Society, 26, 228
Einhard: *Vita Karoli Magni*, 117, 124
Ekkehard IV, 160
eliminatio codicum descriptorum, 214, 215
eliminatio lectionum singularium, 218
Eliot, T. S., 85
Elliot, Tom, 226
emendations, textual, 200–202; *conferre*, 101–102; *emendare*, 101. See also corruption, textual
emendatio ope codicum, 199
emendatio ope ingenii, 199–200, 219
England, cultural growth of, 110
Ennius, 58, 59; *Annales*, 27, 59; "venerable," 88

epic, 197
Epictetus, 197
Epicureanism, 196
EpiDoc, 226
epigraphers, 20–21
epigraphy, 18–24; research tools for, 222–227. See also dedications; epitaphs; graffiti; inscriptions
epistles, composition of, 132
epitaphs, 19, 23
epitomes, 105–106, 109, 140
Erasmus of Rotterdam, 184, 195–196; *Ciceronianus*, 171, 195; *De Libero Arbitrio Diatribe sive Collatio*, 196; *Dialogus de Recta Latini Graecique Sermonis Pronunciatione*, 195; *Praise of Folly*, 195
Eratosthenes of Cyrene, 54
error, polar, 76
errors, scribal, 73–78; abbreviations, 74; anticipation, 75; dittography, 75; haplography, 74–75; interpolative, 77; letter ambiguity, 74; negligence, 73–74; omission, 76; perseveration, 75; preoccupation, 76
Estienne, Henri: *Thesaurus Linguae Graecae*, 201
Estienne, Robert (Stephanus), 197; *Thesaurus Linguae Graecae*, 201, 233; *Thesaurus Linguae Latinae*, 201
Etruscans, 9
Euboea, 7
Euclid, 141
Eugene IV (pope), 203
Eugenius III, 133
Eumenes II, 33, 55
Euripides, 147; *Andromeda*, 40–41
Eustathius, 146
evidence, historical vs. archaeological, 23–34
examinatio, 218
exemplar, 70, 72, 74–76
exiles, Greek, 176
exploration, world, 191
Ezzelino III, 151

Fabius Pictor, 263n6; *Annales*, 89
Fabricius, Johann Albert: *Bibliotheca Graeca*, 207; *Bibliotheca Latina*, 207
Fasti Consulares, 22, 204
Felix, Securus Melior, 101
Feltre, Vittorino da, 179
Ferrara, Pace da, 156
Festus, Pompeius: epitome of *De verborum significatu*, 106, 130, 139
Fichet, Guillaume, 194
Fichte, Johann Gottlieb: *Discourses to the German Nation*, 208
Figulus, Nigidius: *Commentarii grammatici*, 60; *Sacerdotiis Romanorum*, 185
Filelfo, Francesco, 170, 174
Fiocchi, Andrea: *De Magistratibus et Sacerdotiis Romanorum*, 185
Flavianus, Nicomachus, 100
Fleming, Robert, 192
Florence: Greek instruction in, 159, 172–174; as locus of Renaissance culture, 158
Florilegium Gallicum, 135
Florilegium Thuaneum, 128
fora: of Augustus, 19; of Trajan, 39
forgeries, 23, 79–80, 82, 184–185, 289–290n70
Foscolo, Ugo, 207
Fragmente der griechischen Historiker (*FGrHist*), 231
Fragmente der Vorsokrater (Diels), 230–231
France, 200–202; as source of 12th-c. manuscripts, 135–136
Francesca, Piero della, 179, 180
Frederick II, 132
freedmen, 16–17, 245n68
Freeman, Kathleen, 231
Friends of the Herculaneum Society, 228
Froben, Johann, 194
Frontinus: *De aquis*, 136, 160
Fronto, 10
Fulda, monastery of, 128, 160

Gaius (jurist): *Institutiones*, 205
Galen: *De libris propriis*, 82; Latin translations of, 133; library of, 40; *On Grief*, 29
Gaul, 110
Gehrke, Hans-Joachim, 239
George of Trebizond: *Comparatio Philosophorum Aristotelis et Platonis*, 176
Gerard of Cremona, 133
Gerbert of Aurillac (Pope Sylvester II), 129
Germanicus: *Aratea*, 121
Germany, 122, 125, 128; humanism in, 192; Ottonian renaissance in, 129–130
Gildersleeve, Basil, 214
Gilson, Étienne, 181
Giovanni of San Miniato, 181
Giraldi, Giglio Gregorio: *De Historia Poetarum tam Graecorum tam Latinorum*, 171
glossai, Homeric, 52
Gorris, Richard de, 3
gospels, Christian, 36
Goughuenheim, Sylvain, 134
graffiti, 19
Grafton, Anthony, 200
grammarians (*grammatikoi*), 54, 56–57, 59–60, 134, 163
grammata, 15–16
Gratian, Emperor of the West, 97–98, 142
Grattius, *Cynegetica*, 121
Greco-Roman Memoirs series, 295n38
Greece, ancient: book trade in, 40–41; introduction of alphabet to, 6–7; literacy in, 8, 13–16; oral culture of, 64, 211; papyrus in, 26–27
Greek (language), 25; in Byzantine Empire, 110; decline of, in west, 108–109; *digamma*, 9, 202–203; grammars, 55, 173; instruction in, 172–174; in medieval period, 76–77, 156–157; pronunciation of, 195, 286–287n27; in Renais-

sance, 172–176; use of, in Rome, 107; written accents in, 53–54, 57. *See also* literature, Greek *and under* translation
Greek East, 12
Greek-Gothic War, 110
Gregory XII (pope), 159
Gregory XVI (pope), 205
Gregory of Sanok, 192–193
Gregory the Great: *Moralia in Job*, 71; *Regula pastoralis*, 112
Grenfell, Bernard, 26, 228
Gronovius, Johann Friedrich, 202
Grosseteste, Robert, 134
Grottaferrata monastery, 144
Gruter, Janus, 204, 223
Guélis, Germain Vaillant de, 3
Guide de l'épigraphiste, 226
Gullick, Michael, 70
Gutenberg, Johannes, 186
gymnasia, 38, 40

Hadrian, 246n88
Haimo of Auxerre, 124
Han, Ulrich, 186
Handbook for Classical Research (Schaps), 222
Handbuch der klassischen Altertumswissenschaft (*HdA*), 239
handwriting. *See* writing
Harpocration, 57
Hase, Carl Benedict, 233
Haskins, Charles, 130
Hebrew (language), 194
Hegel, Georg, 212
Hegendorff, Christroph, 3
Heinsius, Nicolas, 202, 204
Heiric of Auxerre, 123, 124, 269n128
Henry VIII, 194–195
Henry of Aristippus, 133
Heraclius, 110
Herculaneum: excavations of, 205, 289n70; and papyrus finds, 26, 227–228

heresy, 181–182, 183
Hermann, Johann Gottfried, 212, 213
Herodian, 56, 57. *Katholikes prosodias*, 57
Herodotus, 7
Hersfeld, monastery of, 125
Hesiod, 14
Hesychius, 57
Hewlett-Packard, 232
Heyne, Christian Gottlob, 207
Hippias of Elis, 22
Historia Augusta, 122, 155
histories, annalistic, 87
historiography, 140, 206–207, 208–209; "Livian type," 151
history, ancient: epitomes of, 105; "external," 208; German organization of, 208–209; "internal," 208–209; late antique interest in, 105, 109
Holy Roman Empire, birth of, 117
Homer: *Ataktoi* glossai, 52; Hellenistic editions of, 52–54; *Iliad*, 5, 56, 57; in Latin, 79, 157; place of, in literary tradition, 5–6; *Odyssey*, 5, 57; Romantic appreciation of, 211; ur-text, 207; writing in, 8–9, 242n17
Honorius, 98
Horace: *Ars Poetica*, 33; in Carolingian period, 121; codices, 102–103; emendation to, 203; and early modern satire, 197; "Horatian age," 128; in Middle Ages, 128, 135; on Saturnian verse, 87; on silent reading, 50
Hugo of St. Victor, 132
humanism, renaissance: and Catholic Church, 181–184; and historiography, 206; and idealization of antiquity, 162; and importance of Cicero, 178; and manuscript dating, 164; Petrarch as precursor to, 158; and Protestant Reformation, 195–196; and restoration of Classical Latin, 167–171; and search for manuscripts, 150, 159–161, 164–165; and secularization, 181; spread of,

through Europe, 192–193; and traditionalists vs. modernists, 161; and use of ancient models, 164, 177–178. *See also* culture, humanistic
humanist, definition of, 177
humanitas, 177–178, 282n137
Hunain ibn Ishaq, 143
Hungary, 192, 193
Hunt, Arthur, 26, 228
Hypatia, 142

Iberian Peninsula, 110, 114
Ibycus computer, 232
IconicLIMC, 236
identity, national, 208
Ignatius of Loyola, 196
Ilias Latina, 79
illuminator, 72
imagery, Christian, 76
incunabula, 167
indexing notes (Renaissance), 3
"Index of Prohibited Books," 196
individual, concept of, 149
initial, decorated, 166, *166*
initial, historiated, 166, *167*
ink, 63–64
Inscriptiones Antiquae Totius Orbis Romani, 204
Inscriptiones Graecae, 20, 224–225, 289n66, 294n21
Inscriptiones Latinae Selectae (*ILS*), 224
inscriptions: collecting, 203–204; dating of, 22–23; documentary, 20; forgery of, 23; Greek, 46–47; historical interest in, 20–21; influence of, on handwriting, 64; interpretation of, 21; monumental, 21; "nonsense," 244n50; publication of, 20–21; readership of, 48–49; Roman republican, 64; *stoichedon*, 47–48, *48*. *See also* dedications; epigraphy
instrumenta domestica, 14, 16, 244n50
Internet, 221, 222, 232, 233
interpolation, 77–78, 88, 198–199

Introduction to Manuscript Studies (Clemens and Graham), vii
invasions, barbarian, 110, 143, 162
Ireland, evangelization of, 110
Irenaeus, *Adversus Haereses*, 108
Irenaeus of Lyon, 101–102
Isidorus of Seville, 110
Italy, ancient: book trade in, 41–45, 66; imitation of Greece by, 211; inscriptions in, 19–20; introduction of alphabet in, 9, 16; libraries in, 39–40; oral culture in, 58; papyrus use in, 26; parchment use in, 33; reliance on slaves in, 16
Iucundus, L. Caecilius, 12
ivory, 13

Jacob, Christian, 38
Jacoby, Felix, 231
James of Venice, 133–134
Jeremiah, prophet, 95
Jerome (abbot), 130
Jerome, Saint: on book production, 84; as a "Ciceronian," 96, 152; *Commentarius in Zachariam*, 104; epitome of *Historiae Philippicae*, 106; on luxury codices, 101; and monastic libraries, 68; promotion of pagan literature by, 97; translation of Pachomius by, 67; and vulgate Bible, 108
Jesuits, 196–197
Jews, 211–212
Johannes Scotus Eriugena, 124
John XXIII (anti-pope), 159
John of Damascus, Saint, 133
John of Salisbury, 133–134, 135
John Philagathus, 130
Johnson, William, 49
John the Lydian, 107, 109
Jones, Henry Stuart, 233–234
Judaism, 132, 197
Julian the Apostate, 142
Julius Paris, 155
Justin II, 110

Justinian, 110, 142
Justin Martyr, 94
Juvenal: in Carolingian period, 121; *Satires*, 106

Kahil, Lily, 235
Kaibel, Georg, 230
Kassel, Rudolf, 230
Kirchhoff, Adolf, 8

Lachmann, Karl, 1, 213-214
"Lachmann's Method," 213-220
Lactantius, 161
Laelius Archelaus, 58
Lafranc, archbishop of Canterbury, 71
Lambin, Denis, 201
Lampadio, Gaius Octavius, 58
Lampidio, Octavian, 27
Landriani, Gerardo, 160-161
L'Année épigraphique (*AE*), 225, 294n24
L'Année Philologique (*APh*), 237
Laocoön, 203
lapidaria, 203
lapidary museum of Verona, 205
Lapis Niger, 16
Lascaris, Janus, 200
late antiquity: definition of, 90; transition to Middle Ages, 110
Latin (language), 9, 16; in British culture, 110; Carolingian standardization of, 117-118; composition of, 210; "dead," 169; *De lingua Latina* (Varro), 60; and the *diglossia* debate, 169; earliest use of, in scholarship, 58; in Eastern Empire, 109; Greek words in, 174; Gothic "corruption" of, 162-163; and humanism, 192; instruction in, 179; as liturgical language, 118, 169; in medieval period, 74, 77; in Renaissance, 163-164, 167-169, 179. *See also under* translation
laws, oral publication of, 17
leaden sheets, 10
Leclerc, Jean, 196

lectio difficilior, 218-219
Leiden system, 21-22
lemma, 223
Leo X (pope), 105
Leontius Pilatus, 157
Leopardi, Giacomo: *Ad Angelo Mai*, 205
Lessing, Gotthold Ephraim: *Hamburgische Dramaturgie*, 198
Leto, Pomponio, 183, 190
lexica, 57, 233-234
Lexicon Iconographicum Mythologiae Classicae (*LIMC*), 235-236
Liber linteus zagrabiensis, 10
librarians, of Alexandria, 52-54, 255n56
libraries, ancient, 37-40; Alexandrian, 37-38, 45, 52, 64, 139, 142; Callimachus' role in, 38, 54; in Constantinople, 140-141; censorship in, 89; destruction of, 111, 142; in gymnasia, 38, 40; Palatine, 89; Pergamene, 33, 250n215; private, 37, 39, 42; public, 37-38, 39-40; Roman, 39-40
libraries, post-Antique: Biblioteca Ambrosiana, 153, 205; Biblioteca Capitolare, 153, 155, 204-205; Biblioteca Estense, 207; Biblioteca Medicea Laurenziana, 105, 186; Biblioteca Nazionale of Naples, 91; Carolingian, 119; monastic, 68; Palatine Library of Heidelberg, 202; personal, 185-186; Vatican, 175, 183, 205
Liddell-Scott-Jones Greek Lexicon (LSJ), 233-234
ligatures, 64
Linear A & B, 6
linen, production of, 10
Lipsius, Justus, 200, 202
literacy, 8, 86; Athenian, 15; degrees of, 18; Egyptian (Ptolemaic), 16; and freedmen, 15-17, 245n68; Greek, 13-16; Italian, 16; in monastic life, 66-67; and oral communication, 48-49; and papyrus use, 27; and pedagogical paradigm, 15-16, 45; and Ro-

man religion, 17; and social status, 14, 15; and writing, 14. *See also* reading
literary studies, "theory revolution" in, 2
literature, Arabic, 133
literature, Greek: Arabic translations of, 133, 143; conservation of, 139–141; earliest works of, 5; and fall of Constantinople, 147–148; in late antiquity, 90; Latin translations of, 108–110, 133–134, 173, 174–175; as model for Roman literature, 58–59; in Roman empire, 140. *See also* Greek (language)
literature, heretical, 142
literature, Latin: beginnings of, 58–60; censorship of, 88–89; "decline" of, 163–164; in Imperial period, 86–90; in late antiquity, 90–91; limited circulation of, 87; loss of, 87; and national identity, 208; periodization of, 206, 209; in Republican period, 86–87. *See also* Latin (language)
Livius Andronicus, 58–59
Livy: in late antiquity, 100; on linen texts, 10; and manuscript survival, 104, 105; manuscript tradition, 128–129, 153; in Middle Ages, 128–129; *Periochae*, 105; *volumina*, 27
Lodi, Cathedral of, 160
Lord, Albert Bates, 211
Louis the Pious, 128
Lovati, Lovato, 150–151
Lucilius: *Satires*, 58; *Saturae*, 27
Lucretius: in Carolingian period, 121; censorship of, 196; *De Rerum natura*, 196, 213–214; *Oblongus* codex, 123
Lucullus, 39
Lupus of Ferrières, 123–124
Luther, Martin, 195–196, 201; *On the Bondage of the Will*, 196
Lycophron of Chalcideum, 54

Maas, Paul, 142, 215; *Textkritik*, 217–218
Maas's Law, 31

Mabillon, Jean: *De Re Diplomatica*, 62, 204
Macedonian renaissance, 144
Machiavelli, Niccolò, 1, 193
Macrobius, 77; *Saturnalia*, 96; on *Somnium Scipionis*, 100
Macrus, Pompeius, 89
Madvig, Nicolai, 214
Maffei, Francisco Scipione, 204–205
Magister, Thomas, 214
Mago, 39
Mai, Angelo, 205
manuscripts: appearance of, 49; in Carolingian period, 119–126; dating of, 35; and lexicographers, 56; loss of, in middle ages, 110–113, 138–139, 147–148; original meaning of, 2; photography of, 213; reception of, 85–86; social construction of, 2; valued for antiquity, 125
manuscript studies, 63, 213–220; research tools for, 240
Map, Walter: *De nugis curialium*, 130
Marcus Aurelius, 57; *Meditations*, 144
Marouzeau, Jules, 237
Martial: on book circulation, 44; medieval censorship of, 77; on opisthographic writing, 31, 36
Martianus Capella, *De nuptiis*, 101
Martin V (pope), 159
martyrdom, Christian, 94
Masai, François, 63
Matthias Corvinus (king), 193
Mavortius, Vettius Agorius Basilius, 99
McGann, Jerome, 2
Medici, Lorenzo de', 186
Mehmet II, 193
Mela, Pomponius, 155; *De chorographia*, 100
Melanchthon, Philip, 3, 196
Melania the Elder, 68
Menander, and accusations of plagiarism, 81

Mendoza, Iñigo López de, 192
Merula, Giorgio, 187, 198
Michael II ("the Stammerer"), 124
Middle Ages: cultural decline during, 110–111; ignorance of Greek during, 156–157; imitation of classics during, 152; intellectual rebirth in, 130; pre-humanism in; 150; as Renaissance invention, 162; and scientific texts, 133; Virgilian-Horatian-Ovidian ages, 128
military, Roman, 17
Milton, John: *Paradise Lost*, 197
Minucius Felix: *Octavius*, 95
misattribution, textual, 79
missionaries, 111
Mithridates, 39
Mommsen, Theodor, 20, 223
monasteries: and book production, 67–68, 111; in Carolingian period, 122; Renaissance view of, 165; and scriptoria, 67–71
monastic movement, 99–100
Montagnone, 151
Monte Cassino monastery, 105, 112, 114, 130, 133, 155–156, 160, 277n39
Montefeltro, Federico da, 179, *180*
Montfaucon, Bernard de: *Paleographia Graeca*, 204
Montreuil, John de, 192
More, Thomas, 180–181, 184, 194; *Utopia*, 194
Mount of Olives monastery, 68
Mouseion of Alexandria, 38, 40, 49–50
Müller, Karl Otfried, 212
multispectral imaging, 227
Murethach, 124
Museum der Alterthums-Wissenschaft, 208
Mussato, Albertino, 150; *Ecerinis*, 151

Naevius: *Bellum Poenicum*, 27, 58
Nauck, August, 230
Neleus, 37
neoclassicism, French, 198

neopaganism, 183
Neoplatonism, 97, 197
Nestor's Cup, 16, 47
Netherlands, 202
New Pauly, The, 239
Newton, Isaac, 191, 192
Nicanor, Servius, 56, 81
Niccoli, Niccolò, 105, 125, 156, 158, 161, 173, 185–186
Nicholas V (pope), 175, 176, 183
nicknames, Latin, 119
Nicolaus of Reggio, 133
Niebuhr, Barthold Georg, 206–207
Nietzsche, Friedrich, 214
Nobilior, M. Fulvius, 59
Nonantola, monastery of, 130, 161
Norden, Eduard, 168
notebooks (*pugillares*), 12–13; functionality of, 34; parchment, 33–34
novel, Greek, 145
numbering system, modern, 133
numismatics, 18, 204

Obbink, Dirk, 229
Olsen, Birger Munk, 125–126, 127–128, 138
Olympic games, 22, 142
Omeros (Walcott), 5
onomastics, 22–23
Onos (attr. Lucian), 80
opisthograph, 30–31, 36
oratory, 87
Origen: *Contra Celsum*, 94; translation of Old Testament, 142
Orosius, Paulus, 98
Orsini, Fulvio: *Familiae Romanae*, 204; *Imagines Virorum Illustrum*, 204
Orthodox Church. *See* Church, Orthodox
ostracism, 15
"other," marginalization of, 212
Otto, Walter, 239
Otto III, 129
Ottonian renaissance, 129

Ovid: *aetas Ovidiana*, 126, 128, 135; *Amores*, 84; *Ars Amatoria*, 88; banishment of, 88; in Carolingian period, 121, 126; *Ibis*, 136; manuscript transmission of, 126–127, 136; *Ovid moralisé*, 135
Oxford, 180, 192
Oxyrhynchus Online, 228
Oxyrhynchus Papyri (*P. Oxy.*), 26, 228

Packard, David, 232
Packard Humanities Institute (PHI), 225
Padua, 150, 151
paganism: and Christianity, 94, 99; "revival" of, 96–97
Palaeologus, Michael, 147
Palatine school, 117, 124
paleography, 23, 62–63, 204–206. *See also* writing
palimpsests, 30, 85, 91, 113; methods for reading, 205, 206, 290n73
Panegyrici Latini, 161
Pannartz, Arnold, 186
Pannonius, Janus, 192
Pantheon, 20
Panvinio, Onofrio, 204
paper, introduction of, 132
PapPal, 296n45
papyri: aesthetics of, 31–32, 49; circulation of, 205; documentary, 24; forgeries of, 289–290n70; literary, 24; multispectral imaging of, 227. *See also* papyrus
Papyri.info, 229, 295n43
papyrology, 18, 24–32, 62; and interpretive challenges, 25, 26; methodology of, 24–25; research tools for, 226–230
papyrus (writing material), 10, 18; ancient depictions of, 26; cost of, 27–28; in Egypt, 25; grades of, 28; in Greece, 26–27; history of, 25–26; in Italy, 26; and literacy, 27; and parchment, 32–33; Pliny the Elder on, 24, 28; preservation of, 25–26, 27; production of, 28–29, 60; *recto* and *verso* of, 60; reuse of, 30–31, 249n70. *See also* papyri
papyrus plant, 32
papyrus rolls (*chartae*), 13, 29–30; advantages of, 35–36; as books, 37; pagan use of, 36; *protokollon*, 30; *sillybon*, 30; transition from, to codex, 35, 92, 140; vulnerable surfaces of, 104
parchment (*membrana*), 32–34; cost of, 85; Horace on, 33; in medieval period, 61–62; vs. paper, 131; Quintilian on, 33–34; *recto* and *verso* of, 60–61; ruling of, 61–62; treating of, 61, 257n109. *See also* vellum
Parentucelli, Thomas, 161
Parry, Milman, 211
Parthenon, 203
Passow, Franz, 233
Pattison, Mark, 191
Paul (saint), 93; and Seneca, 96, 127, 182
Paul, Jean, 210–211
Paul II (pope), 183, 193
Paul the Deacon, 117, 123, 139, 268n112
Pauly, August Friedrich, 239
pecia system, 131
pens, ancient, 63–64
performance, dramatic: in Ancient Athens, 84, 221; as Christian tool, 196; humanistic rediscovery of, 172
performance, musical, 15
performance, oral, 58
Pergamum: and invention of parchment, 32–33; library of, 33
Peri antonymias (*On the Pronoun*), 57
Peri epirrematon (*On Adverbs*), 57
Peri syndesmon (*On Conjunctions*), 57
Peri syntaxis (*On Syntax*), 57
persecution, pagan, 98, 142
Perseus Project, 232
Peter Abelard of Bath: *Theologia*, 132
Peter of Pisa, 117
Peter the Deacon, 136
Peter the Venerable, 133
Petrarch, 1, 150; as Ciceronian, 153; and

"earthly things," 152–153; and ignorance of Greek, 157; manuscript collecting by, 153–155; as precursor to humanism, 158; view of, regarding classical antiquity, 151–152; and Virgil codex, 153, *154*
— Works: *Africa*, 162, 171; *Bucolicum Carmen*, 155; *Canzoniere*, 153; *Familiares*, 151
Petronius: *Excerpta vulgaria*, 160; manuscript tradition, 187–188; *Satyricon*, 17, 139, 160, 278n72
Pfeiffer, Rudolf, 208
philhellenism: in 19th c., 211–213; in ancient Rome, 58–59
Philocomus, Vettius, 58
Philodemus, 227
philologists (*philologoi*), 54; Roman republican, 86–87
philology, classical, 62; vs. cultural studies, 212–213; in Early Modern Europe, 200–203; origins of, 123, 188–190; and textual emendation, 200–201
Philostratus: *Vita Apollonii*, 100
Photius: *Bibliotheca*, 144–145, 274n225
photography, 213
Piaggio, Antonio, 227
Pindar, 54
Pithecusae (Ischia), 16
Pius II (pope), 183
Pizan, Christine de, 192
plagiarism, 80–83
Plantin, Christophe, 194
Planudes, Maximus, 147, 148
Platina, Bartolomeo, 183
Plato: and accusations of plagiarism, 81; Latin translations of, 108, 173; in Middle Ages, 146; *Republic*, 15, 173; *Timaeus*, 81, 108
Platonism, 94, 95, 124, 183
Plautus: genuine plays by, 59–60, 86–87; manuscript tradition, 103, 104, 161, 278n77; *Menaechmi*, 172; and renaissance stage production, 172; *Vidularia*, 104

plays, ancient: circulation of, 40–41, 87; and crisis of theatrical production, 87, 263n8; as profit makers, 79; scholarly tools for, 230. See also performance, dramatic
playwrights, canons of, 140
Plethon, Gemistus, 191
Pliny the Elder: in Carolingian period, 268n119; epitomes of, 106; on libraries, 40; multi-volume codices, 103; *Naturalis Historia*, 24, 127, 205; on papyrus, 24, 28; on parchment, 32–33; on Roman writing, 9–10; on writing materials, 10
Pliny the Younger: on bookshops, 44; in Carolingian period, 122; correspondence of, with Trajan, 94; on notebooks and hunting, 12–13; on plagiarism, 80
Plutarch: *Moralia*, 147, 156; and political censorship, 88; translation of, into Latin, 157, 174; *Vitae parallelae*, 174
Poetae Comici Graeci, 230
poetry: Hellenistic, 55; recitation of, 42, 43
Poggio Bracciolini, Gian Francesco, 125, 156, 222; on *diglossia* debate, 170; and pursuit of manuscripts, 159–161, 164–165, 184, 185, 192
Poland, 192
Polenton, Sicco: *Scriptores Illustri Latinae Linguae*, 164, 171
Politian, 1, 187, 193; and mastery of Greek, 176; *Nutricia*, 171; *Oratio in Expositione Homeri*, 176; as philologist, 190, 199
Polybius: *Histories*, 104
Pompeii, 12, 19, 205, 289n70
Pompey the Great, 59
Pomponazzi, Pietro: *De Immortalitate Animae*, 197
Pomposa Abbey, 130, 150
Pope, Alexander: *Imitations of Horace*, 197

Porphyry: *Isagoge*, 108, 109
porticus Octaviae, 39, 40
positivism, 213
Praeconinus, Lucius Aelius Stilo, 59
predestination, 196
"pre-humanism," 150–158
Prelum Ascenianum, 194
printing history, 3
printing press: advent of, 186; and destruction of codices, 189, 198; reception of, 187; spread of, 186–187, 192; and textual corruption, 198–199, 287n39
Priscian: *Institutiones grammaticae*, 123
Priscian of Mauritanian Caesarea, 109
Propertius: on the *Aeneid*, 42; *Elegies*, 159; influence of, on John of Salisbury, 135; manuscript tradition, 153
prosody, 57
Psellus, Michael, 146
Pseudo-Dionysius the Areopagite, 124
pseudo-Theocritus: *Idyll*, 8, 80
Ptolemy II, 37–38
Ptolemy III, 38
Ptolemy VIII, 54–55
publishing houses, 192, 194
punctuation: coronis, 51–52, *51*; *dicola*, 51; in Greek and Latin, 47–48; in inscriptions, 46–49; and interpretation, 254n23; lectional signs, 55–56; *paragraphus*, 50–52, *51*; and reading, 47, 254n23; in texts, 49–52; tricola, 51. *See also* signs, textual

quadrivium, 118, 132, 134, 272n190
Querolus (or *Aulularia*), 201
quill, 64
Quintilian: and assessment of Latin literature, 87–88; and book trade, 44; on codices, 33–34; *Declamationes maiores*, 129; on forgeries of his work, 82; *Institutiones oratoriae*, 160, 164, 253n267
quire, 60
Qur'an, 133

Rabelais, François: *Gargantua et Pantagruel*, 178, 197
Racine, Jean, 198
Ratherius, 128–129
reading: Dionysius' definition of, 55; of inscriptions, 47–48; lectional aids for, 49–50, 55; of plays, 40–41; silent, 50, 254n25; vocalized, 50, 55–56. *See also* literacy
recensio, 214, 215–218. *See also* editions
reception, textual: conditions of, 85; definition of, 86; theory of, 85
record keeping, 15, 245n56
reeds, as writing instruments, 63–64
Reformation, Protestant, 194, 195
Regula magistri, 68–69
Reichenau, monastery of, 160
Reitz, Friedrich Wolfgang, 212
religion, Roman, 17
Remigius of Auxerre, 124, 269n134
Renaissance, the: continuity of, with Middle Ages, 150; Greek instruction in, 172; spread through Europe, 193–194; and use of ancient civilization, 149; and world exploration, 191–192. *See also* humanism, renaissance
research tools, modern: and ancient religion, 236–237, bibliographic, 237–238; dictionaries as, 233–234; encyclopedias as, 239–40; and epigraphy, 222–227; and literature, 230–233; and manuscript studies, 240; and material culture, 240
Reuchlin, Johannes, 194: *De Arte Cabalistica*, 197
Rhetores latini minores, 127
Rhetorica ad Herennium, 80, 106, 127, 135
Ricci, Matteo, 197
Richard of Fournival: *Biblionomia*, 135
Ritschl, Friedrich, 214
Roberts, Colin, 32
Roman Empire: division of, 107; fall of, 163–164
Roman Empire, Eastern, 141–143

Rome: as bilingual culture, 107; cultural authority of, 89–90; decline of, 162–163; early scholarship in, 57–58; philhellenism of, 58, 59; religious texts of, 10, 16; sack of (410), 162; sack of (1527), 193; topographical surveys of, 203. *See also* Italy, ancient
Rossi, Roberto de', 158, 172
rubricator, 71–72
Rudolf of Fulda, 125
Rufinus of Aquileia, 68, 97
Rufus, Q. Curtius: *Historiae*, 104
Rule of Saint Benedict, 66–67

Saint-Germain d'Auxerre, abbey of, 124
Salmasius, Claudius, 202
Salutati, Coluccio, 153, 163, 172,181; *De Tyranno*, 177
Sapegno, Natalino, 151
Sappho, 14
Sassanids, 143
satire, 197
Saturnian verse, 87
Scala, Cangrande della, 151
Scaliger, Joseph Justus, 200, 223; *Thesaurus Temporum*, 201, 202
Scaliger, Julius Caesar: *De Causis Linguae Latinae*, 201; *Poetics*, 201, 206–207
Schedae Rescriptae Veronenses, 86
Schiller, Friedrich, 198
scholars, German, 1
scholia, 56
Schoppe, Caspar (Scioppius), 206
Schow, Niels Iversen, 205
science, ancient, 141, 191–192, 210
Scotus, Sedilius: *Collectaneum*, 123
scribal practice, medieval, 66–73
scribes: in antiquity, 31, 66; as booksellers, 43; Byzantine, 57; medieval, 66–73; professional, 70–71; use of humor by, 66; women, 68. *See also* illuminator; rubricator
Scribes and Scholars (Reynolds and Wilson), 1, 3

scriptio continua, 46, 47–49; and vocalization, 50
scriptoria, monastic, 67–71
scripts: archaic Greek, 7–8; Beneventan, 114; Carolingian miniscule, 65–66, 74, 119–120, *121*, 165; Chancery miniscule, 137–138, *138*; Greek cursive, 64; Greek miniscule, 144; Gothic, 136–137, *137*, 164; half-uncial, 92–93, *94*, 113–114, *114*; humanistic, 166–167, *166*, 279n106; inscriptional cursive, 65, *65*; insular script, 114; Merovingian, 114, *115*; monumental capital, 91–92; New Roman Cursive (miniscule), 65–66, 91, 258n137; Old Roman Cursive (majuscule), 65–66; pre-Caroline letterforms, 119–120, *120*, *121*; rustic capital, 91, *92*, 113; uncial, 92, *93*, 113; Visigothic style, 114, *115*
scriptura monumentalis, 64
"Scuola di Epicuro," 227
senate proceedings, Roman, 17
Seneca: and attribution of *Octavia*, 80; manuscript tradition, 103, 127, 135, 200; and Saint Paul, 79, 127, 182; tragedies of, 127, 130, 135
— Works: *Apocolocyntosis*, 127, 200; *De beneficiis*, 127; *De Clementia*, 127; *Epistulae ad Lucilium*, 127; *Naturales quaestiones*, 127, 135
Sequester, Vibius, 155; *De fluminibus*, 100
Serapeum of Alexandria, 142
Serenus, Septimius: *Opuscula ruralia*, 139
Serenus, Quintus: *Liber medicinalis*, 106
Servius (Maurus S. Honoratus), 78; "Servius Danielis," 201
Shakespeare, William, 1, 197–198
Sibylline books, 17
Siculus, Titus Calpurnius, 135
Siena, Antonio da, 280n122
Sigero, Nicola, 157
Sigismund of Luxembourg (king), 193
signs, textual, 53–54
Sigonio, Carlo, 203–204

INDEX / 341

Silius Italicus: "Additamentum Aldinum," 185; *Punica*, 125, 185
Simon, Richard, 196
slaves, 43, 58–59
Smet, Martin de, 204
social status, and literacy, 14, 15
Socrates: *Apology*, 41; on the ideal curriculum, 15
Solinus: *Collectanea*, 127; *Medicina Plinii*, 106
Sorano: *Gynaecia*, 136
Soranus of Ephesus, 108
soul, mortality of, 197
Spain, 192
sphragis (seal), 81
Stark, Rodney, 132
Statius: *Silvae*, 160
statuary, Greek, 211
St. Catherine at Mount Sinai, monastery of, 114
stemma codicum, 214, 216–219, *216*, *218*
"stemmatic method," 216
St. Gall monastery, 67–68, *68*, *69*, 93, 164
stoichedon inscriptions, 47, 48
stops (*stigmai*), 55–56
Strabo: *Geography*, 191–192; on works of Aristotle, 139
Strada, Zanobi da, 155–156
Studium, Florentine, 172, 173, 174
Studium of Ferrara, 179
Suda, 57, 146
Suetonius: on Augustan censorship, 88–89; in Carolingian period, 122, 125; on misattributed speeches, 79; as model for Einhard, 124; on plagiarism, 81; on Roman Republican scholarship, 57–58; on statue of Apollo Temenites, 40; on uncirculated poetry, 86
— Works: *On Grammarians and Rhetoricians*, 58; *Vitae Caesarum*, 107, 155
Sulla, 39, 42
Supplementum Epigraphicum Graecum (SEG), 225

Sweynheym, Conrad, 186
Symmachus, Quintus Aurelius, 97–98, 100
synizesis, 77

tablets, writing, 8, 242n17; codex form, 11, 243n34; leaf, 11, 13, *25*, 33; stylus (wax), 10–13, *11*, *12*, 33–34. See also notebooks
tabulae dealbatae, 10, 17
Tacitus: manuscript tradition, 104, 125, 156; on Tiberian censorship, 89
Tasso, Torquato: *Gerusalemme Liberata*, 197
Tatian: *Address to the Greeks*, 94
Techne grammatike (attr. Dionysius Thrax), 55, 57, 79
temples: of Hera at Paestum, 19; of Juno Moneta, 10; of Augustus (Rome), 39, 40; of Peace (Rome), 39
Terence, 103
Tertullian: *De praescriptione haereticorum*, 95
texts, literary: Hellenistic contributions to, 52–56
texts, medical, 133
textual transmission, study of, 213–220
Themistius, on library of Constantinople, 140–141
Themistocles Decree, 23–24
Theodoric the Ostragoth, 110
Theodorus Priscianus: *Euporiston*, 108
Theodosius II, 141, 142
Theodulf of Orléans, 117, 118, 123
theologians, 68
Theon of Alexandria, 141
Theophrastus, 37
Theopompus, 140
Thesaurus Cultus et Rituum Antiquorum (ThesCRA), 236–237, 298n76
Thesaurus Linguae Graecae database, 232–233
Thesaurus Linguae Latinae, 234–235
Thomas Aquinas, 134

Tibullus: *Corpus Tibullianum*, 159; spurious works of, 103
Timotheus: *Persians*, 51
Timpanaro, Sebastiano, 214
Tiraboschi, Girolamo: *History of Italian Literature*, 207
Titus Labienus, 88
TOCS-IN, 238
Tortelli, Giovanni: *De orthographia*, 174
Toscanelli, Paolo dal Pozzo, 191–192
trade routes, 7
Tragicorum Graecorum Fragmenta (*TrGF*), 230
training, physical, 15
Trajan, 94
translation: Arabic to Latin, 133; Greek to Arabic, 133–134, 143; Greek to Latin, 108–110, 157, 173, 174–175; methodology, 175; in the middle ages, 109, 157; in the Renaissance, 173–175; into vernacular, 157, 175
transmission, horizontal, 217
transmission, vertical, 217
Traube, Ludwig, 128
Traversari, Ambrogio, 181–182
Trimalchio's banquet, 17, 160, 188
Trinity College, Cambridge, 202
Trismegistos, 229
Trithème, Jean: *De Laude Scriptorum Manualium*, 187
trivium, 118–119, 134, 272n190
Trogus, Pompeius: *Historiae Philippicae*, 106
Turnebus, Adrien, 201
Tuscan dialect, 169; grammar of, 170
Twelve Tables, 16
Tyrtaeus, 14
Tzetzes, John, 146
Tzimiskes, John, 146

Ulpian, definition of book, 36–37
uncial script: i-e confusion, 74
universities, medieval: and development of Gothic script, 137–138; and growth of humanism, 193; impact of, on European culture, 132; role of, in book production, 72, 131
University of Berlin, 208, 212
University of Bologna, 132
University of Halle, 207
University of Montpellier, 132
University of Paris, 132, 134
Ur-text, 207
U.S. Epigraphy Project, 226
usus scribendi, 218, 219

Valerius Flaccus, 125; *Argonautica*, 160, 161
Valerius Maximus: *Epitome*, 100
Valla, Lorenzo: on corruption of Latin language, 162–163; on *diglossia* debate, 170; Greek instruction of, 174; heresy trial of, 182–183; as philologist, 189
— Works: *Elegantiae Latinae Linguae*, 161–163, 167–168, 170; *Emendationes Sex Librorum Titi Livi*, 189; "On the Falsely Believed and Deceptive Donation of Constantine," 195
Vandal kingdom, fall of, 123
Van Groningen, B. A., 42
Vargunteius, Quintus, 58
variants, textual, 218–219
Varro, Marcus Terentius, 59–60; manuscript tradition, 130, 156
— Works: *Antiquitates divinae et humanae*, 256–257n95; in Cicero's *Academia*, 83–84; *De lingua Latina*, 60, 130, 156, 257n97; *De philosophia*, 256–257n95; *De re rustica*, 60, 103; *On Plautine Comedies*, 59–60, 86–87
Vega, Garcilaso de la, 197
Velleius Paterculus, 188
vellum, 32. *See also* parchment
Venice, republic of, 167, 172, 173, 174, 197
Vergerio, Pier Paolo, 158, 193, 277n59
vernacular, 169–170
Veronese, Guarino, 169, 173, 179–80, 206

Verrius Flaccus: *De verborum significatu*, 106
Vesuvius, eruption of, 26, 227
Victorianus, Tascius, 100
Victorinus, Marius, 97, 108
Victory, altar of, 97–98, 142, 266n67
Villa of the Papyri, 26, 205, 227
Vincent of Beauvais: *Speculum maius*, 136
Vindolanda, 13
Virgil: and accusations of plagiarism, 81; "autographs" of, 59, 165; in Carolingian period, 121; commentaries on, 3, 78; editions of, 99, 101, 102; manuscript tradition, 102, 106; "Messianic" eclogue, 96; in Middle Ages, 128, 135; in palimpsest, 86; and printing press, 186; and publication of the *Aeneid*, 42; Renaissance reception of, 3; and scribal error, 76; *sphragis* of, 81; "Virgilian age," 128
Virgil in the Renaissance (Wilson-Okamura), 3, 241n9
Visconti, Gian Galeazzo, 177
Visigoths, 110, 122, 163, 187. *See also under* scripts
Vitruvius: *De Architectura*, 155, 205, 276n36; on plagiarism, 81
Vivarium monastery, 68, 109, 111–112
Vives, Juan Luis, 194–195
Volksgeist, Roman, 208
Voltaire: *Dictionnaire philosophique*, 196
von Eyb, Albrecht, 192
von Hutten, Ulrich, 195
von Wilamowitz-Moellendorff, Ulrich, 188, 190, 289n66
Voss, Johann Gerhard: *De Historis Graecis*, 202; *De Historis Latinis*, 202

Walch, Johann Georg: *Historica Critica Latinae Linguae*, 206
Weber, Karl, 205
Western Roman Empire, unification of, 117
William of Moerbeke, 134, 156
Winckelmann, Johann Joachim, 211
Wolf, Friedrich August, 207–212; "Darstellung Altertumswissenschaft nach der Begriff, Umfang, Zweck und Wert," 207–208, 212; *Geschichte der römischen Literatur*, 209; *Prolegomena ad Homerum*, 207, 209, 210
Woolf, Greg, 16
works, Christian: pagan models for, 95, 96; and textual integrity, 101–102
writing: Etruscan, 9; Greek, 6–9, 46, 47–49; in the *Iliad*, 8–9, 242n17; and literacy, 14; purpose of, 64; Roman, 17–18, 47; *scriptio continua*, 46, 47–49. *See also* alphabets; chirography; paleography; punctuation; scripts
writing materials: ancient, 9–13, 18, 63–64; medieval, 132. *See also* books; linen; notebooks; paper; papyrus; tablets

Xenophon: *Hiero*, 173; *Oeconomicus*, 15

Young, Douglas, 76

Zenodotus, 52–53, 54
Zimmermann, Bernhard, 239
Zonaras, John, 148
Zumpt, Carl, 214

www.ingramcontent.com/pod-product-compliance
Lightning Source LLC
Chambersburg PA
CBHW032016230426
43671CB00005B/107